TOPOLOGICAL METHODS IN GALOIS REPRESENTATION THEORY

Victor P. Snaith

School of Mathematics and Statistics
University of Sheffield

DOVER PUBLICATIONS, INC.
Mineola, New York

Bibliographical Note

This Dover edition, first published in 2013, is an unabridged republication of the work originally published in the series "Canadian Mathematical Society Series of Monographs and Advanced Texts" by John Wiley & Sons, New York, in 1989.

Library of Congress Cataloging-in-Publication Data

Snaith, Victor P. (Victor Percy), 1944–
    Topological methods in Galois representation theory / Victor P. Snaith, School of Mathematics and Statistics, University of Sheffield. — Dover edition.
        pages cm
    Reprint of: New York : Wiley, c1989. — (Canadian Mathematical Society series of monographs and advanced texts).
    Summary: "An advanced monograph on Galois representation theory by one of the world's leading algebraists, this volume is directed at mathematics students who have completed a graduate course in introductory algebraic topology. Topics include Abelian and nonabelian cohomology of groups, characteristic classes of forms and algebras, explicit Brauer induction theory, and much more. 1989 edition"— Provided by publisher.
    Includes bibliographical references and index.
    ISBN-13: 978-0-486-49358-9 (pbk.)
    ISBN-10: 0-486-49358-X (pbk.)
    1. Galois theory. 2. Representations of groups. 3. Invariants. I. Title.
QA174.2.S674 2013
512'.32—dc23

                                                                    2013020627

Manufactured in the United States by Courier Corporation
49358X01   2013
www.doverpublications.com

To Carolyn, Anna, Nina, and Daniel

Preface

This volume studies, from several viewpoints, the representation theory of finite groups which happen to be the Galois groups of finite extensions of fields. In particular, it is concerned with the construction of invariants of such Galois representations.

At the mention of invariants (or characteristic classes, in the topological terminology), an algebraic topologist would at once think of the more-than-adequate theory of Stiefel-Whitney and Chern classes and so might consider the matter closed. True, the methods of algebraic topology are designed for, and have been largely successful in, the process of constructing invariants. However, topology generally presents the seeker with invariants of Galois representations which are natural for all homomorphisms of such groups, which is much more than one insists upon when studying these representations qua Galois representations. It was this novelty which aroused my interest in this subject and, within this volume, I hope to give some bona fide examples in which a modicum of algebraic topology is extremely useful—perhaps even essential.

The first four chapters of this book are concerned with characteristic classes (of Galois representations) whose values lie in mod 2 Galois cohomology. The topic treated is the relationship, first discovered by Jean–Pierre Serre, between algebraic and topological characteristic classes of a Galois representation. That is, as explained in Chapter 2, an orthogonal Galois representation may be considered as giving rise to a bilinear form. The Hasse-Witt classes of this bilinear form turn out to be related to the Stiefel-Whitney classes of the representation. In Chapters 3 and 4 we derive Serre's formula and generalizations of it due to A. Frohlich and B. Kahn. These results we prove by methods that differ from the original ones and which require a modest amount of topology. For example, we develop the Koslowski transfer ab initio, in Chapter 4, in the category of topological spaces. In this setting the process is simpler and more general than the algebraic manner in which Bruno Kahn rediscovered it.

In preparation for later applications, Chapter 1 gives a brief introduction to the abelian cohomology of groups, and Chapter 2 does the same for the nonabelian theory. In those chapters several examples of cohomology rings are calculated. In particular, it is at this point that we collect all the specific cohomological data about dihedral and quaternion groups which will be useful later.

Chapters 6 and 7 are concerned with the construction of invariants of Galois representations in local class field theory. These chapters culminate in a new, essentially local, construction of the local root numbers, which give a local/global factorization of the Artin root number. In addition, as a necessary preliminary step, we construct the orthogonal local root numbers in Chapter 6, section 2, by a new, ad hoc method, involving the Witt group of nonsingular, symmetric, bilinear forms. This feature provides a very satisfactory point of contact between the material of Chapter 3 and that of Chapter 7.

Chapter 6 introduces the canonical form, which I have christened Explicit Brauer Induction, of Brauer's induction theorem. This involves more serious topology, in the form of the Lefschetz Fixed-Point Theorem. This chapter also derives a natural presentation for the representation ring of a finite group in a form which is suitable for the procedure of promoting invariants of abelian Galois representations to give invariants of arbitrary Galois representations. The problem of finding such a presentation is a very natural one and was posed by Jean-Pierre Serre. The construction of the local root number is an excellent example of this procedure in action, and I imagine that the formal nature of the argument will eventually render the technique useful in other contexts.

Finally, I will describe the role of Chapter 5. This chapter treats hard core stable homotopy theory that is not essential to the understanding of the later chapters. If the details are too unfamiliar, I recommend merely the reading of the statements of the main results and the scrutiny of the numerous attendant examples.

In Chapter 5 a result is proved which concerns the stable homotopy classes of maps between classifying spaces of groups. This result leads inexorably to the discovery of the Explicit Brauer Induction formulae, at least in the $I(G)$-adically completed representation ring. All this is described in Chapter 5, with many illustrative examples, and in Chapter 6, section 1. Therefore, Chapter 5 is an example of a result in stable homotopy theory which leads the way to a serious, new result in representation theory and thence to a serious application in number theory. I have included Chapter 5 to emphasize the novelty of this transpiration.

This volume began as lecture notes for a graduate course I gave at the University of Western Ontario during 1985 to 1986. The lecture notes contained Chapters 1–5 in essentially their current form and a far less satisfactory version of Chapters 6–7, in which the representation rings had to be completed and all invariants considered had to be continuous.

Throughout the book I have tried to give sufficient background on the topological prerequisites so that the energetic reader could pursue the details further. On this basis I believe that the reader who has experienced a graduate course on introductory algebraic topology will find this book accessible. On the algebraic and number-theoretic side, I have tried to be more complete, partly because of the constitution of my original audience.

I am very grateful to the University of Western Ontario for granting me a sabbatical year to finish this book. I was fortunate to enjoy the hospitality of the Centre de Recherches Mathématiques, Université de Montréal, and of the Mathematical Sciences Research Institute, Berkeley, during the final stages. I have attempted to embellish the start of each chapter with a quotation of idiosyncratic aptness, in that regard I am very grateful to Nancy Z. Tausky for providing me with the translation of the original lines of Belshazzar to Daniel (from "Cleanness" 1633–1640), and likewise, I would like to thank Mikael Runsten and Udo Zander for the original words of the poem by F. M. Franzén. Finally, I am deeply indebted to Catharine Leggett for typing the manuscript.

Victor Snaith

Hamilton, Ontario
April 1988

Contents

TOPOLOGICAL METHODS
IN GALOIS
REPRESENTATION
THEORY

Chapter One

Abelian Cohomology of Groups

It's like a book, this bloomin' world.
Which you can read and care for just so long,
But presently you feel that you will die
Unless you get the page you're readin' done,
An' turn another—likely not so good;
But what you're after is to turn 'em all.
                    —*RUDYARD KIPLING,*
          "Sestina of the Tramp-Royal" (1896)

In this chapter we first review the basic definitions of group cohomology with abelian coefficients, both continuous and discrete. Then we consider explicit formulae in low dimensions for applications such as products and the transfer (or corestriction) map. We introduce the usual basic concepts, for example, the long exact sequence and the homology/cohomology relationship. Our primary goal is to come away from this chapter with a few specific cohomology rings at our disposal—as well as transfer techniques such as the double coset formula. Transfer techniques are not only technically useful to us at this point; we will need to depend on them when, in Chapter 5, we encounter the double coset formula in stable homotopy theory.

After we have computed the cohomology rings of the cyclic groups, the dihedral group of order eight (with mod 2 coefficients) and (additively) the cohomology of the generalized quaternion groups, certain acts of faith will be required. These take the form of belief in the Stiefel-Whitney classes, Chern classes, and in the properties of spectral sequences. I have taken the view that faith knows no limits! Accordingly, with the briefest review of such things we are able to conclude this chapter with the computation of the mod 2 cohomology rings of dihedral and generalized quaternion groups and the integral cohomology ring of the dihedral group of order eight.

1. BASIC DEFINITIONS

(1.1) Definition

Let G be a group acting upon an abelian group, M. That is, we have a homomorphism

$$\Phi: G \longrightarrow \text{Aut}(M).$$

The associated (left) action

$$\hat{\Phi}: G \times M \to M$$

is defined by $\hat{\Phi}(g, m) = \Phi(g)(m)$ for $g \in G$, $m \in M$. If we abbreviate, as is usual, by writing $\hat{\Phi}(g, m) = g \cdot m$, then

(1.2) $$1 \cdot m = m, \quad g_1 \cdot (g_2 \cdot m) = (g_1 g_2) \cdot m.$$

The set of maps $f: G^n = G \times \cdots \times G \to M$ is denoted by $C^n(G; M)$ and is called the *n-cochains on G with values in M*. Define [Ser, p. I–9]

$$d: C^n(G; M) \to C^{n+1}(G; M)$$

by $(g_1, \ldots, g_{n+1} \in G)$,

(1.3) $$d(f)(g_1, \ldots, g_{n+1}) = g_1 \cdot f(g_2, \ldots, g_{n+1})$$

$$+ \sum_{i=1}^{n} (-1)^i f(g_1, \ldots, g_i g_{i+1}, \ldots, g_{n+1})$$

$$+ (-1)^{n+1} f(g_1, \ldots, g_n).$$

One readily verifies that $dd = 0$ and the *nth cohomology group $(n \geq 0)$ of G with coefficients in M, $H^n(G; M)$*, is defined by

(1.4) $$H^n(G; M) = \frac{(\ker d: C^n(G; M) \to C^{n+1}(G; M))}{(\text{im } d: C^{n-1}(G; M) \to C^n(G; M))},$$

where we set $C^n(G, M) = 0$ if $n < 0$.

(1.5) Remark

I have given the combinatorial definition first because it is the most appropriate one for generalization to continuous cohomology, to which we will turn our attention later in this chapter.

However, for many purposes the homological algebra definition of $H^n(G; M)$ as $\text{Ext}^n_{\mathbb{Z}[G]}(\mathbb{Z}, M)$ is very useful. We will recall that definition before proceeding further.

(1.6)   Define the *bar resolution* [H-S, p. 215]

(1.7) $$0 \leftarrow \mathbb{Z} \xleftarrow{\varepsilon} B_0 G \xleftarrow{d_0} B_1 G \xleftarrow{d_1} B_2 G \xleftarrow{d_2} \cdots$$

as follows. Set $B_n G$ equal to the free left $\mathbb{Z}[G]$-module on the set G^n. Hence, if $g_1, \ldots, g_n \in G$, we may denote the n-tuple $(g_1, \ldots, g_n) \in G^n$ by $[g_1 | \cdots | g_n]$. This is a free generator of $B_n G$ over the integral group-ring $\mathbb{Z}[G]$, and an arbitrary element of $B_n G$ is a sum of elements denoted $X[g_1 | \cdots | g_n]$, where $X \in \mathbb{Z}[G]$. The generator of $B_0 G$ is simply written, $[\ \ ]$.

Define ε and d_i as the $\mathbb{Z}[G]$-module homomorphisms given by

$$\varepsilon[g_1] = 1,$$

$$d_n[g_1 | \cdots | g_{n+1}] = g_1[g_2 | \cdots | g_n]$$

$$+ \sum_{i=1}^{n} (-1)^i [g_1 | \cdots | g_i g_{i+1} | \cdots | g_n]$$

$$+ (-1)^{n+1} [g_1 | \cdots | g_n].$$

The $\mathbb{Z}[G]$-module, $\mathbb{Z}$, is given by the integers with trivial action $g \cdot m = m$ ($g \in G$, $m \in \mathbb{Z}$).

If we define η, S_i in

(1.8) $$0 \to \mathbb{Z} \xrightarrow{\eta} B_0 G \xrightarrow{S_0} B_1 G \xrightarrow{S_1} B_2 G \xrightarrow{S_2} \cdots$$

to be the (abelian group) homomorphisms given by

$$\eta(1) = [\ \ ],$$

(1.9) $$S_n(g_1 [g_2 | \cdots | g_n]) = [g_1 | g_2 | \cdots | g_n] \qquad \text{for } n \geq 0.$$

One easily checks that the following identities are true:

$$1 = \varepsilon \eta,$$

(1.10) $$1 = \eta \varepsilon + d_0 S_0,$$

$$1 = S_{n-1} d_{n-1} + d_n S_n \qquad \text{for } n \geq 1.$$

Without much difficulty, we obtain the following lemma:

(1.11) Lemma

(i) *(1.7) is an exact sequence of free $\mathbb{Z}[G]$-modules. That is, ε is onto, $\ker \varepsilon = \text{image}(d_0)$, and $\ker(d_n) = \text{image}(d_{n+1})$ for all $n \geq 0$.*

(ii) There is a natural isomorphism $\psi_n\colon \mathrm{Hom}_{\mathbb{Z}[G]}(B_n G, M) \to C^n(G, M)$ *given by*

$$\psi(f)(g_1, \ldots, g_n) = f[g_1 | \cdots | g_n].$$

(iii) Furthermore

$$\psi_{n+1}(d_n f) = d(\psi_n(f)).$$

(1.12) Corollary

There is a natural isomorphism

$$\psi^* \colon \frac{\mathrm{Ker}\,(\mathrm{Hom}_{\mathbb{Z}[G]}(d_{n+1}, M))}{\mathrm{Image}\,(\mathrm{Hom}_{\mathbb{Z}[G]}(d_n, M))} \xrightarrow{\;\cong\;} H^n(G; M).$$

(1.13) $Ext^n_{\mathbb{Z}[G]}(\mathbb{Z}, M)$

A summand of a free $\mathbb{Z}[G]$-module is called a *projective $\mathbb{Z}[G]$-module*. If we have an exact sequence of left $\mathbb{Z}[G]$-modules

(1.14)
$$0 \leftarrow \mathbb{Z} \xleftarrow{\;\varepsilon\;} P_0 \xleftarrow{\;d_0\;} P_1 \xleftarrow{\;d_1\;} P_2 \xleftarrow{\;d_2\;} \cdots,$$

in which each $P_i(i \geq 0)$ is projective, then we define $\mathrm{Ext}^n_{\mathbb{Z}[G]}(\mathbb{Z}, M)$ to equal the cohomology of (1.14) with coefficients in M:

(1.15)
$$\frac{\mathrm{Ker}\,(\mathrm{Hom}\,(d_{n+1}, M))}{\mathrm{Image}\,(\mathrm{Hom}\,(d_n, M))}.$$

(1.14) is called a *projective resolution* of the $\mathbb{Z}[G]$-module, $\mathbb{Z}$. Of course, (1.7) is a particular example. However, given a partial commutative diagram in which the lower sequence is exact,

(1.16)
$$
\begin{array}{ccccccccc}
0 \leftarrow & \mathbb{Z} & \leftarrow & P_0 & \leftarrow & P_1 & \leftarrow \cdots \leftarrow & P_n & \leftarrow & P_{n+1} \\
& \downarrow{\scriptstyle f} & & \downarrow{\scriptstyle f_0} & & \downarrow{\scriptstyle f_1} & & \downarrow{\scriptstyle f_n} & & \downarrow \\
0 \leftarrow & \mathbb{Z} & \xleftarrow{\varepsilon'} & Q_0 & \xleftarrow{d_0'} & Q_1 & \leftarrow \cdots \leftarrow & Q_n & \xleftarrow{d_n'} & Q_{n+1} & \leftarrow \cdots,
\end{array}
$$

one can find $f_{n+1}\colon P_{n+1} \to Q_{n+1}$ so that $d_n' f_{n+1} = f_n d_n$. If we have constructed two sequences $\{f_n\}$ and $\{h_n\}$ to make (1.16) commute, we may inductively construct $\{\sigma_n\colon P_n \to Q_{n+1}\}$ so that $(\sigma_{-1} = 0)$

$$f_n - h_n = \sigma_{n-1} d_{n-1} + d_n' \sigma_n.$$

Hence the chain maps $\{f_n\}$ and $\{h_n\}$ induce the same map from the cohomology of the upper sequence of (1.16) to that of the lower sequence. This map is denoted simply by f^*, since it depends only on f. From this, one sees that two choices of resolutions yield isomorphic groups, $\text{Ext}^n_{\mathbb{Z}[G]}(\mathbb{Z}, M)$, which are isomorphic by a canonical isomorphism, namely, 1^*.

Therefore $\text{Ext}^n_{\mathbb{Z}[G]}(\mathbb{Z}, M)$ is a functor of M, and there is a natural isomorphism $(n \geq 0)$,

(1.17) $$H^n(G; M) \cong \text{Ext}^n_{\mathbb{Z}[G]}(\mathbb{Z}, M).$$

(1.18) CONTINUOUS COHOMOLOGY

Suppose that G is a topological group. The example that will occupy us is that of a *profinite group*. That is,

(1.19) $$G = \varprojlim_{\alpha \in \mathscr{A}} G_\alpha$$

where $\{h_{\alpha,\beta} : G_\alpha \to G_\beta\}$ is an inverse system of homomorphisms of finite groups, G_α, as α runs through some partially ordered set $\mathscr{A}$. If $\beta \geq \alpha$, there is one homomorphism $h_{\alpha,\beta}$, and if $\alpha \leq \beta \leq \gamma$, then $h_{\alpha,\gamma} = h_{\beta\gamma} \circ h_{\alpha\beta}$. In these circumstances (1.19) is the set of $\{g_\alpha\} \in \prod_{\alpha \in \mathscr{A}} G_\alpha$ such that $h_{\alpha\beta}(g_\alpha) = g_\beta$. Each G_α has the discrete topology, $\prod_\alpha G_\alpha$ has the product topology, and (1.19) has the resulting subspace topology. It is a compact, totally disconnected group.

(1.20) Example

For $0 < n \in \mathbb{Z}$, set $G_n = \mathbb{Z}/n$, set n less than m if m divides n, and set $h_{n,m} : Z/n \to Z/m$ equal to the canonical surjection, $h_{n,m}(1) = 1$. The resulting group is the *adic-integers*

$$\hat{\mathbb{Z}} = \varprojlim_n \frac{\mathbb{Z}}{n}.$$

Similarly, the *l*-adic integers, for a prime l, are defined by

$$\hat{\mathbb{Z}}_l = \varprojlim_q \frac{\mathbb{Z}}{l^q}.$$

(1.21) Example

Let M be an abelian group, with the discrete topology, on which the topological group, G, acts continuously. This means that for $m \in M$

$$\text{stab}(m) = \{g \in G \mid gm = m\}$$

is open in G. If M^H denotes the subgroup of M fixed by each element of a subgroup, H, then

(1.22)
$$M = \bigcup_{\substack{H < G \\ \text{open}}} M^H.$$

Let $C_{ct}^n(G; M)$ denote the subset of continuous cochains in $C^n(G; M)$. It is clear that d in (1.3) preserves continuous (i.e., locally constant) cochains, and we may define *continuous cohomology* by

(1.23)
$$H_{ct}^n(G; M) = \frac{(\ker d\colon C_{ct}^n(G; M) \to C_{ct}^{n+1}(G; M))}{(\operatorname{im} d\colon C_{ct}^{n-1}(G; M) \to C_{ct}^n(G; M))}.$$

When G is the profinite group given by the absolute Galois group of a field, K:

(1.24)
$$G = \Omega_K = \varprojlim G\left(\frac{N}{K}\right).$$

The resulting continuous cohomology is called *Galois cohomology*. In (1.24) the groups, $G(N/K)$, are the Galois groups of finite Galois extensions N of K, and the inverse limit is taken over such N/K.

In this example we have

(1.25)
$$H_{\text{Gal}}^n(K; M) = H_{ct}^n(\Omega_K; M), \qquad \text{by definition}$$
$$\cong \varinjlim_{N/K} H^n(G(N/K); M^{U(N)}),$$

where $U(N) = \ker(\Omega_K \to G(N/K))$.

2. *BASIC PROPERTIES*

(2.1) Suppose that $f\colon H \to G$ is a homomorphism of groups (or of topological groups in the context of §1.21). Clearly, f induces, by composition, a homomorphism

(2.2)
$$(f^n \cdot \text{—})\colon C^n(G; M) \to C^n(H; M)$$

such that $d(f^n \cdot h) = f^{n+1}(dh)$. Hence (2.2) induces a restriction map on cohomology

(2.3)
$$f^*\colon H^n(G; M) \to H^n(H; M)$$

such that $1^* = 1$ and $f^* \cdot (f')^* = (f'f)^*$.

(2.4) PRODUCTS

Let H and G be groups, and let

$$E_n = \bigoplus_{1 \leq a \leq n} (B_a H) \bigotimes_{\mathbb{Z}} (B_{n-a} G)$$

have the left $\mathbb{Z}[H \times G] \cong \mathbb{Z}[H] \otimes_{\mathbb{Z}} \mathbb{Z}[G]$ module structure given by

$$(x \otimes y) \cdot (b \otimes c) = x \cdot b \otimes y \cdot c$$

$$(x \in \mathbb{Z}[H], y \in \mathbb{Z}[G], b \in B_a H, c \in B_{n-a} G).$$

Define $(d \otimes 1 \pm 1 \otimes d)_n = \partial_n : E_{n+1} \to E_n$ by

(2.5) $$\partial_n(b \otimes c) = d_{a-1}(b) \otimes c + (-1)^a b \otimes d_{n-a-1}(c)$$

and set $\varepsilon = \varepsilon \otimes \varepsilon : B_0 H \otimes B_0 G \to \mathbb{Z} \otimes_{\mathbb{Z}} \mathbb{Z} \cong \mathbb{Z}$.
   Each E_n is a free $\mathbb{Z}[H \times G]$-module, and

(2.6) $$0 \leftarrow \mathbb{Z} \xleftarrow{\varepsilon} E_0 \xleftarrow{\partial_0} E_1 \xleftarrow{\partial_1} E_2 \xleftarrow{\partial_2} \cdots$$

is a free resolution of the $\mathbb{Z}[H \times G]$-module, $\mathbb{Z}$. One sees that (2.6) is exact by means of $(S \otimes 1)_n : E_n \to E_{n+1}$, defined, using the chain homotopy of (1.9).
   The diagonal homomorphism $\Delta : G \to G \times G$ makes (2.6) into an exact sequence of $\mathbb{Z}[G]$-modules, and by the discussion in §1.13 concerning (1.16), we may construct a commutative diagram of $\mathbb{Z}[G]$-modules (here $H = G$):

(2.7)
$$0 \leftarrow \mathbb{Z} \xleftarrow{\varepsilon} B_0 G \xleftarrow{d_0} B_1 G \xleftarrow{d_1} B_2 G \xleftarrow{d_2} \cdots$$
$$\downarrow 1 \qquad \downarrow \Delta_0 \qquad \downarrow \Delta_1 \qquad \downarrow$$
$$0 \leftarrow \mathbb{Z} \xleftarrow{\varepsilon} E_0 \xleftarrow{\partial_0} E_1 \xleftarrow{\partial_1} E_2 \xleftarrow{\partial_2} \cdots$$

An explicit Δ_i is given by the following formula [Mac, p. 296]:

(2.8) $$\Delta_n[g_1|\cdots|g_n] = \sum_{i=0}^{n} [g_1|\cdots|g_i] \otimes g_1 g_2 \cdots g_i [g_{i+1}|\cdots|g_n].$$

If $f \in \mathrm{Hom}_{\mathbb{Z}[H]}(B_n H, M)$ and $f' \in \mathrm{Hom}_{\mathbb{Z}[G]}(B_m G, M)$, then we may project from E_{n+m} onto the factor $B_n H \otimes B_m G$ and compose with $f \otimes f'$ to yield

(2.9) $$f \otimes f' \in \mathrm{Hom}_{\mathbb{Z}[H \times G]}(E_{n+m}, M \otimes M').$$

From (2.5) and (2.7) we see easily that if f and f' are cocycles, so is $f \otimes f'$, and its cohomology class depends only on that of f and f'. In this manner we obtain the *external cup-product*.

(2.10) $H^n(H; M) \otimes H^m(G; M') \to H^{n+m}(H \times G; M \otimes M')$,

given by $[f] \cup [f']$, where $[-]$ denotes a cohomology class. The cup-product is associative.

The *internal product* is

(2.11)
$$H^n(G; M) \otimes H^m(G; M') \to H^{n+m}(G; M \otimes M') \quad \text{given by}$$
$$[f][f'] = \Delta^*([f] \cup [f']).$$

If $T_*: H^q(G; M \otimes M') \to H^q(G; M' \otimes M)$ is induced by interchanging M and M', then the product is commutative in the sense that

(2.12) $T_*([f][f']) = (-1)^{mn}[f'][f]$.

(2.13) By virtue of the explicit formula (2.8), one finds that the products of (2.10) and (2.11) make sense for continuous cochains and induce analogous products on continuous cohomology groups.

(2.14) Example

Suppose that R is a ring with trivial G action. We have a product $m: R \times R \to R$ so that we may form

$$H^n(G; R) \otimes H^m(G; R) \to H^{n+m}(G; R \otimes R) \xrightarrow{m_*} H^{n+m}(G; R).$$

Explicitly, by (2.8), this product is given by

$$[f][f'][g_1 | \cdots | g_{n+m}] = (f[g_1 | \cdots | g_n])(f'[g_{n+1} | \cdots | g_{n+m}]).$$

This product makes $\oplus_{0 \leqslant n} H^n(G; R)$ into a graded ring, $H^*(G; R)$. Similarly, $H_{ct}^*(G; R)$ is a graded ring.

(2.15) HOMOLOGY

The nth homology of a discrete group, G, with coefficients in a $\mathbb{Z}[G]$-module, M, is denoted by $H_n(G; M)$. It is defined by

(2.16) $H_n(G; M) = \dfrac{(\ker d_{n-1} \otimes_G 1: B_n G \otimes_G M \to B_{n-1} G \otimes_G M)}{(\operatorname{im} d_n \otimes_G 1: B_{n+1} G \otimes_G M \to B_n G \otimes_G M)}$,

where $\{B_n G, d_n\}$ is the bar resolution of (1.7). As in §1.13, one can show that (up to canonical isomorphism) $\{B_n G, d_n\}$ may be replaced in (2.16) by any projective resolution of $\mathbb{Z}$, as a $\mathbb{Z}[G]$-module.

(2.17) Suppose that $0 \to L \xrightarrow{\alpha} M \xrightarrow{\beta} N \to 0$ is a short exact sequence of $\mathbb{Z}[G]$-modules. Then we are entitled to two long exact sequences:

$$\cdots \to H^n(G; L) \xrightarrow{\alpha_*} H^n(G; M) \xrightarrow{\beta_*} H^n(G; N) \xrightarrow{\delta} H^{n+1}(G; L) \to \cdots$$

(2.18)

$$\cdots \to H_n(G; L) \xrightarrow{\alpha_*} H_n(G; M) \xrightarrow{\beta_*} H_n(G; N) \xrightarrow{\delta} H_{n-1}(G; L) \to \cdots$$

In the upper sequence $\alpha_*[f] = [\alpha \cdot f]$, $\beta_*[h] = [\beta \cdot h]$, and in the lower sequence $\alpha_*[a] = [(1 \otimes \alpha)(a)]$.

To construct each of the sequences in (2.18), we first produce a commutative diagram of the form

(2.19)

$$
\begin{array}{ccccccccc}
& & \vdots & & \vdots & & \vdots & & \\
& & \downarrow & & \downarrow & & \downarrow & & \\
0 & \to & A_{n+1} & \xrightarrow{\alpha} & B_{n+1} & \xrightarrow{\beta} & C_{n+1} & \to & 0 \\
& & \downarrow{\scriptstyle d} & & \downarrow{\scriptstyle d} & & \downarrow{\scriptstyle d} & & \\
0 & \to & A_n & \xrightarrow{\alpha} & B_n & \xrightarrow{\beta} & C_n & \to & 0 \\
& & \downarrow & & \downarrow & & \downarrow & & \\
& & \vdots & & \vdots & & \vdots & &
\end{array}
$$

in which the vertical columns are chain complexes (i.e., $dd = 0$) and the horizontal rows are short exact. The cohomology sequence arises from the case

(2.20) $$\left.\begin{array}{c} A_n \\ B_n \\ C_n \end{array}\right\} = \mathrm{Hom}_G\left(B_{-n}G, \left\{\begin{array}{c} L \\ M \\ N \end{array}\right\}\right),$$

whereas the homology sequence comes from the case

(2.21) $$\left.\begin{array}{c} A_n \\ B_n \\ C_n \end{array}\right\} = B_n G \otimes_{\mathbb{Z}[G]} \left\{\begin{array}{c} L \\ M \\ N \end{array}\right\}.$$

If $x \in \ker d \subset C_{n+1}$, we define

(2.22) $$\partial'[x] = [\alpha^{-1} d \beta^{-1}(x)] \in H_n(A_*, d).$$

This is a homology class for the chain complex (A_*, d) which depends only on the homology class $[x] \in H_{n+1}(C_*, d)$.

It is straightforward to prove the next lemma, from which (2.18) follows, being special cases (2.20) and (2.21).

(2.23) Lemma

In the situation of (2.19),

$$\cdots \to H_n(A) \xrightarrow{\alpha} H_n(G) \xrightarrow{\beta} H_n(C) \xrightarrow{\partial'} H_{n-1}(A) \xrightarrow{\alpha} \cdots$$

is an exact sequence.

(2.24) Application

Suppose that $\{\cdots \to C_n \xrightarrow{d} C_{n-1} \xrightarrow{d} \cdots\}$ is a chain complex of free modules over a principal ideal domain, R. Suppose also that M is a left R-module. We can form the cohomology groups $H^n(C_*; M)$ given by the homology of the chain complex $\{\mathrm{Hom}_R(C_n, M), d^*\}$, or we can form the R-module given by $H_n(C_*, d)$, the homology of $\{C_*, d\}$. These are related by the *universal coefficient* exact sequence:

(2.25) $0 \to \mathrm{Ext}^1_R(H_{n-1}(C_*), M) \to H^n(C_*; M) \to \mathrm{Hom}_R(H_n(C_*), M) \to 0.$

Briefly, I will recall the derivation of (2.25). Firstly $Z_n = \ker d$ and $B_n = \mathrm{im}\, d$ are *free* sub-R-modules of C_n so that the exact sequence

(2.26) $$0 \to B_n \xrightarrow{i} Z_n \to H_n(C_*) \to 0$$

is a free R-module resolution of $H_n(C_*)$. Hence, by definition, the map

(2.27) $$\mathrm{Hom}_R(Z_n, M) \xrightarrow{i^*} \mathrm{Hom}_R(B_n, M)$$

satisfies

(2.28) $$\mathrm{Ker}\, i^* \cong \mathrm{Hom}_R(H_n(C_*), M)$$

$$\mathrm{Coker}\, i^* \cong \mathrm{Ext}^1_R(H_n(C_*), M).$$

Now take the exact sequence

(2.29) $$0 \to Z_n \to C_n \xrightarrow{d} B_{n-1} \to 0,$$

and form a diagram like (2.19) by applying $\mathrm{Hom}_R(-, M)$ to (2.29), where $d: Z_n \to Z_{n-1}$ and $d: B_n \to B_{n-1}$ are defined to be zero. By §2.23, there results a long exact sequence

(2.30) $\cdots \to \mathrm{Hom}_R(Z_{n-1}, M) \xrightarrow{\partial} \mathrm{Hom}_R(B_{n-1}, M) \to H^n(C_*; M) \to \cdots,$

and (2.25) follows from (2.28) and the verification that ∂ in (2.30) equals $i*$ of (2.27).

(2.31)   If $\text{Tor}_R^1(H_n(C), M)$ is defined as

$$\text{Ker}(B_n \otimes_R M \to Z_n \otimes_R M),$$

a similar argument yields the following exact sequence:

(2.32) $0 \to H_n(C_*) \otimes_R M \to H_n(C_* \otimes M, d \otimes 1) \to \text{Tor}_R^1(H_{n-1}(C_*), M) \to 0.$

(2.33) Particularly useful special cases of (2.25) and (2.32) are the following exact sequence and isomorphism: Let G be a discrete group, and let p be a prime. We have exact sequences ($\mathbb{Z}$ and $\mathbb{Z}/p$ having trivial G-action)

$$0 \to H_n(G; \mathbb{Z}) \otimes \frac{\mathbb{Z}}{p} \to H_n\left(G; \frac{\mathbb{Z}}{p}\right) \to \text{Tor}_\mathbb{Z}^1\left(H_{n-1}(G; \mathbb{Z}), \frac{\mathbb{Z}}{p}\right) \to 0,$$

$$H^n\left(G; \frac{\mathbb{Z}}{p}\right) \cong \text{Hom}\left(H_n\left(G; \frac{\mathbb{Z}}{p}\right), \frac{\mathbb{Z}}{p}\right).$$

(2.34)   TRANSFER OR CORESTRICTION

Suppose that $i: H \subset G$ are discrete groups with index $[G:H] = m$. Let $\{x_i : 1 \le i \le m\}$ be a set of coset representatives for G/H. If we are given a free $\mathbb{Z}[H]$-resolution

$$0 \leftarrow \mathbb{Z} \xleftarrow{\varepsilon} Q_0 \xleftarrow{d} Q_1 \xleftarrow{d} Q_2 \xleftarrow{d} \cdots,$$

then, because $\mathbb{Z}[G]$ is a free $\mathbb{Z}[H]$-module,

$$0 \leftarrow \mathbb{Z}[G] \otimes_{\mathbb{Z}[H]} \mathbb{Z} \xleftarrow{1 \otimes \varepsilon} \mathbb{Z}[G] \otimes_{\mathbb{Z}[H]} Q_0 \xleftarrow{1 \otimes d} \mathbb{Z}[G] \otimes_{\mathbb{Z}[H]} Q_1 \leftarrow \cdots,$$

is an exact $\mathbb{Z}[G]$-module resolution of $\mathbb{Z}[G] \otimes_{\mathbb{Z}[H]} \mathbb{Z}$. Hence we have an isomorphism, in which M is a $\mathbb{Z}[G]$-module,

(2.35) $H^n(H; M) \cong \text{Ext}_{\mathbb{Z}[G]}^n(\mathbb{Z}[G] \otimes_{\mathbb{Z}[H]} \mathbb{Z}, M).$

However, there is a $\mathbb{Z}[G]$-module map $\tau: \mathbb{Z} \to \mathbb{Z}[G] \otimes_{\mathbb{Z}[H]} \mathbb{Z}$ defined by $\tau(z) = \sum_{i=1}^m x_i \otimes z$. For, if $g \in G$, then $g x_i \in x_{\sigma(i)} H$ for some permutation, $\sigma \in \Sigma_m$, so that $g(\tau(z)) = \tau(z) = \tau(g \cdot z)$. The induced map

(2.36) $\tau^*: \text{Ext}_{\mathbb{Z}[G]}^n(\mathbb{Z}[G] \otimes_{\mathbb{Z}[H]} \mathbb{Z}, M) \to \text{Ext}_{\mathbb{Z}[G]}^n(\mathbb{Z}, M) \cong H^n(G; M)$

composes with (2.33) to yield the *transfer homomorphism*

(2.37) $\text{Tr} = i_*: H^n(H; M) \to H^n(G; M).$

In an entirely analogous manner we obtain the homology transfer map:

(2.38) $$\mathrm{Tr} = i^*: H_n(G; M) \to H_n(H; M).$$

(2.39) Lemma

In §2.24 each composite homomorphism $i_* i^*: H^n(G; M) \to H^n(G; M)$ and $i^* i_*: H_n(G; M) \to H_n(G; M)$ is multiplication by $m = [G:H]$.

Proof. We will prove only the cohomology case, leaving the similar homology result to the interested reader. Suppose that we have a $\mathbb{Z}[G]$-resolution

$$0 \leftarrow \mathbb{Z} \xleftarrow{\ \varepsilon\ } P_0 \xleftarrow{\ d\ } P_1 \xleftarrow{\ d\ } P_2 \leftarrow \cdots,$$

then the homology of $\{\mathrm{Hom}_G(P_i, M), d^*\}$ is $H^*(G; M)$. Since a free $\mathbb{Z}[G]$-module is a free $\mathbb{Z}[H]$-module, the homology of $\{\mathrm{Hom}_H(P_i, M), d^*\}$ is $H^*(H; M)$. The natural map

$$\mathrm{Hom}_G(P_i, M) \to \mathrm{Hom}_H(P_i, M) \cong \mathrm{Hom}_G(\mathbb{Z}[G] \otimes_{\mathbb{Z}[H]} P_i, M)$$

is induced by $f: \mathbb{Z}[G] \otimes_{\mathbb{Z}[H]} \mathbb{Z} \to \mathbb{Z}$, $f(g \otimes \omega) = \omega$. Therefore, in cohomology, $i_* i^*$ is induced by $f \cdot \tau: \mathbb{Z} \to \mathbb{Z}[G] \otimes_{\mathbb{Z}[H]} \mathbb{Z} \to \mathbb{Z}$, but $f(\tau(z)) = f(\sum_{i=1}^m x_i \otimes z) = mz$. However, from §1.13, it is clear that $(m \cdot -): \mathbb{Z} \to \mathbb{Z}$ induces multiplication by m on $\mathrm{Ext}^n_{\mathbb{Z}[G]}(\mathbb{Z}, M)$. $\qquad\square$

(2.40) Remark

Returning to the definition of the transfer in cohomology, observe that if

$$0 \leftarrow \mathbb{Z} \xleftarrow{\ \varepsilon\ } Q_0 \xleftarrow{\ d\ } Q_1 \xleftarrow{\ d\ } Q_2 \leftarrow \cdots$$

is a $\mathbb{Z}[G]$-module resolution, then there is an adjunction isomorphism

$$\Phi: \mathrm{Hom}_G(Q_i, \mathrm{Hom}_H(\mathbb{Z}[G], M)) \xrightarrow{\ \cong\ } \mathrm{Hom}_G(\mathbb{Z}[G] \otimes_{\mathbb{Z}[H]} Q_i, M)$$

given by $\Phi(f)(a \otimes b) = f(b)(a)$. This induces an isomorphism

(2.41) $$H^n(G; \mathrm{Hom}_H(\mathbb{Z}[G], M)) \cong H^n(H; M)$$

The transfer is induced by the $\mathbb{Z}[G]$-module homomorphism [H-S, p. 266]:

(2.42) $$\begin{cases} \theta: \mathrm{Hom}_H(\mathbb{Z}[G], M) \to M, \\ \theta(\Phi) = \displaystyle\sum_{i=1}^m x_i^{-1} \Phi(x_i). \end{cases}$$

Here G acts on the left of $\text{Hom}_H(\mathbb{Z}[G], M)$ *by* $(g\Phi)(x) = \Phi(xg)$ *so that*

$$\theta(g(\Phi)) = \sum_i x_i^{-1} \Phi(x_i g) = \sum_i g(x_i g)^{-1} \Phi(x_i g) = g\theta(\Phi).$$

(2.43) Theorem

Let H and K be subgroups of finite index in G, and let M be a $\mathbb{Z}[G]$-module. Then the composite

$$H^n(H; M) \xrightarrow{\text{Tr}_H^G} H^n(G; M) \xrightarrow{i^*} H^n(K; M)$$

is equal to the sum

$$\sum_{g \in K \backslash G / H} \psi_g,$$

where ψ_g is the composite,

$$H^n(H; M) \xrightarrow{(g-g^{-1})^*} H^n(gHg^{-1}; M) \xrightarrow{i^*} H^n(K \cap (gHg^{-1}); M) \xrightarrow{\text{Tr}} H^n(K; M).$$

Proof. $H^*(H; M)$ is the homology of the chain complex, $\text{Hom}_H(Q_i, M)$, where $\{Q_i, d\}$ is a free $\mathbb{Z}[G]$-resolution of $\mathbb{Z}$, as in §2.40. By §2.40, the transfer, Tr_H^G, is induced by $\hat{\theta}: \text{Hom}_H(Q_i, M) \to \text{Hom}_G(Q_i, M)$,

$$\hat{\theta}(f) = \sum_{i=1}^m x_i(f(x_i^{-1}\cdot -)).$$

However, as a K-map in $\text{Hom}_K(Q_i, M)$, $\hat{\theta}(f)$ is the sum of maps such as

$$\hat{f}_i = \sum_{x_j \in Kx_iH} x_j(f(x_j^{-1}\cdot -)): Q_i \to M,$$

one for each double coset of $K \backslash G / H$.

However, the bijection

$$Kx_iH/H \leftrightarrow K/(K \cap (x_iHx_i^{-1})),$$

$$kx_iH \leftrightarrow k(K \cap (x_iHx_i^{-1})),$$

shows that $\hat{f}_i$ equals the sum as g runs over coset representatives of $K/K \cap (x_iHx_i^{-1})$ of the maps $g(f(g^{-1}\cdot -))$. However, this sum is the representative of the composition, ψ_g which completes the proof. $\square$

(2.44) Remark

I leave to the interested reader the task of proving the analogous double coset formula in homology.

(2.45) Corollary

If $H \lhd G$ has finite index and M is a $\mathbb{Z}[G]$-module, then

$$H^*(H; M) \xrightarrow{\mathrm{Tr}} H^*(G; M) \xrightarrow{i^*} H^*(H; M)$$

equals $\sum_g (g - g^{-1})^*$, where g runs through a set of representatives of G/H.

Proof. When $H = K \lhd G$, then $H = K \cap (x_i H x_i^{-1})$ and $\mathrm{Tr}_H^H = 1$ so that ψ_g becomes $(g - g^{-1})^*$. □

(2.46) I will close this section with the explicit description of the transfer on $H^1(G; M)$ when M is a trivial G-module. We use the notation of §2.34.

From the exact sequence associated to

$$0 \to I(G) \to \mathbb{Z}[G] \xrightarrow{\varepsilon} \mathbb{Z} \to 0,$$

we find [H-S, p. 193] that, in general,

(2.47) $H^1(G; M) \cong \mathrm{Hom}_G(I(G), M)/\{(g - 1) \mapsto gm - m; m \in M\}.$

When G acts trivially on M, (2.47) yields

(2.48) $H^1(G; M) \cong \mathrm{Hom}_G(I(G), M) \cong \mathrm{Hom}(G, M),$

where the second isomorphism sends $f : I(G) \to M$ to the homomorphism $\mu(f)(g) = f(g - 1)$.

For $g \in G$, we have a permutation $\pi \in \Sigma_m$ and $h(i, g) \in H (1 \le i \le m)$, satisfying

(2.49) $x_i g = h(i, g) x_{\pi(i)}.$

We will derive the following formula:

(2.50) Proposition

Let M be a trivial G-module. Then the following diagram commutes

$$
\begin{array}{ccc}
H^1(H; M) & \xrightarrow{\ i_* \ } & H^1(G; M) \\
\Big\downarrow{\scriptstyle \cong} & & \Big\downarrow{\scriptstyle \cong} \\
\mathrm{Hom}(H, M) & \xrightarrow{\ I \ } & \mathrm{Hom}(G, M),
\end{array}
$$

where $I(\alpha)(g) = \sum_{i=1}^m \alpha(h(i, g)).$

Proof. By (2.42), we must unravel the isomorphism between $H^1(H; M)$ and

(2.51) $H^1(G; \operatorname{Hom}_H(\mathbb{Z}[G], M)) \cong \dfrac{\operatorname{Hom}_G(I(G), \operatorname{Hom}_H(\mathbb{Z}[G], M))}{\{(g-1) \mapsto gz - z; z \in \operatorname{Hom}_H(I(G), M)\}}.$

On the other hand, the diagram of exact sequences

$$
\begin{array}{ccccccccc}
0 & \longrightarrow & I(H) & \longrightarrow & \mathbb{Z}[H] & \longrightarrow & \mathbb{Z} & \longrightarrow & 0 \\
 & & \downarrow & & \downarrow & & \downarrow{\scriptstyle 1} & & \\
0 & \longrightarrow & I(G) & \longrightarrow & \mathbb{Z}[G] & \longrightarrow & \mathbb{Z} & \longrightarrow & 0
\end{array}
$$

gives isomorphisms

(2.52)
$$
\begin{aligned}
H^1(H; M) &\cong \frac{\operatorname{Hom}_H(I(G), M)}{\operatorname{Hom}_H(\mathbb{Z}[G], M)} \\
&\cong \operatorname{Hom}_H(I(H), M),
\end{aligned}
$$

where the second isomorphism is induced by restriction to $I(H)$. The isomorphism connecting (2.51) and (2.52) is given by

(2.53)
$$
\begin{array}{ccc}
\operatorname{Hom}_G(I(G), \operatorname{Hom}_H(\mathbb{Z}[G], M)) & \xrightarrow[{\cong}]{\lambda} & \operatorname{Hom}_H(\mathbb{Z}[G] \otimes_G I(G), M) \\
 & & \downarrow{\scriptstyle \cong} \\
 & & \operatorname{Hom}_H(I(G), M)
\end{array}
$$

given by $\lambda(f)(a \otimes b) = f(b)(a)$.

Now suppose $f \in \operatorname{Hom}(H, M)$, and define $\hat{f} \in \operatorname{Hom}_H(I(G), M)$ by

(2.54) $\hat{f}(hx_i - 1) = f(h - 1)$ $(h \in H; 1 \le i \le m)$

so that $\hat{f}$ corresponds to f in (2.52). If $\lambda(F) = \hat{f}$ in (2.53), then

(2.55) $F(1 \otimes (hx_i - 1)) = f(h - 1),$

and $i_*(f)$ is represented, by (2.42), by the map $J: I(G) \to M$,

(2.56) $J(g - 1) = \displaystyle\sum_{i=1}^{m} F(x_i \otimes (g - 1)) = \sum_{i=1}^{m} \hat{f}(x_i g - x_i),$

since x_i^{-1} acts trivially on M. However,

$$\sum_{i=1}^{m} \hat{f}(x_i g - x_i) = \sum_{i=1}^{m} \hat{f}(x_i g - 1) - \sum_{i=1}^{m} \hat{f}(x_i - 1)$$

$$= \sum_{i=1}^{m} \hat{f}(x_i g - 1)$$

$$= \sum_{i=1}^{m} \hat{f}(h(i, g) - 1)$$

$$= \sum_{i=1}^{m} f(h(i, g)), \qquad \text{by (2.54).}$$

This establishes the formula of §2.50.   □

3. EXAMPLES OF COHOMOLOGY RINGS

In this section I will evaluate some rings, $H^*(G; \Lambda)$ for suitable finite groups, G, and rings, Λ.

(3.1) THE CYCLIC GROUPS

Let $G = \mathbb{Z}/n$, the cyclic group of order n, acting trivially on $\mathbb{Z}$, the integers. Hence we have a free resolution of $\mathbb{Z}[G]$-modules:

$$0 \leftarrow \mathbb{Z} \xleftarrow{\varepsilon} \mathbb{Z}[G] \xleftarrow{d} \mathbb{Z}[G] \xleftarrow{\Delta} \mathbb{Z}[G] \xleftarrow{d} z[G] \xleftarrow{\Delta} \cdots$$

given by $\varepsilon(g) = 1$ for $g \in G$, $d(z) = (x - 1)z$, where x generates G and $\Delta(z) = \sum_{i=1}^{n} x^i(z)$. At once one sees that $H^*(\mathbb{Z}/n; \mathbb{Z})$ is the cohomology of the complex

$$\mathbb{Z} \xrightarrow{0} \mathbb{Z} \xrightarrow{n} \mathbb{Z} \xrightarrow{0} \mathbb{Z} \xrightarrow{n} \mathbb{Z} \xrightarrow{0} \cdots$$

so that

(3.2) $H^i(\mathbb{Z}/n; \mathbb{Z}) \cong \begin{cases} \mathbb{Z} & \text{if } i = 0, \\ 0 & \text{if } 0 < i \text{ odd}, \\ \mathbb{Z}/n & \text{if } 0 < i \text{ even}. \end{cases}$

From the universal coefficient theorem and Künneth formula (2.25), and (2.32), we can evaluate $H^*(\mathbb{Z}/n: M)$ and $H_*(\mathbb{Z}/n; M)$ for any trivial module, M.

Suppose that the ith copy of $\mathbb{Z}[G]$, in the preceding resolution, has generator e_i ($i \geq 0$). Then the differential is given by

(3.3) $\partial(e_j) = \begin{cases} xe_{j-1} - e_{j-1}, & j \quad \text{odd}, \\ \sum_{i=1}^{n} x^i e_{j-1}, & 0 < j \text{ even}, \\ 0, & j = 0. \end{cases}$

If $\{B_i(\mathbb{Z}/n), d_i\}$ is the bar resolution (see §1.6), define, for $i \geq 0$, a $\mathbb{Z}/n$-module homomorphism $\Phi: P_i \to B_i(\mathbb{Z}/n)$ by

$$(3.4) \quad \phi(e_i) = \begin{cases} \sum_I (x^{i_1}|x|x^{i_2}|x|\cdots|x^{i_s}|x] & \text{if } i = 2s > 0, \quad I = (i_1, i_2, \ldots, i_s), \\ \sum_I [x|x^{i_1}|x|x^{i_2}|\cdots|x^{i_s}|x] & \text{if } i = 2s+1, \quad I = (i_1, i_2, \ldots, i_s), \\ [-] & \text{if } i = 0. \end{cases}$$

(3.5) Lemma

Φ is a chain map in (3.4).

Proof. We must show that $\Phi \partial = d\Phi$. However,

$$\Phi(\delta e_{2s}) = \sum_{i_1} x^{i_1} \Phi(e_{2s-1})$$

$$= \sum_{i_1, i_2, \ldots, i_s} x^{i_1}[x|x^{i_2}|x|x^{i_3}|\cdots],$$

whereas

$$d_{2s-1}\Phi(e_{2s}) = d_{2s-1}\left(\sum_{i_1, \ldots, i_s} [x^{i_1}|x|\cdots-]\right)$$

$$= \sum x^{i_1}[x|x^{i_2}|x|\cdots] - \sum [x^{i_1+1}|x^{i_2}|x|\cdots]$$

$$+ \sum [x^{i_1}|x^{i_2+1}|x|\cdots]\cdots$$

$$\vdots$$

$$= \sum x^{i_1}[x|x^{i_2}|x|\cdots]$$

$$= \Phi(\partial e_{2s}).$$

Similarly, $\Phi\partial(e_{2s+1}) = d_{2s}\Phi(e_{2s+1})$, which completes the proof. $\square$

(3.6) If $G = \mathbb{Z}/n$, we have a chain map

$$\phi_n: P_n \xrightarrow{\Phi} B_n G \xrightarrow{\Delta_n} \sum_{a=0}^{n} B_a G \otimes B_{n-a} G,$$

where Δ is defined in (2.8).

If $f \in \text{Hom}_G(B_2 G, \mathbb{Z})$ and $h \in \text{Hom}_G(B_n G, \mathbb{Z})$ are cocycles, then the product, $[f][h]$, is represented by $(f \times h) \cdot \phi_{n+2}$. If $n = 2s$, we obtain

$$[f][h](e_{2s+2}) = (f \otimes h)(\Delta_{2s+2})(\sum [x^{i_1}|x|\cdots|x^{i_{s+1}}|x])$$

$$= \sum f[x^{i_1}|x]h[x^{i_2}|x|\cdots|x^{i_{s+1}}|x],$$

$$= f(\Phi_2(e_2))h(\Phi_n(e_n)).$$

The generator of $H^2(\mathbb{Z}/n; \mathbb{Z})$ is given by $[f]$, where $f(e_2) = 1$ so that we have deduced the following result, since we have shown that $[f]^s$ generates $H^{2s}(\mathbb{Z}/n; \mathbb{Z})$:

(3.7) Theorem

As a graded ring $H^*(\mathbb{Z}/n; \mathbb{Z}) = \mathbb{Z}[f]/(nf)$, where $deg(f) = 2$.

(3.8) Let p be a prime; then, if $H \subset \mathbb{Z}/n$ is its p-Sylow subgroup, we have isomorphisms

$$H^*(\mathbb{Z}/n; \mathbb{Z}/p) \xrightarrow{\cong} H^*(H; \mathbb{Z}/p)$$

and

$$H_*(H; \mathbb{Z}/p) \xrightarrow{\cong} H_*(\mathbb{Z}/n; \mathbb{Z}/p).$$

Hence the mod p cohomology algebra of a cyclic group is determined by the following result:

(3.9) Theorem

Let p be a prime; then, for $\alpha \geq 1$,

$$H^*(\mathbb{Z}/p^\alpha; \mathbb{Z}/p) \cong \begin{cases} E(v) \otimes P[f] & \text{if } p \neq 2 \quad \text{or} \quad p = 2, \alpha \geq 2, \\ P[v] & \text{if } p = 2, \alpha = 1, \end{cases}$$

where $deg(v) = 1$, $deg(f) = 2$, $E(z)$ and $P[z]$ are, respectively, the $\mathbb{Z}/p$-exterior or polynomial algebra on one generator, z.

Proof. From the cohomology universal coefficient theorem—see (2.25) and (2.33)—the theorem is correct additively. In addition, since $H^{2n}(\mathbb{Z}/p^\alpha; \mathbb{Z}/p) \cong H^{2n}(\mathbb{Z}/p^\alpha; \mathbb{Z}) \otimes \mathbb{Z}/p$, the generator is the nth power of the two-dimensional generator, by Theorem 3.7. When $p \neq 2$, the anticommutativity, $x \cdot y = (-1)^{\deg(x)\deg(y)} y \cdot x$, implies $v^2 = -v^2 = 0$, and the result follows.

When $p = 2$, we must show that $v^2 \neq 0$ if $\alpha = 1$, and that $v^2 = 0$ if $\alpha \geq 2$. This follows from the explicit formula of §2.14. For v is represented by the canonical epimorphism $v: \mathbb{Z}/2^\alpha \to \mathbb{Z}/2$ so that $v^2[g_1|g_2] = v(g_1)v(g_2)$.

Composing this with $\Phi: P_2 \to B_2 \mathbb{Z}/2^\alpha$ of (3.4) gives a representation $\lambda = \Phi v^2$: $P_2 \to \mathbb{Z}/2$,

$$\lambda(e_2) \equiv \sum_{i=1}^{2^\alpha - 1} v(x^i)v(x) \equiv 2^{\alpha - 1} \quad (\text{mod } 2).$$

Hence $v^2 = f \neq 0$ if $p = 2$, $\alpha = 1$, and $v^2 = 0$ if $p = 2$, $\alpha \geq 2$, as required. □

(3.10)   THE GENERALIZED QUATERNION GROUPS

Let $Q_{4n} = \{x, y \mid x^n = y^2, y^4 = 1, xyx = y\}$ which is the *generalized quaternion group of order 4n*. From [C-E, p. 253] we have a periodic free $\mathbb{Z}[Q_{4n}]$-resolution

$$0 \leftarrow \mathbb{Z} \xleftarrow{\varepsilon} X_0 \xleftarrow{d_0} X_1 \xleftarrow{d_1} X_2 \xleftarrow{d_2} \cdots,$$

in which

(i) $X_{4j} = \mathbb{Z}[Q_{4n}]\langle a_j \rangle$, $d(a_j) = \displaystyle\sum_{i=1}^{2n} (x^i e_{j-1} + x^i y e_{j-1})$,

(ii) $X_{4j+1} = \mathbb{Z}[Q_{4n}]\langle b_j, b_j' \rangle$, $d(b_j) = (x-1)a_j$,
$$d(b_j') = (y-1)a_j,$$

(iii) $X_{4j+2} = \mathbb{Z}[Q_{4n}]\langle c_j, c_j' \rangle$, $d(c_j) = \displaystyle\sum_{i=0}^{n-1} x^i b_j - (y+1)b_j'$,
$$d(c_j') = (xy+1)b_j + (x-1)b_j',$$

(iv) $X_{4j+3} = \mathbb{Z}[Q_{4n}]\langle e_j \rangle$, $d(e_j) = (x-1)c_j - (xy-1)c_j'$.

Hence, if $\wedge$ is a trivial Q_{4n}-module, then $H^*(Q_{4n}; \wedge)$ is the cohomology of the (periodic) cochain complex

$$\wedge \xrightarrow{0} \wedge \oplus \wedge \xrightarrow{d} \wedge \oplus \wedge \xrightarrow{0} \wedge \xrightarrow{4n} \wedge \xrightarrow{0} \cdots,$$

where $d : \mathrm{Hom}_{Q_{4n}}(X_{4j+1}, \wedge) \to \mathrm{Hom}_{Q_{4n}}(X_{4j+2}, \wedge)$ is composition with d_{4j+1}. Therefore

$$d(1, 0) = (n, 2)$$

$$d(0, 1) = (-2, 0).$$

Hence if n is even,

$$\ker d = \{(a, b) \in \wedge \oplus \wedge \mid 2a = 2b = 0\},$$

but if n is odd,

$$\ker d = \{(a, b) \in \wedge \oplus \wedge \mid 2a = 0, 2b = a\}$$

$$\cong \{b \in \wedge \mid 4b = 0\}.$$

If n is even,

$$\mathrm{coker}\, d \cong (\wedge \oplus \wedge) \otimes \mathbb{Z}/2,$$

but if n is odd,

$$\operatorname{coker} d \cong \wedge \otimes \mathbb{Z}/4.$$

We recapitulate this result as follows:

(3.11) Theorem

If Q_{4n} acts trivially on $\wedge$, then

$$H^s(Q_{4n}; \wedge) \cong \begin{cases} \wedge & \text{if } s = 0, \\ \wedge \otimes \mathbb{Z}/4n & \text{if } s = 4j > 0, \\ \operatorname{Tor}(\mathbb{Z}/4n, \wedge) & \text{if } s = 4j + 3, \\ \operatorname{Tor}(\mathbb{Z}/2, \wedge \oplus \wedge) & \text{if } s = 4j + 1, n \text{ even}, \\ \operatorname{Tor}(\mathbb{Z}/4, \wedge) & \text{if } s = 4j + 1, n \text{ odd}, \\ (\wedge \oplus \wedge) \otimes \mathbb{Z}/2 & \text{if } s = 4j + 2, n \text{ even}, \\ \wedge \otimes \mathbb{Z}/4 & \text{if } s = 4j + 2, n \text{ odd}. \end{cases}$$

(3.12)   THE DIHEDRAL GROUPS

The *dihedral group of order* $2n$, D_{2n}, is given by

$$D_{2n} = \{x, y \,|\, x^n = 1 = y^2, xyx = y\}.$$

In the next section we will compute the algebra, $H^*(D_{2n}; \mathbb{Z}/2)$, of course, where D_{2n} acts trivially on $\mathbb{Z}/2$. However, for the present we will content ourselves with a brief consideration of $H^*(D_8; M)$. For this we will rewrite D_8 as

$$D_8 = \{s_1, s_2, t \,|\, s_1^2 = s_2^2 = t^2 = 1, ts_1 t = s_2, s_1 s_2 = s_2 s_1\}.$$

Let $P_*^1 = \{0 \leftarrow \mathbb{Z} \leftarrow P_0^1 \leftarrow \cdots\}$ denote the $\mathbb{Z}[\langle s_1 \rangle]$-resolution of the trivial module, $\mathbb{Z}$, given in §3.1. Similarly, let P_*^2 and P_* be the corresponding resolutions when we replace s_1 by s_2 and t, respectively. Hence, for example, $P_* = \{\mathbb{Z} \leftarrow P_0 \leftarrow P_1 \leftarrow \cdots\}$, where P_i is the free $\mathbb{Z}[\mathbb{Z}/2]$ $(= \mathbb{Z}[\langle t \rangle])$ module on one generator, e_i.

Form the tensor product chain complex,

$$(P_* \otimes P_*^1 \otimes P_*^2, d \otimes 1 \otimes 1 \pm 1 \otimes d \otimes 1 \pm 1 \otimes 1 \otimes d)$$

(3.13) $$0 \leftarrow \mathbb{Z} = \mathbb{Z} \otimes \mathbb{Z} \otimes \mathbb{Z} \leftarrow P_0 \otimes P_0^1 \otimes P_0^2 \leftarrow \cdots,$$

as described in §2.4 and (2.5). By the Künneth formula [H-S, p. 166], (3.13) is an exact complex. Furthermore the nth group, $\oplus_{a+b+c=n} P_a \otimes P_b^1 \otimes P_c^2$ is a free

$\mathbb{Z}[D_8]$-module if we endow it with the action

$$t(e_a \otimes e_b^1 \otimes e_c^2) = (-1)^{bc}(te_a) \otimes e_c^1 \otimes e_b^2,$$

$$s_1(e_a \otimes e_b^1 \otimes e_c^2) = e_a \otimes (s_1 e_b^1) \otimes e_c^2,$$

$$s_2(e_a \otimes e_b^1 \otimes e_c^2) = e_a \otimes e_b^1 \otimes (s_2 e_c^2).$$

This D_8-action commutes with the differentials and turns (3.13) into a free $\mathbb{Z}[D_8]$-resolution of $\mathbb{Z}$.

Applying $\mathrm{Hom}_{D_8}(-, M)$ to the chain complex (3.13) (with the $\mathrm{Hom}_{D_8}(\mathbb{Z}, M)$ group omitted) gives the chain complex from which to compute $H^*(D_8; M)$.

For example, let $M = \mathbb{Z}/2$ with trivial action, of course. Set $E_n = \bigoplus_{a+b+c=n} P_a \otimes P_b^1 \otimes P_c^2$, and let $f_{a,b,c} \in \mathrm{Hom}_{D_8}(E_n, \mathbb{Z}/2)$ be given by $(a+b+c=n)$

$$f_{a,b,c}(e_a, \otimes e_b, \otimes e_{c'}) = (\delta_{a,a'})(\delta_{b,b'})(\delta_{c,c'}).$$

Therefore $H^n(D_8; \mathbb{Z}/2)$ is generated by the cohomology classes of $\{[f_{a,b,b}]; a+2b=n\}$ and of $\{[f_{0,b,c} + f_{0,c,b}]; b+c=n, b<c\}$. Hence, if

$$\varepsilon_n = \#\{(a,b) | a+2b=n, a,b \geq 0\}$$

and

$$\varepsilon_n' = \#\{(x,y) | x+2y=n, y \geq 0, x \geq 1\},$$

we have shown the following result (as is seen by setting $b=y, c=x+y$):

(3.14) Proposition

$$\dim_{\mathbb{F}_2}(H^n(D_8; \mathbb{Z}/2)) = \varepsilon_n + \varepsilon_n'.$$

(3.15) Remark

If Q is the graded $\mathbb{Z}/2$-algebra $\mathbb{Z}/2[x, y, w]/(x^2 + xy)$ $(deg\ x = deg\ y = 1, deg\ w = 2)$, then $\dim_{\mathbb{F}_2} Q_n = \varepsilon_n + \varepsilon_n'$ also.

(3.16) STIEFEL-WHITNEY CLASSES

If G is a discrete group, then the topological manner in which to construct $H^*(G; M)$ (for convenience, let us assume G acts trivially on M) is the following: Choose any contractible space, EG, on which G acts freely. Form the orbit space $BG = (EG)/G$, and then

(3.17) $$H^*(G; M) \cong H^*(BG; M),$$

where the latter is singular cohomology of the space, BG [Spa]. The isomorphism arises from the fact that any two models for EG define the same right-hand side of (3.17), but one model exists whose singular chain complex $(C_*(EG), d)$ is isomorphic, as G-chain complexes, to the bar resolution of G.

On the other hand, if G is any topological group, we may define the right-hand side of (3.17). For example, let $O(\mathbb{R}) = \bigcup_n O_n(\mathbb{R})$ be the infinite orthogonal group obtained by including

$$O_n(\mathbb{R}) = \{X \in GL_n\mathbb{R} \mid XX^t = I_n\}$$

into $O_{n+1}(\mathbb{R})$, by sending X to

$$\begin{bmatrix} X & 0 \\ 0 & 1 \end{bmatrix} \in O_{n+1}(\mathbb{R}),$$

and taking the union over all $n \geq 1$.

Write $O(\mathbb{R})$ for the orthogonal group with its classical topology [H] and $O_\delta(\mathbb{R})$ for the orthogonal group as a discrete group. The homomorphism of topological groups, given by the identity, induces

(3.18) $$H^*(BO(\mathbb{R}); \mathbb{Z}/n) \to H^*(O_\delta(\mathbb{R}); \mathbb{Z}/n).$$

The following is an easy corollary of results of A. A. Suslin:

(3.19) Theorem ([Su], see also [Ka])

For any $1 \leq n$, (3.18) is an isomorphism.

When $n = 2$ the left hand of (3.18) is a polynomial ring on the Stiefel-Whitney classes, $w_i \in H^i(BO(\mathbb{R}); \mathbb{Z}/2)$ [H] so that

(3.20) $$H^*(O_\delta(\mathbb{R}); \mathbb{Z}/2) \cong \mathbb{Z}/2[w_1, w_2, \ldots].$$

In addition the w_i are characterized by their restriction to the diagonal matrices $\{\pm 1\}^n = O_1(\mathbb{R})^n \subset O_n(\mathbb{R})$. If $H^*(O_1(\mathbb{R}); \mathbb{Z}/2) = \mathbb{Z}/2[u]$, then $H^*(O_1(\mathbb{R})^n$; $\mathbb{Z}/2) \cong \mathbb{Z}/2[u_1, u_2, \ldots, u_n]$, where $\deg u_i = 1$ and u_i comes from the ith $O_1(\mathbb{R})$-factor. The w_i are characterized by

(3.21) $$w_i = \sigma_i(u_1, u_2, \ldots, u_n) \in H^*(O_1(\mathbb{R})^n; \mathbb{Z}/2)$$

for all $i \leq n$, where σ_i is the ith elementary symmetric function.

(3.22) Definition

If $p: G \to O_n(\mathbb{R})$ is a homomorphism, then we define the ith Stiefel-Whitney class of p,

$$SW_i(p) \in H^i(G; \mathbb{Z}/2)$$

to be $p^*(w_i)$, where w_i is as in (3.21).

It is not necessary to use Suslin's result (§3.19) to define the classes, $SW_i(p)$. However, the alternative involves going into details about classifying spaces—a digression which I would prefer to minimize for the purposes of these notes.

4. SPECTRAL SEQUENCE CALCULATIONS OF SOME COHOMOLOGY RINGS

(4.1)   Suppose that we have an extension of groups

$$(4.2) \qquad N \to G \to B.$$

The most useful method to calculate the algebra $H^*(G;R)$, where R is a commutative ring on which G acts trivially, is via the Hochschild-Serre spectral sequence

$$(4.3) \qquad E_2^{s,t} = H^s(B;H^t(N;R)) \Rightarrow H^{s+t}(G;R).$$

I am not intending to derive the spectral sequences which we will use in this section. For derivations, see [H-S. Ch. 8] and [Mac].

Here $E_2^{s,t}$ is the cohomology of B acting via conjugation (on the left) on N and thereby on $H^t(N;R)$. The spectral sequence consists of a series of differential algebras

$$\{E_r^{s,t}, d_r; r \geq 2\}$$

such that

(i) $d_r d_r = 0$.

(ii) $E_{r+1}^{s,t} = \dfrac{\ker(d_r: E^{s,t} \to E^{s+r,t-r+1})}{\operatorname{im}(d_r: E_r^{s-r,t+r-1} \to E_r^{s,t})}.$

(iii) If $a \in E_r^{s,t}, b \in E_r^{s',t'}$, then $d_r(ab) = d_r(a)b + (-1)^{s+t}a d_r(b)$.

(iv) If $E_\infty^{s,t}$ denotes $E_N^{s,t}$ for $N \gg 0$, there is a filtration of $H^n(G;R)$,

$$H^n(G;R) = F^0 H^n \supset \cdots \supset F^s H^n \supset F^{s+1} H^n \supset \cdots \supset F^n H^n \supset 0$$

such that

$$E_\infty^{s,t} \cong \frac{F_s H^{s+t}(G;R)}{F_{s+1} H^{s+t}(G;R)}.$$

(v) The spectral sequence is natural with respect to homomorphisms of group extensions.

Let D_{2n} denote the dihedral group of order $2n$, as in §3.12. We have an extension

$$\mathbb{Z}/n \xrightarrow{i} D_{2n} \xrightarrow{\pi} \mathbb{Z}/2$$

in which $\mathbb{Z}/n = \langle x \rangle$. We will use the spectral sequence to calculate $H^*(D_{2n}; \mathbb{Z}/2)$. First, suppose that $n = 2^j(2v + 1)$, then we have $\mathbb{Z}/2^j = \langle x^{2v+1} \rangle \lhd D_{2n}$ and a map of extensions

$$
\begin{array}{ccccc}
\mathbb{Z}/2^j & \longrightarrow & D_{2^{j+1}} & \longrightarrow & \mathbb{Z}/2 \\
\downarrow{\scriptstyle\Phi'} & & \downarrow{\scriptstyle\Phi} & & \downarrow{\scriptstyle=} \\
\mathbb{Z}/n & \longrightarrow & D_{2n} & \longrightarrow & \mathbb{Z}/2
\end{array}
$$

In addition we have two homomorphisms $x_i \colon D_{2^{j+1}} \to \mathbb{Z}/2$ given by

(4.4)
$$x_i((x^{2v+1})^\varepsilon y^\delta) = \begin{cases} \delta \,(\mathrm{mod}\,2) & \text{if } i = 1, \\ \varepsilon \,(\mathrm{mod}\,2) & \text{if } i = 2. \end{cases}$$

These homomorphisms define $x_1, x_2 \in H^1(D_{2^{j+1}}; \mathbb{Z}/2)$ by (2.48). In addition the inclusion of $\mathbb{Z}/n$ into the circle, $S^1 = SO(2) \subset O_2(\mathbb{R})$ and setting $y = \begin{bmatrix} 0 & 1 \\ 1 & 0 \end{bmatrix}$ defines a canonical inclusion $k \colon D_{2n} \subset O_2(\mathbb{R})$ so that we may define

(4.5)
$$w = k^*(w_i) = SW_2(k) \in H^2(D_{2n}; \mathbb{Z}/2),$$

the second Stiefel-Whitney class of k (see §3.22).

We will prove the following:

(4.6) Theorem

(i) If $n = (2v + 1)2^j$, then $\Phi^* \colon H^*(D_{2n}; \mathbb{Z}/2) \to H^*(D_{2^{j+1}}; \mathbb{Z}/2)$ is an isomorphism.

(ii) As an algebra, if $n \geq 2$,

$$H^*(D_{2n}; \mathbb{Z}/2) \cong \mathbb{Z}/2[x_1, x_2, w]/(x_2^2 + x_1 x_2).$$

(iii) If $n = 1$, $D_4 = \mathbb{Z}/2 \times \mathbb{Z}/2$ and $H^*(D_4; \mathbb{Z}/2) \cong \mathbb{Z}/2[x_1, x_2]$.

Proof. Part (iii) follows from the Künneth formula and the calculations of §3.9.

Also from §3.9 we see that the map $\Phi \colon D_{2^{j+1}} \to D_{2n}$ induces a map of spectral sequences, by §4.1(v) which is an isomorphism of $E_2^{s,t}$-terms. By induction Φ^* is an isomorphism on $E_r^{s,t}$-terms, using §4.1(ii), and so on each quotient $F^s H^n / F^{s+1} H^n$. Therefore, by downward induction s, part (i) follows.

Now consider part (ii) when $n = 2^j$ and $j \geq 2$. From §3.9, we know that $H^*(\mathbb{Z}/2^j; \mathbb{Z}/2) \cong E(\alpha) \otimes P(\beta)$, where $\deg \alpha = 1$, $\deg \beta = 2$. Also α is represented by the canonical homomorphism onto $\mathbb{Z}/2$ so that the following diagram commutes:

(4.7)

$$\begin{array}{ccc} \mathbb{Z}/2^j & \longrightarrow & D_{2^{j+1}} \\ & \alpha \searrow & \downarrow x_2 \\ & \mathbb{Z}/2 & \end{array}$$

Also β is the restriction of the second Stiefel-Whitney class via the homomorphism $\mathbb{Z}/2^j \xrightarrow{i} D_{2^{j+1}} \xrightarrow{k} O_2(\mathbb{R})$. This can be seen by observing that the element of order two goes under ki to the matrix

$$\tau = \begin{bmatrix} -1 & 0 \\ 0 & -1 \end{bmatrix}.$$

Since w_2 restricts to $u_1 u_2 \in H^2((\pm 1)^2; \mathbb{Z}/2)$, by (3.21) it restricts to

$$0 \neq u^2 = \Delta^*(u_1 u_2) \in H^2(\langle \tau \rangle; \mathbb{Z}/2).$$

An inner automorphism acts like the identity on cohomology so, since $i^*: H^*(D_{2^{j+1}}; \mathbb{Z}/2) \to H^*(\mathbb{Z}/2^j; \mathbb{Z}/2)$ is into, we conclude that

(4.8) $$E_2^{*,*} = H^*(\mathbb{Z}/2; \mathbb{Z}/2) \otimes H^*(\mathbb{Z}/2^j; \mathbb{Z}/2)$$

as an algebra. The first factor in (4.8) is $\mathbb{Z}/2[\gamma]$, where $\gamma \in E_2^{1,0}$. This class is represented by x_1 in $E_\infty^{1,0} = F^1 H^1 / F^2 H^1$ by standard properties of $E_2^{s,0}$ in the spectral sequence. This means that $d_r(\gamma) = 0$ for all $r \geq 2$.

Also, since $\alpha, \beta \in \text{im}(i^*)$, standard properties of the spectral sequence imply that $d_r(\alpha) = 0 = d_r(\beta)$ for all $r \geq 2$ and that

$$\alpha \in E_\infty^{0,1} = F^0 H^1 / F^1 H^1$$

represents x_2, whereas

$$\beta \in E_\infty^{0,2} = F^0 H^2 / F^1 H^2$$

represents w.

Hence we find that

(4.9) $$E_2^{*,*} = E_\infty^{*,*} \cong \mathbb{Z}/2[\alpha, \gamma, w]/(\alpha^2).$$

From (4.9) and the preceding remarks about representatives, we see that $H^*(D_{2^{j+1}}; \mathbb{Z}/2)$ has a basis

$$\{x_1^a x_2^\varepsilon w^b; a, b \geq 0, \varepsilon = 0 \text{ or } 1\}.$$

To determine the algebra structure, we must find the coefficients in the relation

(4.10) $$x_2^2 = z_1 x_1^2 + z_2 x_1 x_2 + z_3 w \qquad (z_i \in \mathbb{Z}/2).$$ □

Theorem (4.6) will be completed by the next result.

(4.11) Proposition

In (4.10), $z_3 \equiv z_1 \equiv 0 \pmod 2$, and $z_2 \equiv 1 \pmod 2$.

Proof. We observe at once that restriction of (4.10) to $\mathbb{Z}/2^j$ yields $0 = z_3 \beta \in H^2(\mathbb{Z}/2^j; \mathbb{Z}/2)$ so that $z_3 \equiv 0 \pmod 2$.

Secondly, restriction of x_2 to $D_4 = \mathbb{Z}/2 \times \mathbb{Z}/2$ is trivial since it factors through $\mathbb{Z}/2^j$ and $j \geq 2$. Hence, in $H^2(\mathbb{Z}/2 \times \mathbb{Z}/2; \mathbb{Z}/2)$, (4.10) becomes $0 = x_2^2 = z_1 x_1^2$, which is only zero if $z_1 = 0$.

Hence we are required to prove that $x_2^2 \neq 0$ to finish the proof.

Let Φ denote the automorphism of D_8 (corresponding to rotation in $O_2(\mathbb{R})$) given by

(4.12) $$\Phi(x) = x, \quad \Phi(y) = xy.$$

We check that $\Phi(y)^2 = xyxy = y^2 = 1$ and $\Phi(xyx) = \Phi(y) = xy = x(xy)x = \Phi(x)\Phi(y)\Phi(x)$. Also

$$\Phi^*(x_i)(x^\varepsilon y^\delta) = x_i(x^\varepsilon(xy^\delta))$$

so that $\Phi^*(x_1) = x_1, \Phi^*(x_2) = x_1 + x_2$. Applying this to what remains of (4.10) yields $z_2 x_1(x_1 + x_2) = x_1^2 + x_2^2$, which shows that $z_2 \equiv 1 \pmod 2$. The relation $x_2^2 = x_1 x_2$ in D_8 implies it, by restriction, for all D_{2^j} for $j \geq 2$.   □

(4.13) Remark

The transfer, $i_*: H^*(\mathbb{Z}/2^j; \mathbb{Z}/2) \to H^*(D_{2^{j+1}}; \mathbb{Z}/2)$, is an $H^*(D_{2^{j+1}}; \mathbb{Z}/2)$-module homomorphism. For example, if $0 \neq \alpha \in H^1(\mathbb{Z}/2^{j+1}; \mathbb{Z}/2)$ and $x_j \in H^1(D_{2^{j+1}}; \mathbb{Z}/2)$, then

$$i_*(\alpha i^*(x_j)) = i_*(\alpha)x_j.$$

When $j = 1$, $i^*(x_1) = 0$, so $i_*(\alpha)x_1 = 0$. When $j = 2$, $i^*(x_2) = \alpha$, and $\alpha^2 = 0$ if $j \geq 2$ so that $i_*(\alpha)x_2 = 0$ also. Since $i_*(\alpha) = ax_1 + bx_2$, we must have $0 = ax_1^2 + bx_1 x_2$ so that $i_*(\alpha) = 0$ for $j \geq 1$.

Exercise Verify this by use of §2.50.

(4.14) Let us consider more closely the example of $D_8 = \{x, y | x^4 = y^2 = 1, xyx = y\}$. As in §3.12, we may write $D_8 = \{s_1, s_2, t | s_1^2 = s_2^2 = t^2 = 1, ts_1 t = s_2, s_1 s_2 = s_2 s_1\}$, where $x = s_1 t$, $y = t$. Since $\langle s_1, s_2 \rangle \cong \mathbb{Z}/2 \times \mathbb{Z}/2 \lhd D_8$, we have a homomorphism

(4.15) $$d: D_8 \to \mathbb{Z}/2,$$

which is trivial on $\langle s_1, s_2 \rangle$. In terms of (4.14) we see easily that

(4.16) $$d = x_1 + x_2 \in H^1(D_8; \mathbb{Z}/2).$$

Now define $z \in H^1(\langle s_1, s_2 \rangle; \mathbb{Z}/2)$ by $z(s_1^a s_2^b) \equiv a \pmod{2}$.
From §2.50, one finds that the transfer $i_*(z) \in H^1(D_8; \mathbb{Z}/2)$ is given by

(4.17) $$i_*(z)(g) = \begin{cases} z(g) + z(tgt) & \text{if } g \in \langle s_1, s_2 \rangle, \\ z(gt) + z(tg) & \text{otherwise.} \end{cases}$$

Hence

$$i_*(z)(x) = i_*(z)(s_1 t) = 1$$

$$i_*(z)(y) = z(1) + z(t) = 0,$$

$$i_*(z)(xy) = z(s_1) + z(s_2) = 1$$

so that

(4.18) $$i_*(z) = x_2.$$

However, i_* is an $H^*(D_*; \mathbb{Z}/2)$-module homomorphism so that

(4.19) $$x_2(x_1 + x_2) = i_*(zi^*(d)) = 0$$

since $i^*(d) = 0$. This gives a second proof of Proposition 4.11 in this case.

(4.20)   We will now calculate the integral cohomology ring, $H^*(D_8; \mathbb{Z})$. To do this, we will use the (mod 2) Bockstein spectral sequence (there exists a similar spectral sequence at each odd prime).

Let G be a finite group. The spectral sequence

(4.21) $$E_1^s = H^s(G; \mathbb{Z}/2) \Rightarrow (\mathbb{Z}/2)$$

is a sequence of differential graded algebras $\{E_r^s, d_r : E_r^s \to E_r^{s+1}; r \geq 1\}$ such that

(i) $d_r d_r = 0$.

(ii) $E_{r+1}^s = \dfrac{(\ker d_r : E_r^s \to E_r^{s+1})}{(\operatorname{im} d_r : E_r^{s-1} \to E_r^s)}.$

(iii) If $a \in E_r^s$, $b \in E_r^a$, then $d_r(ab) = d_r(a)b + (-1)^s a d_r(b)$.

(iv) $O = E_\infty^s = E_r^s$ for $r \gg 0$, for all $s > 0$, and $E_\infty^0 \cong \mathbb{Z}/2$.

(v) $d_1 = Sq^1$, the Bockstein associated to the coefficient sequence $\mathbb{Z}/2 \to \mathbb{Z}/4 \to \mathbb{Z}/2$ [S-E].

(vi) An image under d_r is represented (by a coset of an element in E_1^s) by the image an element of order 2^r in $H^s(G; \mathbb{Z})$ under $H^s(G; \mathbb{Z}) \to H^s(G; \mathbb{Z}/2)$.

(4.22) Theorem

In the Bockstein spectral sequence

$$E_1^* = H^*(D_8; \mathbb{Z}/2) = \mathbb{Z}/2[x_1, x_2, w]/(x_2^2 + x_1 x_2) \Rightarrow (\mathbb{Z}/2).$$

(i) $d_1(x_2^\varepsilon x_1^a w^b) = (a + b + \varepsilon)x_2^\varepsilon x_1^{a+1} w^b$ ($\varepsilon = 0$ or $1; 0 \le a, b$).

(ii) $E_2^* \cong \mathbb{Z}/2[\xi, \lambda]/(\lambda^2)$, where $\xi = [w^2]$ and $\lambda = [x_2 w]$.

(iii) $d_2(\lambda) = \xi$ and $E_3^* = 0$ for $* > 0$.

Proof. From [S-E] we know that $d_1(x_i) = Sq^1(x_i) = x_i^2$ for $i = 1, 2$. In addition $x_1 = SW_1(k)$, where (see §4.2) $k: D_8 \to O_2(\mathbb{R})$ is the canonical inclusion, since SW_1 is just the determinant. However, by the Wu formula, $Sq^1(SW_2) = SW_3 + (SW_1)(SW_2)$ [H; M-St], we have

$$d_1(w) = Sq^1(SW_2(k))$$
$$= SW_3(k) + SW_1(k)SW_2(k)$$
$$= x_1 w,$$

since SW_3 is zero on a two-dimensional representation.

Parts (i) and (ii) follow from the preceding computation, together with properties (iii) and (ii) of the spectral sequence.

We will conclude the proof by showing that $\xi = [w^2] \in E_2^4$ represents a boundary under some d_r. This will mean that $d_r(\lambda) = \xi$ since ξ must be a d_r-boundary for some r, by property (vi). Finally, we show that ξ is represented by the image of an integral class of order four which, by property (vi), means $d_2(\lambda) = \xi$:

$$H^*(BU(n); \mathbb{Z}) \cong \mathbb{Z}[c_1, c_2, \ldots, c_n],$$

where $U(n)$ is the unitary group of complex $n \times n$ matrices, with the classical topology. By analogy with Stiefel-Whitney classes, a complex representation $l: D_8 \to U(2)$ has Chern classes $c_1(l) = l^*(c_1)$, $c_2(l) = l^*(c_2)$. Setting $l = \mathrm{Ind}_{\langle x \rangle}^{D_8}(y)$, where $y(x) = \sqrt{(-1)}$, we have $c_2(l) \in H^4(D_8; \mathbb{Z})$. However, l is the complexification of the orthogonal representation, k. That is, $l \cong k \otimes_\mathbb{R} \mathbb{C}$, which implies that the image of $c_2(l)$ in $H^4(D_8; \mathbb{Z}/2)$ satisfies [H; M-St]:

(4.23) $c_2(l) = SW_1(k)^4 + SW_2(k)^2 = x_1^4 + w^2 \in H^4(D_8; \mathbb{Z}/2).$

Since $[w^2] = [w^2 + x_1^4] \in E_2^4$ we have shown that $d_r(\lambda) = \xi$ for some $r \geq 2$. Also, since $H^*((1); \mathbb{Z}) = 0$ for $* > 0$, $8c_2(l) = [D_8 : (1)]c_2(l) = 0$ by §2.39. Hence either $d_2(\lambda) = \xi$ or $d_3(\lambda) = \xi$.

If $d_3(\lambda) = \xi$, we would have (since $d_1(x_1^3) = x_1^4$, $d_1(x_1^2 x_2) = x_1^3 x_2$) $H^4(D_8; \mathbb{Z}) = \mathbb{Z}/8 \oplus \mathbb{Z}/2 \oplus \mathbb{Z}/2$, by property (vi) again. However, if we look at the Hochschild-Serre spectral sequence (cf., §4.2)

$$E_2^{s,t} = H^s(\mathbb{Z}/2; H^t(\mathbb{Z}/4; \mathbb{Z})) \Rightarrow H^{s+t}(D_8; \mathbb{Z}),$$

one finds that $t \in \mathbb{Z}/2$ acts as (-1) on $H^2(\mathbb{Z}/4; \mathbb{Z})$ and $(-1)^2 = 1$ on $H^4(\mathbb{Z}/4; \mathbb{Z})$ so that

(4.24)
$$\begin{cases} E_2^{0,4} \cong \mathbb{Z}/4, \quad E_2^{2,2} \cong \mathbb{Z}/2, \quad E_2^{4,0} \, \mathbb{Z}/2, \\ E_2^{1,3} = 0 = E_2^{3,1}. \end{cases}$$

This means $\#(H^4(D_8; \mathbb{Z})) \leq 16$, so it is impossible that $H^4(D_8; \mathbb{Z}) \cong \mathbb{Z}/8 \oplus \mathbb{Z}/2 \oplus \mathbb{Z}/2$. Therefore this group must be $\mathbb{Z}/4 \oplus \mathbb{Z}/2 \oplus \mathbb{Z}/2$ and so $d_2(\lambda) = \xi$, which completes the proof. □

(4.25) In $H^*(D_8; \mathbb{Z})$, by Theorem 4.22 and property (vi), there are unique classes—*each of order two*—$\alpha, \beta \in H^2(D_8; \mathbb{Z})$ and $v \in H^3(D_8; \mathbb{Z})$, reducing mod 2 to $x_2^2 = x_1 x_2$, x_1^2, and $x_1 w$, respectively. Let $\xi = c_2(\lambda) \in H^4(D_8; \mathbb{Z})$ be the element of order four given by the second Chern class introduced in the proof of Theorem 4.22.

From §4.22 (see end of proof),

$$H^4(D_8; \mathbb{Z}) \cong \mathbb{Z}/4\langle \xi \rangle \oplus \mathbb{Z}/2\langle \beta^2 \rangle \oplus \mathbb{Z}/2\langle \alpha^2 \rangle.$$

Also $\alpha\beta$ and α^2 map to $x_1^3 x_2 \in H^4(D_8; \mathbb{Z}/2)$:

(4.26)
$$\alpha\beta + \alpha^2 + 2v\xi = 0 \in H^4(D_8; \mathbb{Z})$$

(some $v \in \mathbb{Z}$) as we see from the long exact sequence

$$\cdots \longrightarrow H^4(D_8; \mathbb{Z}) \xrightarrow{2} H^4(D_8; \mathbb{Z}) \longrightarrow H^4(D_8; \mathbb{Z}/2) \longrightarrow \cdots$$

We will now show that $2v\xi = 0$ in (4.26). Let x_i be the complexification of $x_i : D_8 \to O_1(\mathbb{R})$. Hence

$$x_i^2 = SW_1(x_i)^2 = c_1(x_i) \in H^2(D_8; \mathbb{Z}/2).$$

This means $\alpha = c_1(x_1)$ and $\beta = c_1(x_2)$. Therefore

$$\alpha + \beta = c_1(x_1 \otimes x_2) = c_1(\det),$$

which means that $\alpha + \beta$ goes to zero in $H^2(\langle x \rangle; \mathbb{Z}) \cong \mathbb{Z}/4$, since $\langle x \rangle = \ker(\det: D_8 \to \mathbb{Z}/2)$. However, l restricts to $y + y^3$ on $\mathbb{Z}/4$ (y is the canonical one-dimensional representation) so that $c_2(l) = \xi$ restricts to $c_1(y)c_1(y^3) = -c_1(y)^2$, which generates $H^4(\langle x \rangle; \mathbb{Z}) \cong \mathbb{Z}/4$. Thus, as $2v\xi$ goes to zero, we must have $2v\xi = 0$. That is,

$$(4.27) \qquad\qquad\qquad \alpha\beta = \alpha^2 \in H^4(D_8; \mathbb{Z}).$$

From §4.22, $H^6(D_8; \mathbb{Z})$ is all two-torsion and so injects into $H^6(D_8; \mathbb{Z}/2)$. Since $\beta\xi$ and v^2 both go to $x_1^2 w^2$, we obtain

$$(4.28) \qquad\qquad\qquad v^2 = \beta\xi \in H^6(D_8; \mathbb{Z}).$$

Examination of the d_1- and d_2-boundaries in §4.22 shows that $H^*(D_8; \mathbb{Z})$ is generated by monomials

$$\{\alpha^\varepsilon v^{\varepsilon'} \beta^a \xi^b \,|\, \varepsilon, \varepsilon' = 0 \quad \text{or} \quad 1; a, b \geq 0\}.$$

Therefore we have proved the following result:

(4.29) Theorem

With the notation of §4.25,

$$H^*(D_8; \mathbb{Z}) \cong \mathbb{Z}[\alpha, \beta, v, \xi]/(2\alpha, 2\beta, 2v, 4\xi, \alpha^2 + \alpha, v^2 + \beta\xi).$$

(4.30)   We will conclude this section by computing the mod 2 cohomology ring of the generalized quaternion groups of (3.10):

$$Q_{4n} = \{x, y \,|\, x^n = y^2, \ y^4 = 1, \ xyx = y\}.$$

Let D_{2n} denote the dihedral group of (3.12),

$$D_{2n} = \{x, y \,|\, x^n = y^2 = 1, \ xyx = y\},$$

then we have a central extension, given by sending x and y to their namesakes

$$\mathbb{Z}/2 \rightarrowtail Q_{4n} \twoheadrightarrow D_{2n}.$$

When $n = 1$, then Q_4 is cyclic of order four, and from Theorem 3.9, we have

$$H^*(Q_4; \mathbb{Z}/2) \cong E(v) \otimes P[f].$$

Now suppose that $n = 2^\alpha(2s + 1)$, then the Sylow 2-subgroup of Q_{4n} is $Q_{2^{\alpha+2}}$. In a similar manner to §4.6(i), we may show that the inclusion induces an isomorphism

$$H^*(Q_{4n}; \mathbb{Z}/2) \xrightarrow{\cong} H^*(Q_{2^{\alpha+2}}; \mathbb{Z}/2).$$

Henceforth let us assume that $n = 2^\alpha(2s + 1)$ with $\alpha \geq 1$.

It will be useful to know, in advance, the dimensions of these cohomology groups. From the Künneth formula combined with Theorem 3.11, we know that

$$H^{4j+i}(Q_{4n}; \mathbb{Z}/2) = \begin{bmatrix} \mathbb{Z}/2 & \text{if } i \equiv 0, 3 \pmod 4, \\ \mathbb{Z}/2 \oplus \mathbb{Z}/2 & \text{if } i \equiv 1, 2 \pmod 4. \end{bmatrix}$$

If $x_1, x_2 \in H^1(D_{2^{\alpha+1}}; \mathbb{Z}/2)$ are the classes defined in (4.4), then it is simple to see that π induces an isomorphism

$$\pi^*: H^1(D_{2^{\alpha+1}}; \mathbb{Z}/2) \xrightarrow{\cong} H^1(Q_{2^{\alpha+2}}; \mathbb{Z}/2).$$

In addition, from §3.11, the Bockstein, Sq^1 (cf., §4.21(v)) is an isomorphism

$$Sq^1: H^1(Q_{2^{\alpha+2}}; \mathbb{Z}/2) \to H^2(Q_{2^{\alpha+2}}; \mathbb{Z}/2).$$

The latter group is therefore generated by $\pi^*(x_1)^2$ and $\pi^*(x_2)^2$.

The periodicity in $H^*(Q_{4n}; \Lambda)$ is induced by cup-product with the generator of $H^4(Q_{4n}; \mathbb{Z})$. This is because Q_{4n} is a subgroup of the group of quaternions of unit norm, which is the 3-sphere, S^3 (or alternatively the first symplectic group, $Sp(1)$) and any group which acts freely on S^3 must have cohomological periodicity of this type [Br, pp. 153–154].

Henceforth denote by p the generator

$$0 \neq p \in H^4(Q_{4n}; \mathbb{Z}/2).$$

This class is the mod 2 reduction of the *first Pontrjagin class* of the symplectic representation afforded by the inclusion of Q_{4n} into $Sp(1)$ [H; M-St].

Let us consider first the case of Q_8:

$$\mathbb{Z}/2 \rightarrowtail Q_8 \xrightarrow{\pi} \mathbb{Z}/2 \times \mathbb{Z}/2.$$

If $H^*(\mathbb{Z}/2 \times \mathbb{Z}/2; \mathbb{Z}/2) \cong \mathbb{Z}/2[x_1, x_2]$, then the class, δ, representing this central extension is, according to [Q], given by the 2-cocycle:

(4.31)
$$\begin{cases} \delta[x|x] = x^2 \in Q_8, \\ \delta[y|y] = y^2 \in Q_8, \\ \\ \delta[x|y] = xy(xy)^{-1} = 1, \\ \delta[y|x] = yx(xy)^{-1} = x^{-2} = x^2. \end{cases}$$

By Example 2.14, the class of (4.31) is given by

$$\delta = x_1^2 + x_2^2 + x_1 x_2.$$

The Hochschild-Serre spectral sequence of π is

(4.32) $E_2^{*,*} = H^*(\mathbb{Z}/2 \times \mathbb{Z}/2; \mathbb{Z}/2) \otimes H^*(\mathbb{Z}/2; \mathbb{Z}/2) \Rightarrow H^*(Q_8; \mathbb{Z}/2)$

so that $E_2^{*,*} \cong \mathbb{Z}/2[x_1, x_2, v]$ with bideg $(x_i) = (1, 0)$ and bideg $(v) = (0, 1)$.

As in the proof of Theorem 4.6 the classes x_1 and x_2 are d_2-cycles, and in addition we must have (see Chapter 2 §§1.28b/1.30)

$$d_2(v) = x_1^2 + x_2^2 + x_1 x_2.$$

One readily computes the homology and obtains

(4.33) $E_3^{*,*} \cong \mathbb{Z}/2[x_1, x_2, u]/(x_1^2 + x_2^2 + x_1 x_2),$

where $u = [v^2]$.

A particular case of the Kudo Transgression Theorem (which I leave to the assiduous reader for verification) relates d_2, d_3 and the Bockstein, Sq^1, in the following manner:

$$\begin{aligned}
d_3(u) &= d_3[v^2] \\
&= d_3[Sq^1(v)] \\
&= Sq^1 d_2(v) \\
&= Sq^1(x_1^2 + x_2^2 + x_1 x_2) \\
&= x_1^2 x_2 + x_1 x_2^2 \\
&= x_1^3 \in E_3^{3,0}.
\end{aligned}$$

Therefore one may compute the homology and obtain

(4.34) $E_4^{*,*} \cong \mathbb{Z}/2[x_1, x_2, p]/(x_1^2 + x_2^2 + x_1 x_2, x_1^3),$

where $p = [u^2] \in E_4^{0,4}$. By counting dimensions, we see that there can be no further differentials in the spectral sequence, since taking cohomology groups in a nontrivial manner can only reduce the dimension in each total degree (where the *total degree* of $E_r^{s,t}$ is $s + t$). Therefore $E_4^{*,*} = E_\infty^{*,*}$. Furthermore the relations in (4.34) imply that

$$\pi^*(x_1^2 + x_2^2 + x_1 x_2) = 0 = \pi^*(x_1^3) \in H^*(Q_8; \mathbb{Z}/2).$$

Therefore, if we write x_i for $\pi^*(x_i)$, then

(4.35) $$H^*(Q_8; \mathbb{Z}/2) \cong \mathbb{Z}/2[x_1, x_2, p]/(x_1^2 + x_2^2 + x_1 x_2, x_1^3),$$

where $\deg(x_i) = 1$ and $\deg(p) = 4$.

Now we turn our attention to $Q_{2^{\alpha+2}}$ with $\alpha \geq 2$. Consider the following commutative diagram of central extensions:

(4.36)

$$
\begin{array}{ccc}
\mathbb{Z}/2 \rightarrowtail Q_8 & \xrightarrow{\pi} & \mathbb{Z}/2 \times \mathbb{Z}/2 \\
= \downarrow \text{\scriptsize 1} \quad \downarrow i_1 & & \downarrow i_2 \\
\mathbb{Z}/2 \rightarrowtail Q_{2^{\alpha+2}} & \xrightarrow{\pi} & D_{2^{\alpha+1}}
\end{array}
$$

Since $i_2^*(x_1) = x_1$ and $i_2^*(x_2) = 0$, the lower extension in (4.36) corresponds to a cohomology class

(4.37) $$\delta = \alpha_1 w + \alpha_2 x_1^2 + \alpha_3 x_1 x_2 \in H^2(D_{2^{\alpha+1}}; \mathbb{Z}/2),$$

with $\delta = i_2^*(\delta) = x_1^2 + x_2^2 + x_1 x_2 \in H^2(\mathbb{Z}/2 \times \mathbb{Z}/2; \mathbb{Z}/2)$. However, $i_2^*(w)$ is the second Stiefel-Whitney class of the inclusion of $D_{2^{\alpha+1}}$ into $O_2(\mathbb{R})$, restricted to $\mathbb{Z}/2 \times \mathbb{Z}/2$. For $i = 1, 2$, let $L_i: \mathbb{Z}/2 \times \mathbb{Z}/2 \to O_1(\mathbb{R}) = \{\pm 1\}$ denote the ith projection, then

$$
\begin{aligned}
i_2^*(w) &= SW_2(L_1 L_2 + L_2) \\
&= SW_1(L_1 L_2) SW_1(L_2) \\
&= (x_1 + x_2) x_2 \in H^2(\mathbb{Z}/2 \times \mathbb{Z}/2; \mathbb{Z}/2).
\end{aligned}
$$

Therefore $x_1^2 + x_2^2 + x_1 x_2 = \alpha_1(x_1 x_2 + x_2^2) + \alpha_2 x_1^2$ implies that $\alpha_1 \equiv \alpha_2 \equiv 1$ (mod 2).

The spectral sequence for the lower extension in (4.36) takes the form

(4.38) $$E_2^{*,*} \cong \mathbb{Z}/2[x_1, x_2, w, v]/(x_1 x_2 + x_2^2) \Rightarrow H^*(Q_{2^{\alpha+2}}; \mathbb{Z}/2)$$

in which $\operatorname{bideg}(x_i) = (1, 0)$, $\operatorname{bideg}(w) = (2, 0)$, $\operatorname{bideg}(v) = (0, 1)$, and

$$d_2(v) = w + x_1^2 + \alpha_3 x_1 x_2 \in E_2^{2,0}.$$

One easily finds that if $u = [v^2] \in E_3^{0,2}$, then

$$
\begin{aligned}
E_3^{*,*} &\cong \mathbb{Z}/2[x_1, x_2, w, u]/(x_1 x_2 + x_2^2, w + x_1^2 + \alpha_3 x_1 x_2) \\
&\cong \mathbb{Z}/2[x_1, x_2, u]/(x_1 x_2 + x_2^2).
\end{aligned}
$$

Also, by the Kudo Transgression Theorem, we have

$$d_3[u] = d_3[Sq^1(v)]$$
$$= Sq^1(d_2(v))$$
$$= Sq^1(w + x_1^2 + \alpha_3 x_1 x_2), \qquad \text{by §4.22(i)},$$
$$= x_1 w + \alpha_3(x_1^2 x_2 + x_1 x_2^2)$$
$$= x_1^3 + 2\alpha_3 x_1^2 x_2 + \alpha_3 x_1 x_2^2$$
$$= x_1^3 + \alpha_3 x_1 x_2^2 \in E_3^{3,0}.$$

Thus, if $p = [u^2] \in E_4^{0,4}$, then

(4.39) $$E_4^{*,*} \cong \mathbb{Z}/2[x_1, x_2, p]/(x_1 x_2 + x_2^2, x_1^3 + \alpha_3 x_1 x_2^2).$$

We now have to decide whether α_3 is even or odd in (4.39). If $\alpha_3 \equiv 1 \pmod 2$, then $0 \neq x_1^3 \in E_4^{3,0} = E_\infty^{3,0}$ so that

$$0 \neq x_1^3 \in H^3(Q_{2^{\alpha+2}}; \mathbb{Z}/2),$$

and similarly $0 \neq x_1^4 \in E_4^{4,0} = E_\infty^{4,0}$, since $E_4^{0,3} = 0$, which means that

$$Sq^1 : H^3(Q_{2^{\alpha+2}}; \mathbb{Z}/2) \to H^4(Q_{2^{\alpha+2}}; \mathbb{Z}/2)$$

is nonzero, since $Sq^1(x_1^3) = x_1^4$. However, this is impossible since $H^4(Q_{2^{\alpha+2}}; \mathbb{Z}/2) \cong \mathbb{Z}/2^{\alpha+2}$, by §3.11. Therefore we find that $\alpha_3 \equiv 0 \pmod 2$ from which, as in the case of Q_8, we find that

$$H^*(Q_{2^{\alpha+2}}; \mathbb{Z}/2) \cong \mathbb{Z}/2[x_1, x_2, p]/(x_2^2 + x_1 x_2, x_1^3),$$

which completes the calculation.

To recapitulate, in the foregoing discussion we have proved the following result:

(4.40) Theorem

If $n = 2^\alpha(2s + 1)$, then

(i) $H^*(Q_{4n}; \mathbb{Z}/2) \cong H^*(Q_{2^{\alpha+2}}; \mathbb{Z}/2)$ *and*

(ii) $H^*(Q_{2^{\alpha+2}}; \mathbb{Z}/2) \cong \begin{cases} \mathbb{Z}/2[v, f]/(v^2) & \text{if } \alpha = 0, \\ \mathbb{Z}/2[x_1, x_2, p]/(x_1^2 + x_2^2 + x_1 x_2, x_1^3) & \text{if } \alpha = 1, \\ \mathbb{Z}/2[x_1, x_2, p]/(x_2^2 + x_1 x_2, x_1^3) & \text{if } \alpha \geq 2, \end{cases}$

where $deg(v) = 1 = deg(x_i)$ and $deg(p) = 4$.

Chapter Two

Nonabelian Cohomology of Groups

Lord of the Two Ways, these are the foreigners,
They come out of nowhere.
Sometimes they come to tell us things,
Mostly they are the greedy ones.
What then do they want?

—*D. H. LAWRENCE,*
"The Fourth Hymn of Ramon" (1926)

In this chapter we introduce nonabelian cohomology of groups. Group cohomology with nonabelian coefficients is ungainly at first sight. In contrast with abelian cohomology, it requires case-by-case analysis of its low dimensional exact sequence of pointed sets and nonabelian groups. Initially, one may wonder what its purpose could possibly be.

For us, nonabelian cohomology will be used in the context of Galois cohomology, where the groups are Galois groups of field extensions. Having set up the exact sequence, we consider various examples of the manner in which Galois cohomology classifies algebraic objects by the technique of Galois descent. The examples in which we are interested are bilinear forms, equivariant bilinear forms, central simple k-algebras, and central simple k[G]-algebras. We go through all these from first principles—once again with the intent of obtaining explicit formulae. In addition we establish the connection between central, simple algebras and the Brauer group.

In §3 we look at specific representatives of bilinear forms and for operations, such as the Scharlau transfer, upon them. Similarly, for central, simple algebras we derive the Brauer group representative for a cyclic algebra. We close the chapter by proving the basic symbol relation (from algebraic K-theory) in the Brauer group. We will need this later, in Chapter 3, together with our dihedral group cohomology calculations, to derive relations between products in mod 2 Galois cohomology.

1. BASIC DEFINITIONS

Suppose that X is a topological G-module. By this we mean that X is a (not necessarily abelian) topological group endowed with a (left) G-action, which is continuous. We have in mind particularly the cases where G is a discrete group or a profinite group (as in Chapter 1, section 1.18).

In the case of abelian cohomology, $H^0(G; M)$ is isomorphic to the invariants

$$M^G = \{m \in M \mid gm = m \text{ for all } g \in G\}.$$

Accordingly, we make the following definition:

(1.1) Definition

With G, X as given earlier, the 0th cohomology of G with coefficients in X is given by

$$H^0(G; X) = X^G = \{x \in X \mid gx = x \text{ for all } g \in G\}.$$

Thus $H^0(G; X)$ is a subgroup of X.

(1.2) If G, X are as in §1.1, we may form the *semidirect product* group, $G \ltimes X$. As a set $G \ltimes X$ is the product, $G \times X$. The product on $G \ltimes X$ is defined by $(g, g_1 \in G; \, x, x_1 \in X)$:

(1.3) $$(g, x)(g_1, x_1) = (gg_1, xg(x_1)).$$

Note that $X \lhd G \ltimes X$ is a subgroup.

Suppose that we have a (continuous) homomorphism

$$\Phi: G \to G \ltimes X \qquad \text{of the form}$$

(1.4) $$\Phi(g) = (g, f(g)),$$

then $f: G \to X$ is a continuous map satisfying

(1.5) $$f(gg_1) = f(g)g(f(g_1)) \qquad (g, g_1 \in G).$$

Conversely, if f satisfies (1.5), then $\Phi(g) = (g, f(g))$ defines a (continuous) homomorphism, $\Phi: G \to G \ltimes X$.

(1.6) Definition

The set of 1-cocycles of G with coefficients in X, $Z^1(G; X)$, consists of the (continuous) functions f satisfying (1.5). There is a bijection between $Z^1(G; X)$ and the continuous homomorphisms $\Phi: G \to G \ltimes X$ of the form (1.4). The bijection associates Φ to f as in (1.4).

(1.7) If $f_1, f_2 \in Z^1(G, X)$, define an equivalence relation, $\approx$, by $f_1 \approx f_2$ if and only if there exists $b \in X$ such that

$$(1.8) \qquad\qquad f_1(g) = b^{-1} f_2(g) g(b).$$

Observe that $(1, b^{-1}) = (1, b)^{-1} \in G \ltimes X$ and that, in $G \ltimes X$,

$$(1, b^{-1})(g, f_2(g))(1, b) = (g, b^{-1} f_2(g))(1, b)$$
$$= (g, b^{-1} f_2(g) g(b)).$$

Hence we have shown the following.

(1.9) Lemma

$f_1 \approx f_2$ if and only if $\Phi_1 = 1 \times f_1$ and $\Phi_2 = 1 \times f_2 : G \to G \ltimes X$ differ by conjugation by an element of X.

(1.10) Definition

With G and X as in §1.1, define $H^1(G; X)$, the first cohomology of G with coefficients in X to be

$$H^1(G; X) = Z^1(G; X)/\approx.$$

This is a pointed set with base-point corresponding to $f(g) = b^{-1} g(b)$ for any $b \in X$.

(1.11) Proposition

Let M be an abelian, topological, trivial $(G \ltimes X)$-module. Then there is a well-defined homomorphism

$$z^*: H^i(G \ltimes X; M) \to H^i(G; M)$$

associated to each $z \in H^1(G; X)$. If $z = [f]$, then $z^* = \Phi^*$, where $\Phi = (1 \times f)$ as in (1.4) and §1.6.

If z is the base-point, then z^* is the map induced by the natural map $\lambda: G \to G \ltimes X$ given by $\lambda(g) = (g, 1)$.

Proof. This result follows immediately from the preceding discussion (particularly §1.9), together with the fact that inner automorphisms (of $G \ltimes X$ in this case) induce the identity on cohomology [Ho-S].

(1.12) Let $i: X \to Y$ be an inclusion of G-groups, as in §1.1, and let Y/X denote the *G-set* of left cosets. We have a sequence of maps

(1.13) $$\{1\} \longrightarrow H^0(G; X) \xrightarrow{i_*} H^0(G; Y) \xrightarrow{j_*} (Y/X)^G$$

$$\xrightarrow{\delta} H^1(G; X) \xrightarrow{i_*} H^1(G; Y).$$

In (1.13) i_*, j_* are induced by composition with the canonical maps. To define δ, let $yX = gyX$ for $y \in Y$ and for all $g \in G$. Hence we may define a 1-cocycle by

(1.14) $$\delta(yX)(g) = y^{-1}g(y) \in X.$$

(1.15) Proposition [Ser, I-64]

(1.13) *is an exact sequence of pointed sets, as explained in the proof.*

Proof. Clearly, $i_*: X^G \to Y^G$ is injective, being the restriction of the inclusion homomorphism.

In addition, if $y \in Y^G$ satisfies $j_*(y) = 1X$, then $y \in Y^G \cap X = X^G$, and if $j_*(y_1) = j_*(y_2)$, then $x = y_1^{-1}y_2 \in X^G$ so that y_1, y_2 differ by the action of $x \in X^G$. If $\delta(y_1 X) = \delta(y_2 X)$, then there exists $b \in X$ so that for all $g \in G$,

$$y_1^{-1}g(y_1) = b^{-1}y_2^{-1}g(y_2)g(b)$$

so that $y = y_2 b y_1^{-1} \in Y^G$ and $y_1 X$, $y_2 X$ differ by the action of $y \in Y^G$.

If $f_1, f_2: G \to X$ are two 1-cocycles and $i_*(f_1) = i_*(f_2)$, then there exists $y \in Y$ such that $f_1(g) = y^{-1}f_2(g)g(y)$ for all $g \in G$. Therefore $i_*^{-1}(1)$ consists of all $f(g) = y^{-1}g(y)$, where $y \in (Y/X)^G$, since $f(g) \in X$. Also Y^G acts trivially on $i_*^{-1}(1)$ so

$$i_*^{-1}(1) \cong \{(Y/X)^G \text{ modulo the action } Y^G\}. \qquad \square$$

(1.16) Proposition [Ser, I-66]

If $X \rightarrowtail Y \twoheadrightarrow W$ is a short exact sequence of G-groups, as in §1.1, in which $Y \twoheadrightarrow W$ admits a continuous section, then

$$H^0(G; X) \xrightarrow{i_*} H^0(G; Y) \xrightarrow{j_*} H^0(G; W)$$

$$\xrightarrow{\delta} H^1(G; X) \xrightarrow{i_*} H^1(G; Y) \xrightarrow{j_*} H^1(G; W)$$

is an exact sequence of sets.

Proof. We have dealt with most of this in §1.15. If $f: G \to Y$ is a 1-cocycle and $j_*(f) = [1]$, then there exists $y \in Y$ such that $h(g) = yf(g)g(y)^{-1} \in X$ for each $g \in G$. Thus $[f] = i_*[h]$, as required. $\qquad \square$

(1.17) Suppose now that

$$X \rightarrowtail Y \twoheadrightarrow W$$

is an extension of G-groups in which X *is abelian.* Therefore we have

$$\lambda: W^G \to \operatorname{Aut}_G(X)$$

given by $\lambda(\hat{w})(x) = \hat{w}x(\hat{w})^{-1}$, since $\lambda(w)(gx) = g(\hat{w}x\hat{w}^{-1})$, when $\hat{w} \in Y$ is a continuous lifting of $w \in W$.

(1.18) Proposition

  (i) *For* $w \in W^G$, $\alpha \in H^1(G; X)$

$$w(\alpha) = \lambda(w^{-1})(\alpha) + \delta(w).$$

  (ii) *If* X *is central in* Y, *then*

$$w(\alpha) = \alpha + \delta(w)$$

and $\delta: H^0(G; W) \to H^1(G; A)$ *is a homomorphism. Here* $w(\alpha)$ *is the action of*

$$W^G = (Y/X)^G \quad \text{on } H^1(G; X),$$

which was introduced in §1.15 (proof).

Proof. Represent α by $f: G \to X$, a 1-cocycle, and lift w to $\hat{w} \in Y$, as in §1.16.
  By definition, $\delta(w)(g) = \hat{w}^{-1}g(\hat{w}) \in X$, while $\lambda(w^{-1})(\alpha)(g) = \hat{w}^{-1}\alpha(g)\hat{w}$ so that if we write the "product" in X additively,

$$[\lambda(w^{-1})(\alpha) + \delta(w)](g) = \hat{w}^{-1}\alpha(g)\hat{w}(\hat{w})^{-1}g(\hat{w})$$

$$= \hat{w}^{-1}\alpha(g)g(\hat{w})$$

$$= w(\alpha)(g),$$

as required for part (i).
  Part (ii) is easy since $\lambda(w) = 1_X$ when X is central and

$$\delta(ww') = ww'(\alpha) - \alpha$$

$$= w(w'(\alpha)) - w'(\alpha) - \alpha + w'(\alpha)$$

$$= \delta(c) + \delta(c'), \quad \text{as required.} \qquad \square$$

(1.19) Consider further the group extension $X \rightarrowtail Y \twoheadrightarrow W$ in which X is abelian. If $f: G \to W$ is a 1-cocycle, we define a new G-action on X, *denoted by* $_f X$, given by

(1.20)
$$\begin{cases} \#: G \times_f X \to {}_f X, \\ g \# x = f(g)(g(x)), \ (g \in G, x \in X = {}_f X). \end{cases}$$

Here $f(g) \in W$ acts on $g(x) \in X$ by conjugation, as explained later. Observe that if $\hat{w} \in Y$ is a lift of $w \in W$, then

$$g\#(g'\#x) = g\#(f(g')g'(x)(f(g'))^{-1})$$
$$= f(g)[g(f(g')g'(x)(f(g'))^{-1})](f(g))^{-1}$$
$$= (f(g)gf(g'))(gg'(x))$$
$$= f(gg')gg'(x)$$
$$= (gg')\#(x).$$

Also if $X \lhd Y$ is *central*, then $_f X = X$.

(1.21) Definition

In the situation of §1.19, define for $[f] \in H^1(G; W)$ the class

$$\Delta[f] \in H^2(G; {}_f X)$$

by the formula $(g, g' \in G)$,

(1.22)
$$\Delta[f][g|g'] = h(g)g(h(g'))[h(gg')]^{-1} \in {}_f X,$$

where $h: G \to Y$ is a lift of the representing cocycle, $f: G \to W$.

We must verify that $\Delta(f)$ is a 2-cocycle whose cohomology class is well-defined for each 1-cocycle, f.

However, if we change the representing 1-cocycle from $f: G \to W$ to $(f': G \mapsto w^{-1}f(g)g(w))$ for some $w \in W$, the effect is to change $_f X$ to $_{f'} X$ which are related as follows. The new action $\#': G \times_{f'} X \to {}_{f'} X$ is given by

(1.23)
$$g\#'(x) = \hat{w}^{-1}h(g)(g(\hat{w})g(x)g(\hat{w}^{-1}))h(g)^{-1}\hat{w},$$

where $\hat{w} \in Y$ and $h: G \to Y$ lift w and f, respectively. Hence we have a commutative diagram

(1.24)
$$\begin{array}{ccc} G \times_{f'} X & \xrightarrow{\#'} & {}_{f'} X = X \\ {\scriptstyle 1 \times (w \cdot -)} \downarrow & & \downarrow {\scriptstyle (w \cdot -)} \\ G \times_f X & \xrightarrow{\#} & {}_f X = X \end{array}$$

To verify the cocycle condition, we must show that

$$\Delta[f](\delta[g|g'|g'']) = 1,$$

which, written multiplicatively, is equivalent to the identity

$$ABCD = 1,$$

where

$$A = g\#\Delta(f)[g'|g''] = h(g)g(h(g'))gg'(h(g''))(gh(gg''))^{-1}h(g)^{-1},$$

$$B = \Delta(f)[gg'|g'']^{-1} = h(gg'g'')(gg'(h(g'')))^{-1}h(gg')^{-1},$$

$$C = \Delta(f)[g|g'g''] = h(g)(gh(g'g''))(h(gg'g''))^{-1},$$

$$D = \Delta(f)[g|g']^{-1} = h(gg')(gh(g'))^{-1}(h(g))^{-1}.$$

Therefore, since A, B, C, $D \in X$, we have

$$ABCD = ACBD = Ah(g)(gh(g'\,g''))((gg'h(g'')))^{-1}(gh(g'))^{-1}h(g)^{-1},$$

after cancelling two pairs of terms. However, this last expression is identically zero, as required.

Suppose we use a different lifting, h', in the formula (1.22). This lifting will have the form

$$h'(g) = x(g)h(g),$$

where $x: G \to X$ is a continuous function. This changes $\Delta(f)\{g|g'\}$ to

$$\Delta'(f)[g|g'] = x(g)h(g)(g(x(g'))g(h(g')))[x(gg')h(gg')]^{-1}.$$

Therefore we have

$$\Delta'(f)[g|g'](\Delta(f)[g|g'])^{-1} = x(g)h(g)(g(x(g')))(gh(g'))h(gg')^{-1}$$
$$\cdot x(gg')^{-1}h(gg')(gh(g'))^{-1}h(g)^{-1},$$

as the product of the last three factors lies in X, which is abelian, and so does $x(gg')$. We obtain

$$= x(g)h(g)g(x(g'))h(g)^{-1}x(gg')^{-1}$$

$$= x(g)(g\#x(g'))x(gg')^{-1},$$

which is the boundary $\delta(x) \in C^2(G, {}_f X)$ evaluated on $[g|g']$. Hence the cohomology classes satisfy

$$[\Delta'(f)] = [\Delta(f)] \in H^2(G; {}_fX).$$

If we change f to a cohomologous 1-cocycle, we have, from (1.23), an *isomorphism*

$$H^2(G; {}_{f'}X) \xrightarrow{(w \cdot -)^*} H^2(G; {}_fX),$$

which carries $\Delta(f')$ to $\Delta(f)$. In this sense $\Delta(f)$ is well-defined and depends only on the cohomology class of $f \in Z^1(G; W)$.

(1.25) Proposition

With the notation of §1.21, the cohomology class $[f] \in \mathrm{im}\,(H^1(G; Y) \to H^1(G; W))$ if and only if $\Delta(f) = 1 \in H^2(G; {}_fX)$. [By the discussion of §1.21, $\Delta(f) = 1$ for one cocycle representative implies $\Delta(f) = 1$ for all of them.]

Proof. Clearly, if f lifts to a 1-cocycle, $h: G \to Y$, then (1.22) shows that $\Delta(f)[g|g'] = \delta(h)[g|g'] = 1$. Conversely, if

$$h(g)(g(h(g')))(h(gg'))^{-1} = (g\#u)(g')(u(gg'))^{-1}(ug))$$

$$= h(g)(g(u(g')))h(g)^{-1}u(g)u(gg)^{-1},$$

as X is abelian, so that

$$g((u^{-1}h)(g'))[(u^{-1}h)(gg')] = [(u^{-1}h)(g)]^{-1},$$

which means that $u^{-1}h: G \to Y$ is a *1-cocycle* covering $f: G \to W$, as required. $\square$

Combining §§1.13–1.25

(1.26) Proposition

Let $X \overset{i}{\rightarrowtail} Y \overset{j}{\twoheadrightarrow} W$ be a central extension of groups upon which G acts continuously and for which the map, j, has a continuous section. Then the following sequence of pointed sets (and groups) is exact:

$$H^0(G; X) \xrightarrow{i_*} H^0(G; Y) \xrightarrow{j_*} H^0(G; W) \xrightarrow{\delta} H^1(G; X)$$

$$\xrightarrow{i_*} H^1(G; Y) \xrightarrow{j_*} H^1(G; W) \xrightarrow{\Delta} H^2(G; X).$$

(1.27) Suppose now that

$$H \overset{i}{\rightarrowtail} G \overset{\pi}{\twoheadrightarrow} Q$$

is an extension of *discrete* groups and that M is an abelian (left) Q-module. There are exact sequences of the following form [H-S, p. 202]:

(1.28a)
$$H_2(G; M) \overset{\pi_*}{\to} H_2(Q; M) \overset{\delta}{\to} H_{ab} \otimes_Q M$$
$$\overset{i_*}{\to} H_1(G; M) \overset{\pi_*}{\to} H_1(Q; M) \to 0,$$

where $H_{ab} = H_1(H; \mathbb{Z})$, the abelianization of H, and

(1.28b)
$$0 \longrightarrow H^1(Q; M) \overset{\pi^*}{\to} H^1(G; M)$$
$$\overset{i^*}{\to} \operatorname{Hom}_Q(H_{ab}, M) \overset{\delta}{\to} H^2(Q; M) \overset{\pi^*}{\to} H^2(G; M).$$

I will sketch the derivation of (1.28b), using the Hochschild-Serre spectral sequence

(1.29)
$$E_2^{s,t} = H^s(Q; H^t(H; M)) \Rightarrow H^{s+t}(G; M).$$

In low dimensions this spectral sequence looks as follows:
(1.30)

$\overset{t}{\uparrow}$ 3	$H^3(H; M)^Q$			
2	$H^2(H; M)^Q$	$H^1(Q; H^2(H; M))$		
1	$H^1(H; M)^Q$	$H^1(Q; H^1(H; M))$	$H^2(Q; H^1(H; M))$	
0	M^G	$H^1(Q; M^H)$	$H^2(Q; M^H)$	$H^3(Q; M^H)$
	0	1	2	$\overset{3}{\underset{\to s}{}}$

Now we have an exact sequence resulting from the filtration of $H^1(G; M)$ associated to (1.30):

$$F^1 H^1 = H^1(Q; M) \rightarrowtail H^1(G; M) \rightarrow \frac{F^0 H^1}{F^1 H^1} = E_\infty^{0,1} = \ker(d_2),$$

since $M^H = M$, which yields an exact sequence

$$H^1(Q; M) \overset{\pi^*}{\rightarrowtail} H^1(G; M) \overset{i^*}{\longrightarrow} \mathrm{Hom}_Q(H, M) \overset{d_2}{\longrightarrow} H^2(Q; M).$$

The result now follows since, as M is abelian, $\mathrm{Hom}_Q(H, M) \cong \mathrm{Hom}_Q(H_{ab}, M)$, and

$$\frac{H^2(Q; M)}{(\mathrm{im}(d_2))} = F^2 H^2 \subset H^2(G; M).$$

For a more explicit derivation of this exact sequence, see [H-S, p. 202]. With the evaluation of d_2 in [Ho-S], one can verify that the preceding exact sequence coincides with that of [H-S, p. 202].

(1.31)   GALOIS COHOMOLOGY EXAMPLES

We close this section with some nonabelian, continuous cohomology examples which we will need later.

Let K be a field with separable closure, $\bar{K}$. If L/K is a finite Galois extension, $K \subset L \subset \bar{K}$, we denote its *Galois group* by $G(L/K)$. If $L \subset F$ is another finite Galois extension, we have an extension

(1.32) $$G(F/L) \rightarrowtail G(F/K) \twoheadrightarrow G(L/K),$$

and we may form the *absolute Galois* group

(1.33) $$\Omega_K = \varprojlim_{K \subset L \subset \bar{K}} G(L/K)$$

(the limit is taken over finite Galois extensions within a fixed algebraic closure, $\bar{K}$), giving (1.33) the profinite topology (as in §1.18). Any matrix, $X \in GL_n \bar{K}$, lies in some $GL_n L$ so that Ω_K acts continuously (i.e., in a "locally constant" manner) on $GL_n \bar{K}$. Similarly, Ω_K acts continuously on the subgroups of $GL_n \bar{K}$ given by

$$SL_n \bar{K} = \{X \in GL_n \bar{K} \,|\, \det(X) = 1\} \text{ and } O_n \bar{K} = \{X \in GL_n \bar{K} \,|\, X X^t = I_n\}$$

(where X^t is the *transpose* of X).

Hence we may form *continuous cohomology* groups (or *Galois cohomology* groups), $G(L/K)$, such as $H^1(G(L/K); O_n L)$.

(1.34)   Proposition [Ser 2, p. 151]

Let L/K be a finite Galois extension. Then $H^1(G(L/K)); GL_n(L)) = \{1\}$, and *consequently, taking limits over L,*

$$H^1(K; GL_n\bar{K}) = \{1\}$$

as well.

(1.35) Corollary

Let L/K be a finite Galois extension. Then

(i) $H^1(G(L/K); SL_n(L)) = \{1\}$,

(ii) $H^1(K; SL_n\bar{K}) = \{1\}$.

Proof. Clearly, (ii) follows from (i) by taking the limit over L. However, (i) follows from §2.26 applied to the extension $SL_n(L) \rightarrowtail GL_n(L) \xrightarrow{\det} L^*$, bearing in mind that $GL_1(L) = L^*$ and $H^0(G(L/K); GL_n(L)) = GL_nK$ so that $j_* = \det: GL_nK \to K^*$ is onto. $\qquad\square$

2. EXAMPLES OF GALOIS DESCENT

In this section we will recall how nonabelian Galois cohomology can be used to classify bilinear forms, Hermitian forms with a group action, central simple K-algebras and central simple $K[\Gamma]$-algebras.

The general idea—namely, that $H^1(K; \text{Aut}_{\bar{K}}(X))$ classifies "objects over K" which become isomorphic with X when we "extend scalars to $\bar{K}$"—is succinctly described in [Ser 2, pp. 152–153]. Nevertheless, I have chosen to give the complete derivation from first principles in each case, in order to have access to the explicit cocycle formulae when necessary.

(2.1) BILINEAR FORMS

Let L/K be a finite Galois extension, and set $G = G(L/K)$, temporarily. Let V be a finite dimensional K-vector space, and let $\Phi: V \times V \to K$ be a fixed, nondegenerate, K-bilinear form. Write V_L for $V \otimes_K L$ and Φ_L for $\Phi \otimes_K L$: $V_L \times V_L \to L$. Hence, if $v, w \in V$ and $a, b \in L$, then

(2.2) $$\Phi_L(v \otimes a, w \otimes b) = \Phi(v, w) \otimes ab.$$

Let $O(\Phi_L)$ be the (discrete) orthogonal groups of (V_L, Φ_L)

(2.3) $$O(\Phi_L) = \{\alpha \in \text{Aut}_L(V_L) | \Phi_L(\alpha(x), \alpha(y)) = \Phi_L(x, y) \text{ for all } x, y \in V_L\}.$$

Hence G acts on V_L, $\text{Aut}_L(V_L)$ and $O(\Phi_L)$ by the formulae

$$(2.4) \quad \begin{cases} g(v \otimes a) = v \otimes g(a) & (v \in V, a \in L), \\ g(\alpha)(w) = g(\alpha(g^{-1}(z))) & (\alpha \in \mathrm{Aut}_L(V_L), z \in V_L, g \in G). \end{cases}$$

Now suppose that $\psi: V \times V \to K$ is a second such bilinear form such that $(V_L, \Phi_L) \cong (V_L, \psi_L)$. That is, there exists $A \in \mathrm{Aut}_L(V_L)$ so that $(u_1, u_2 \in V_L)$,

$$(2.5) \qquad\qquad \Phi_L(u_1, u_2) = \psi_L(A(u_1), A(u_2)).$$

Observe that

$$g(\Phi_L(v \otimes a, w \otimes b)) = \Phi(v, w) g(ab)$$
$$= \Phi(v, w) g(a) g(b)$$
$$= \phi_L(g(v \otimes a), g(w \otimes b))$$

so that, in general,

$$(2.6) \quad \begin{cases} g(\Phi_L(u_1, u_2)) = \Phi_L(gu_1, gu_2) \\ g(\psi_L(u_1, u_2)) = \psi_L(gu_1, gu_2). \end{cases}$$

For $g \in G = G(L/K)$, set

$$(2.7) \qquad\qquad f(g) = A^{-1} g(A) \in \mathrm{Aut}_L(V_L).$$

(2.8) Lemma

(2.7) defines a map $f: G \to O(\Phi_L)$.

Proof. We have

$$\Phi_L(A^{-1}g(A)(u_1), A^{-1}g(A)(u_2)) = \psi_L(g(A)(u_1), g(A)(u_2)) \qquad \text{by (2.5)}$$
$$= \psi_L(g(A(g^{-1}u_1)), g(A(g^{-1}u_2))) \quad \text{by (2.4)}$$
$$= g(\psi_L(A(g^{-1}u_1), A(g^{-1}u_2))) \qquad \text{by (2.6)}$$
$$= g(\Phi_L(g^{-1}u_1, g^{-1}u_2)) \qquad\qquad \text{by (2.5)}$$
$$= \Phi_L(u_1, u_2)) \qquad\qquad\qquad\qquad \text{by (2.6).}$$

(2.9) Proposition

In (2.7) f *is a 1-cocycle and the association* $\psi \mapsto [f]$ *induces a bijection*

$$\begin{Bmatrix} \text{K-isomorphism classes of} \\ \text{bilinear forms, } \psi, \text{ L-isomorphic} \\ \text{to } \Phi_L \end{Bmatrix} \longleftrightarrow H^1(G(L/K); O(\Phi_L)).$$

Proof. Clearly,

$$f(gh) = A^{-1}gh(A)$$

$$= A^{-1}g(A)g(A^{-1})g(h(A))$$

$$= f(g)g(f(h)),$$

so f is a 1-cocycle. If $B \in \mathrm{Aut}_L(V_L)$ also satisfies (2.5) in place of A, then

$$\Phi_L(B^{-1}(A(x)), B^{-1}(A(y))) = \psi_L(A(x), A(y))$$

$$= \Phi_L(x, y)$$

so that $B^{-1}A \in O(\Phi_L)$. However,

$$(A^{-1}B)(B^{-1}g(B))(g(A^{-1}B)^{-1}) = A^{-1}g(A)$$

shows that the 1-cocycle defined using B in (2.7) is cohomologous to that defined using A. Hence sending ψ to $[f]$ is well-defined.

Conversely, given a 1-cocycle $f: G \to O(\Phi_L)$, there exists, by §1.34, an $A \in \mathrm{Aut}_L(V_L)$ such that $f(g) = A^{-1}g(A)$, and we may define $\psi: V_L \times V_L \to L$ by (2.5). However, if $\psi_L(x, y) = \Phi_L(A^{-1}x, A^{-1}y)$, then, for $g \in G$,

$$\psi_L(gx, gy) = \Phi_L(A^{-1}g(x), A^{-1}g(y))$$

$$= \Phi_L(f(g)^{-1}A^{-1}g(x), f(g)^{-1}A^{-1}g(y))$$

$$= \Phi_L(g(A)^{-1}g(x), g(A)^{-1}g(y))$$

$$= \Phi_L(g(A^{-1}(x)), g(A^{-1}(y)))$$

$$= g\Phi_L(A^{-1}(x), A^{-1}(y))$$

$$= g\psi_L(x, y)$$

so that ψ_L restricts to a bilinear form, ψ, (non-degenerate because $\Phi_L \cong \psi_L$ is) $\psi: V \times V \to K$, on the G-invariant subset, $(V_L \otimes V_L)^G = V \otimes V$.

Furthermore, if $B \in \mathrm{Aut}_L(V_L)$ and

$$A^{-1}g(A) = f(g) = B^{-1}g(B),$$

then $AB^{-1} \in \mathrm{Aut}_K(V) \subset \mathrm{Aut}_L(V_L)$, which means that the bilinear forms resulting from choosing B and from A are K-isomorphic.

Clearly, the two constructions given here are mutually inverse, which completes the proof. $\quad\square$

(2.10) Corollary

Let K be a field of characteristic different from two. Then $H^1(K; O_n\bar{K})$ corresponds bijectively, via the construction of §2.9, to the set of K-isomorphism classes of symmetric, nonsingular K-bilinear forms of rank n.

Proof. Any symmetric, K-bilinear form, $\psi: K^n \times K^n \to K$, can be diagonalized. That is, K^n has a basis $\{v_i\}$ such that $\psi(v_i, v_j) = \delta_{ij}a_j \ (a_j \in K)$. Hence ψ is isomorphic to the standard bilinear form on K^n.

$$\Phi((x_1, x_2, \ldots), (y_1, y_2, \ldots)) = \sum x_i y_i$$

over $K(\sqrt{a_1}, \sqrt{a_2}, \ldots, \sqrt{a_n}) = L$ and $O_n(\Phi_R) = O_n\bar{K}$. $\square$

(2.11) HERMITIAN CLASS GROUPS

The hermitian class group is studied, for example, in [CN-T1/2/3; F2; T]. Among other uses it is the means to relate local constants of L-functions to the Galois-hermitian structure of rings of algebraic integers.

   The setting is the following: Let Γ be a finite group. The category of (right) hermitian $\mathbb{Z}[\Gamma]$-modules has objects (M, h), where M is a finitely generated, locally free (i.e., free over each localized group ring, $\mathbb{Z}_{(p)}[\Gamma]$) and

(2.12) $h: (M \otimes_{\mathbb{Z}} \mathbb{Q}) \times (M \otimes_{\mathbb{Z}} \mathbb{Q}) \to \mathbb{Q}[\Gamma]$ ($\mathbb{Q}$ is the rationals)

is a nondegenerate hermitian form which is

(2.13)
   (i) $\mathbb{Q}$-bilinear,
   (ii) satisfying $h(m, n\alpha) = h(m, n)\alpha$ $(m, n \in M \otimes_{\mathbb{Z}} \mathbb{Q}, \ \alpha \in \mathbb{Q}[\Gamma])$. Hence, if $n = x \otimes y \ (x \in M, y \in \mathbb{Q})$ and $\alpha = \sum_\gamma \alpha_\gamma \gamma \ (\gamma \in \Gamma, \alpha_\gamma \in \mathbb{Q})$, then $n\alpha = \sum_\gamma (x)\gamma \otimes y\alpha_\gamma$.
   (iii) $\overline{h(m, n)} = h(n, m)$, where $\bar{\alpha} = \overline{(\sum_\gamma \alpha_\gamma \gamma)} = \sum_\gamma \alpha_\gamma \gamma^{-1}$, is the canonical anti-involution.

Being nondegenerate means that there is an adjunction isomorphism

(2.14) $M \otimes_{\mathbb{Z}} \mathbb{Q} \xrightarrow{\cong} \mathrm{Hom}_{\mathbb{Q}[\Gamma]}(M \otimes_{\mathbb{Z}} \mathbb{Q}, \mathbb{Q}[\Gamma])$

given by $(m \mapsto h(m, -))$.

(2.15)   Here is an equivalent formulation of §2.13. Write $h(m, n) = \sum_{\gamma \in \Gamma} h_\gamma(m, n)\gamma$, where h is as in (2.12). If $e \in \Gamma$ is the identity, then $h_e: (M \otimes_{\mathbb{Z}} \mathbb{Q}) \times (M \otimes_{\mathbb{Z}} \mathbb{Q}) \to \mathbb{Q}$ determines h by the identity

(2.16) $$h_e(m, (n)\gamma^{-1}) = h_\gamma(m, n)$$

which follows from §2.13 (ii).

If h is as in §2.13, then

(2.17) $$h_e: (M \otimes_Z \mathbb{Q}) \times (M \otimes_Z \mathbb{Q}) \to \mathbb{Q}$$

is nondegenerate,

  (i) $\mathbb{Q}$-bilinear,
  (ii) satisfying $h_e(m, n) = h_e((m)\alpha, (n)\alpha)$ $(m, n \in M \otimes_Z \mathbb{Q}, \alpha \in \Gamma)$,
  (iii) $h_e(m, n) = h_e(n, m)$.

The parts (i), (iii), and nondegeneracy are clear, while (ii) follows from

$$h((m)\alpha, m) = \overline{\overline{h((m)\alpha, m)}}$$

$$= \overline{h(n, (m)\alpha)} \qquad \text{by §2.13 (iii),}$$

$$= \overline{(h(n, m)\alpha)} \qquad \text{by §2.13 (ii),}$$

$$= \alpha^{-1} h(m, n)$$

so that $h((m)\alpha, (n)\alpha) = \alpha^{-1} h(m, n)\alpha \in \mathbb{Q}[\Gamma]$.

(2.18) Examples

(a) Let N/K be a tamely ramified Galois extension of number fields. Set $\Gamma = G(N/K)$. The integers, O_N, in N are locally free over $\mathbb{Z}[\Gamma]$. Also $O_N \otimes_Z \mathbb{Q} \cong N$ as a (right) $\mathbb{Q}[\Gamma]$-module, so we may consider (O_N, h), where

$$h: N \times N \to \mathbb{Q}[\Gamma]$$

is given by

(2.19) $$h(m, n) = \sum_{\gamma \in \Gamma} \mathrm{Tr}_{N/\mathbb{Q}}(m, (n)\gamma)\gamma^{-1}$$

and $h_e(m, n) = \mathrm{Tr}_{N/\mathbb{Q}}(mn)$, the *trace form* of $N/\mathbb{Q}$.

(b) With N, K, Γ as in (a), set $M = O_K[\Gamma]$ so that $M \otimes_Z \mathbb{Q} \cong K[\Gamma]$ as a (right) $\mathbb{Q}[\Gamma]$-module, and set

(2.20) $$H: K[\Gamma] \times K[\Gamma] \to \mathbb{Q}[\Gamma]$$

equal to $H(m, n) = \mathrm{Tr}_{K/\mathbb{Q}}((\bar{m})n)$.

Hence, if $m = \sum k_\gamma \gamma$, $n = \sum k'_\gamma \gamma$, then

$$(2.21) \qquad H_e(m, n) = \left(\sum_{\gamma, \delta} \mathrm{Tr}_{K/Q}(k_\gamma k'_\delta) \gamma^{-1} \delta \right)_e$$

$$(2.22) \qquad \qquad = \sum_{g \in \Gamma} \mathrm{Tr}_{K/Q}(k_{\gamma^{-1}} k'_\gamma).$$

(2.23) Lemma

Let $\bar{Q}$ be the separable closure of Q. In (2.19) and (2.21) the map $\Phi: N \otimes_Q \bar{Q} \xrightarrow{\cong} \bar{Q}[\Gamma]$ given by

$$\Phi(n \otimes \alpha) = \alpha \sum_{\gamma \in \Gamma} ((n)\gamma) \gamma^{-1}$$

induces an isomorphism of Γ-invariant, bilinear forms

$$(O_N, h)_{\bar{Q}} \cong (O_K[\Gamma], H)_{\bar{Q}}.$$

Proof. Firstly Galois theory (the normal basis theorem) implies that Φ is an isomorphism. Also, if $z \in \Gamma$,

$$\Phi((n)z \otimes \alpha) = \alpha \sum_{\gamma \in \Gamma} ((n)z\gamma) \gamma^{-1}$$

$$= \alpha \sum_{\gamma} ((n)z\gamma) \gamma^{-1} z^{-1} z$$

$$= \alpha \sum_{\gamma} ((n)\gamma) \gamma^{-1} z$$

$$= \Phi(n \otimes \alpha) z$$

so that Φ is a Γ-equivalent isomorphism. Also we have $(m, n \in N, \alpha, \beta \in \bar{Q})$,

$$H_{\bar{Q}}(\Phi(n \otimes \alpha), \Phi(m \otimes \beta)) = \alpha\beta \sum_{\gamma, \delta} \mathrm{Tr}_{K/Q}(((n)\gamma \cdot (m)\delta) \gamma^{-1} \delta^{-1}),$$

whereas $h_{\bar{Q}}(n \otimes \alpha, m \otimes \beta) = \sum_\gamma \alpha\beta \, \mathrm{Tr}_{N/Q}(n((m))\gamma^{-1}$ so that, comparing e-components,

$$H_{\bar{Q}}(\Phi(n \otimes \alpha), \Phi(m \otimes \beta))_e = \alpha\beta \sum_\gamma \mathrm{Tr}_{K/Q}((mn)\gamma)$$

$$= \alpha\beta \, \mathrm{Tr}_{K/Q}(\mathrm{Tr}_{N/K}(mn))$$

$$= \alpha\beta \, \mathrm{Tr}_{N/Q}(mn)$$

$$= h_{\bar{Q}}(n \otimes \alpha, m \otimes \beta)_e$$

as required. $\square$

(2.24)   Motivated by the example of the previous result, we consider the Galois cohomology classification of Γ-invariant, nondegenerate, symmetric bilinear forms.

Let Γ be a finite group which, for my convenience, acts on the *left* of the K-vector space, V. Suppose that $\Phi: V \times V \to K$ is a fixed, Γ-invariant, nonsingular bilinear form. Let $\psi: V \times V \to K$ be another such form, and suppose that there exists

$$A \in \operatorname{Aut}_L(V_L)$$

that is *not necessarily Γ-equivariant*, satisfying (2.5) (i.e., $\Phi_L \cong \psi_L$).

As in (2.7) form ($g \in G(L/K)$),

(2.25) $$f(g) = A^{-1}(g \cdot (A(-\!\!\!-))) \in O(\Phi_L).$$

For $\gamma \in \Gamma$, set

(2.26) $$q(\gamma) = A^{-1}(\gamma(A(-\!\!\!-))) \in O(\Phi_L).$$

To see that $q(\gamma)$ is orthogonal, observe that

$$\Phi_L(A^{-1}\gamma(A(x)), A^{-1}\gamma(A(y))) = \psi_L(\gamma(A(x)), \gamma(A(y))) \qquad \text{by (2.5)}$$

$$= \psi_L(A(x), A(y)) \qquad \text{by } \Gamma\text{-invariance}$$

$$= \Phi_L(x, y) \qquad \text{by (2.5).}$$

Clearly, q is a homomorphism. Also we have

$$q(\gamma)f(g)(z) = q(\gamma)(A^{-1}(g(A(g^{-1}z))))$$

$$= A^{-1}\gamma(A(A^{-1}g(A(g^{-1}z))))$$

$$= A^{-1}(\gamma g(A(g^{-1}(z))))$$

$$= A^{-1}(g\gamma(A(g^{-1}(z))))$$

$$= A^{-1}(g(\gamma A)(z))$$

$$= A^{-1}g(A)g(A^{-1})g(\gamma A)(z)$$

$$= f(g)g(q(\gamma))(z).$$

Hence

(2.27) $$q(\gamma)f(g) = f(g)g(q(\gamma)) \qquad (g \in G(L/K), \ \gamma \in \Gamma).$$

Now let $\Gamma \times G(L/K)$ act on $O(\Phi_L)$ by the projection onto $G(L/K)$. Define

(2.28)
$$\begin{cases} F: \Gamma \times G(L/K) \to O(\Phi_L), \\ F(\gamma, g) = q(\gamma)f(g). \end{cases}$$

(2.29) Proposition

F in (2.28) is a 1-cocycle, and $\{\psi \mapsto [F]\}$ establishes a bijection

$$\left\{ \begin{array}{l} \Gamma\text{-invariant, symmetric} \\ \text{nonsingular, } K\text{-bilinear} \\ \text{forms } \psi: V \times V \to K \text{ which} \\ \text{are } L\text{-equivalent to } \Phi \end{array} \right\} \leftrightarrow H^1(\Gamma \times G((L/K); O(\Phi_L)).$$

Proof. Firstly, F is a 1-cocycle because

$$F(\gamma\gamma', gg') = q(\gamma)q(\gamma')f(g)g(f(g'))$$

$$= q(\gamma)f(g)g(q(\gamma')f(g')) \qquad \text{by (2.27)}$$

$$= F(\gamma, g)((\gamma, g)F(\gamma', g')), \qquad \text{as required.}$$

If we choose $\beta \in \mathrm{Aut}_L(V_L)$, instead of A, to define $q'(\gamma)$ and $f'(g)$, then $(A^{-1}B)f'(g)g(A^{-1}B)^{-1} = f(g)$ from §2.9. Whereas

$$q'(\gamma) = B^{-1}\gamma(B(-\!-\!))$$

$$= (B^{-1}A)A^{-1}\gamma(A(A^{-1}B)-\!\!)$$

$$= (B^{-1}A)q(\gamma)((A^{-1}B)-\!\!)$$

so that

$$F'(\gamma, g) = q'(\gamma)f'(g)$$

$$= (B^{-1}A)q(\gamma)f(g)g(A^{-1}B)$$

$$= (B^{-1}A)F(\gamma, g)(\gamma, g)(A^{-1}B)$$

and $[F] = [F'] \in H^1(\Gamma \times G(L/K), O(\Phi_L))$.

Conversely, given $[F] \in H^1(\Gamma \times G(L/K), O(\Phi_L))$, we can reconstruct $\psi: V \times V \to K$ by writing $F(1, g) = f(g) = A^{-1}g(A)$ and $\psi = \Phi_L(A^{-1}(-\!-\!), A^{-1}(-\!-\!))$ restricted to $(V_L \otimes V_L)^{G(L/K)} = V \otimes V$. Also ψ is Γ-invariant with respect to f, the (left) Γ-action on V_L via $F(\gamma, 1) = q(\gamma) \in O(\Phi_L)$, where we note that $F(-\!-\!, 1): \Gamma \to O(\Phi_L)$ is a homomorphism. Also, if $F(1, g) = B^{-1}g(B)$, then, from §2.9, we know that the resulting form on V differs by $AB^{-1} \in \mathrm{Aut}_K(V)$ from the form defined using A. However, we have

$$q(\gamma) = B^{-1}BA^{-1}\gamma(AB^{-1}(B\text{---}))$$
$$= B^{-1}((AB^{-1})^{-1}\gamma(AB^{-1})(B\text{---}))$$

so that $AB^{-1} \in \mathrm{Aut}_K(V)$ transports the $q(\gamma)$-action to the $q'(\gamma)$-action, so that the Γ-isomorphism class of (V, ψ) depends only on $[F] \in H^1(\Gamma \times G(L/K); O(\Phi_L))$.

These two constructions are evidently mutually inverse. $\qquad\square$

Taking a limit over L, we obtain the following:

(2.30) Corollary

Let Γ be a finite group. Let K be a field of characteristic different from two. Then $H^1(\Gamma \times K; O_n\bar{K})$ corresponds bijectively, via the construction of §2.28, to the set of Γ-isomorphism classes of symmetric, Γ-invariant, nonsingular K-bilinear forms of rank n.

(2.31) CENTRAL SIMPLE ALGEBRAS

Suppose that K is a field and that A is a finite dimensional K-algebra, simple and having centre equal to K. By Wedderburn's theorem [Ser 2, p. 157] there exists a finite Galois extension, F/K, such that there is an isomorphism of F-algebras:

$$(2.32) \qquad\qquad B: M_n F \xrightarrow{\cong} A \otimes_K F,$$

where $\dim_K A = n^2$ and $M_n F$ denotes the $n \times n$ matrices with entries in F.

We have (left) $G(F/K)$ actions on these algebras

$$(2.33) \qquad\qquad \begin{cases} g(a \otimes f) = a \otimes g(f), \\ g(X) = g((x_{ij})) = (g(x_{ij})), \end{cases}$$

where $g \in G(F/K)$; $a \in A$; $f, x_{ij} \in F$.

Define

$$(2.34) \qquad\qquad g(B): M_n F \xrightarrow{\cong} A \otimes_K F$$

by $g(B)(X) = g(B(g^{-1}(X)))$.

(2.35) Lemma

In (2.34), $g(B)$ is an isomorphism of F-algebras.

Proof. Clearly, $g(B)$ is a ring homomorphism. However, if $f \in F$ and $X \in M_n F$, then

$$g(B)(fX) = g(B(g^{-1}(f)g^{-1}(X)))$$
$$= g(g^{-1}(f)B(g^{-1}(X)))$$
$$= (gg^{-1}(f))g(B)(X)$$
$$= fg(B)(X),$$

as required. $\qquad\qquad\qquad\qquad\qquad\qquad\qquad\qquad\qquad\qquad\qquad\qquad$ $\square$

Any matrix $Y \in GL_n F$ acts on $M_n F$ by conjugation, $(X \mapsto YXY^{-1})$, and this F-algebra automorphism depends only on Y modulo factors, $F^* I_n$. This induces an isomorphism

(2.36) $\qquad \begin{cases} PGL_n F = (GL_n F)/F^* \cong \text{Alg Aut}_F (M_n F), \\ Y(\text{mod } F^*) \mapsto (Y - Y^{-1}). \end{cases}$

Therefore we may construct a 1-cocycle by the formula

(2.37) $\qquad \begin{cases} \psi_A : G(F/K) \to PGL_n F, \\ \psi_A(g) = B^{-1}g(B) \in \text{Alg Aut}_F (M_n F). \end{cases}$

If we choose a second isomorphism, $C : M_n F \xrightarrow{\cong} A \otimes_K F$, then the resulting 1-cocycle is

$$\psi'_A(g) = C^{-1}g(C)$$
$$= (C^{-1}B)\psi_A(g)g(C^{-1}B)^{-1}$$

so that we have shown the following:

(2.38) Lemma

The construction of (2.37) *gives a well-defined map*

$$\begin{cases} \text{central simple } K\text{-algebras,} \\ \text{which are } F\text{-isomorphic to} \\ M_n F \end{cases} \xrightarrow{\psi} H^1(G(F/K); PGL_n F).$$

(2.39) Next observe that if ψ_A is a 1-coboundary in (2.37), then there exists $C \in \text{Alg Aut}_F (M_n F) \cong PGL_n F$ such that $B^{-1}g(B) = C^{-1}g(C)$ so that the F-algebra isomorphism

$$CB^{-1} : A \otimes_K F \xrightarrow{\cong} M_n F$$

commutes with the $G(F/K)$-actions. Therefore taking CB^{-1} restricted to $G(F/K)$-invariants, we obtain the following:

(2.40) Lemma

If ψ_A is a 1-coboundary then $A \cong M_n K$, as K-algebras.

(2.41) Now suppose that $f: G(F/K) \to PGL_n F$ is a cocycle. We now proceed to extract an n^2-dimensional, central, simple K-algebra from f.

(2.42)   Set $A = \{X \in M_n F \mid g(X) = (f(g)^{-1}(X)$ for all $g \in G(F/K)\}$.

A is a K-algebra since $g(-)$ and $f^{-1}(g)$ are K-algebra automorphisms of $M_n F$.

If we change $f(g)$ to $C^{-1} f(g) g(C)$ for $C \in PGL_n F = \text{Alg Aut}_F(M_n F)$, then

$$g(X) = g(C)^{-1} f(g)^{-1} C(X)$$

so that

$$g(C(X)) = f(g)^{-1} C(X).$$

Therefore C induces an isomorphism of K-algebras between the algebra associated to $f(-)$ and that associated to the (cohomologous) $C^{-1} f(-)((-)(C))$.

(2.43) Proposition

The construction of (2.37) yields a bijection, ψ, in §2.38. The base-point in $H^1(G(F/K); PGL_n F)$ corresponds to $\text{End}_K(K^n) \cong M_n K$.

Proof. In §2.38, we checked that ψ was well-defined, and we gave, in (2.42), a well-defined construction (λ, say) which we must now verify to be ψ^{-1}.
   Suppose that we have B, as in (2.32), then we form $f(g) = B^{-1} g(B)$ and consider matrices $X \in M_n F$ satisfying

$$g(X) = f(g)^{-1}(X)$$
$$= g(B)^{-1}(B(X))$$
$$= g(B^{-1}(g^{-1}(B(X)))).$$

This happens if and only if $B(X) \in (A \otimes_K F)^{G(F/K)} = A$. Hence up to K-isomorphism, $\lambda(\psi(A)) \cong A$, as required.
   Suppose, conversely, that $f \in Z^1(G(F/K); PGL_n F)$. The map $(\rho: g \mapsto f(g) g(-) \in \text{Alg Aut}_F(M_n F))$ is a homomorphism, since

$$f(gh)(gh(X)) = f(g)g(f(h))(gh(X))$$
$$= \rho(g)(f(h)h(X))$$
$$= \rho(g)(\rho(h)(X)).$$

Therefore, by a standard Galois theory argument [A, pp. 36–37], $\dim_K((M_nF)^{\rho(G(F/K))}) = \dim_F M_nF = n^2$. Hence we may consider $\psi(\lambda(f))$. The isomorphism

$$B^{-1} : (M_nF)^\rho \otimes_K F \to M_nF$$

is given by scalar multiplication, where $(M_nF)^\rho$ denotes the $\rho(G(F/K))$-invariants.

Now suppose $V \in (M_nF)^\rho$ and $k \in F$, then

$$B^{-1}g(B(vk)) = B^{-1}(v \otimes gk)$$
$$= v \cdot g(k)$$
$$= g(g^{-1}(v)k)$$
$$= g(f(g^{-1})^{-1}(vk)).$$

Hence

$$B^{-1}g(B)(vk) = B^{-1}g(B(g^{-1}vk))$$
$$= g(f(g^{-1})^{-1}(g^{-1}vk))$$
$$= g(f(g^{-1})^{-1}(vk)).$$

However, such elements, $z = vk$, generate M_nF so that $B^{-1}g(B) = g(f(g^{-1})^{-1})$. One may verify that $(\hat{f} : g \mapsto g(f(g^{-1})^{-1}))$ is a 1-cocycle if and only if f is. Also $f \mapsto \hat{f}$ is an involution on 1-cocycles which passes to $H^1(G(F/K); PGL_nF)$ so that $\psi\lambda[f] = [\hat{f}]$, which completes the proof.   □

(2.44)   Consider the central extension

$$F^* \rightarrowtail GL_nF \twoheadrightarrow PGL_nF,$$

and the resulting exact sequence of sets and groups given by §1.26:

$$\cdots \longrightarrow \{1\} \longrightarrow H^1(G(F/K); PGL_nF) \xrightarrow{\Delta} H^2(G(F/K); F^*).$$

Here we have used §1.34 to replace $H^1(G(F/K); GL_nF)$ by $\{1\}$.

(2.45) Lemma [Ser 2, p. 158–159]

If $n = [F : K]$, then Δ is onto in §2.44.

Proof. We will show that every 2-cocycle $a: G(F/K)^2 \to F^*$ is writable (cf., (1.22)) as

$$a(g, g') = p(g)g(p(g'))(p(gg'))^{-1},$$

where $p: G(F/K) \to GL_n F$ reduces to a 1-cocycle in $PGL_n F$.

Let V be the F-vector space with basis $\{e_g | g \in G(F/K)\}$. Let

$$p(g) \in \mathrm{End}_F(V)$$

be given by $p(g)(e_{g'}) = a(g, g')e_{gg'}$. Hence $p(g) \in GL(V) \cong GL_n F$. Also

$$p(g)g(p(g'))(e_{g''}) = p(g)(g(a(g', g''))e_{g'g''})$$

$$= a(g, g'g'')g(a(g', g''))e_{gg'g''},$$

$$a(g, g')p(gg')e_{g''} = a(g, g')a(gg', g'')e_{gg'g''}.$$

Since a is a 2-cocycle, we see that

$$p(g)g(p(g')) = a(g, g')p(gg'),$$

as required. Note that this last equation shows that p is a 1-cocycle *in $PGL_n F$* since $a(g, g') \in F^*$. □

(2.46) A finite-dimensional K-algebra is central simple if and only if it is isomorphic to a matrix ring with entries in a division algebra, D, with centre K. See [Ser 2, p. 157, Prop. 7].

The *Brauer group* of K is the set of isomorphism classes of central simple K-algebras subject to the relation that $M_n D$ is equivalent to D. The operation $(\_ \otimes_K \_)$ makes this set into a group, denoted $\mathrm{Br}(K)$. This relation is equivalent to that of making $\mathrm{End}_K(K^n) \cong M_n K$ equivalent to K, the unit of the group.

Combining §§2.38, 2.40, 2.43, and 2.45, and taking F/K to the limit in $\bar{K}/K$, we obtain the following:

(2.47) Theorem

Δ of §2.44 *induces an isomorphism*

$$\Delta: \mathrm{Br}(K) \xrightarrow{\cong} H^2(K; \bar{K}^*).$$

Proof. Certainly we have shown that Δ is a bijection. It remains to check that tensor product of algebras corresponds to the sum in $H^2(K; \bar{K}^*)$. However, this is easy. If $B^{-1}: A \otimes_K F \xrightarrow{\cong} M_n F$ and $C^{-1}: A' \otimes_K F \xrightarrow{\cong} M_m F$ are F-algebra isomorphisms, then $A \otimes_K A'$ is represented by the 1-cocycle

$(g \mapsto B^{-1}g(B) \otimes_F C^{-1}g(C))$. We have a commutative diagram of group extensions

$$
\begin{array}{ccccc}
F^* \times F^* & \longrightarrow & GL_nF \times GL_mF & \longrightarrow & PGL_nF \times PGL_mF \\
\downarrow\text{\footnotesize mult} & & \downarrow{\otimes} & & \downarrow \\
F^* & \longrightarrow & GL_{nm}F & \longrightarrow & PGL_{nm}F
\end{array}
$$

which, by naturality of the coboundary, Δ, shows that the class $[A \otimes A']$ is given by the image of $([A], [A'])$ under the map

$$(\text{mult})_* : H^2(G(F/K), F^* \times F^*) \to H^2(G(F; K); F^*).$$

This image represents the $H^2(K; \bar{K}^*)$-sum, $[A] + [A']$, as required.   □

(2.48)   THE EQUIVARIANT BRAUER GROUP

Let Γ be a finite group, and let K be a field. By analogy with §2.46, we may define the Brauer group of central, simple $K[\Gamma]$-algebras, $\mathrm{Br}(\Gamma, K)$, in the following manner [F3].

A (left) central, simple $K[\Gamma]$-algebra, R, is a finite dimensional, central simple K-algebra together with a left action by Γ on R in such a way that $(\gamma \cdot -): R \to R$ is a K-*algebra* automorphism for all $\gamma \in \Gamma$.

$\mathrm{Br}(\Gamma, K)$, the *equivariant Brauer group*, is the group of isomorphism classes of central, simple $K[\Gamma]$-algebras modulo the following relation:

$$(2.49) \qquad\qquad \mathrm{End}_K(V) \sim 1$$

for all finite dimensional $K[\Gamma]$-modules, V. The product in $\mathrm{Br}(\Gamma, K)$ is induced by tensor product (over K).

Let A be such an algebra, and let $B: M_nF \xrightarrow{\cong} A \otimes_K F$ be the isomorphism of (2.32). Define a homomorphism

$$(2.50) \qquad \begin{cases} \Phi_A : \Gamma \to PGL_nF = \mathrm{Alg\,Aut}_F(M_nF) & \text{by} \\ \Phi_A(\gamma) = B^{-1}((\gamma \cdot -) \otimes_K 1)B. \end{cases}$$

Note that $\Phi_A(\gamma)$ is an F-algebra map because B is and $(\gamma \cdot -)$ is a K-algebra map.

Let $\psi_A(g) = B^{-1}g(B) \in PGL_nF$ be the 1-cocycle of (2.37). In PGL_nF we have $(\gamma \in \Gamma, g \in G(F/K))$,

$$\Phi_A(\gamma)\psi_A(g) = [B^{-1}(\gamma \otimes 1)B]B^{-1}g(B)$$

$$= B^{-1}(\gamma \otimes 1)g(B),$$

while

$$\psi_A(g)g(\Phi_A(\gamma)) = B^{-1}g(B)g(B^{-1}(\gamma \otimes 1)B)$$

$$= B^{-1}g(\gamma \otimes 1)g(B)$$

$$= B^{-1}(\gamma \otimes 1)g(B),$$

since γ is a K-algebra map. Hence $\Phi_A(\gamma)\psi_A(g) = \psi_A(g)g(\Phi_A(g))$, and as in §§2.27–2.29, this gives a 1-cocycle on $\Gamma \times G(F/K)$.

(2.51) Proposition

With the preceding notation,

$$\{A \mapsto ((\gamma, g) \mapsto \Phi_A(\gamma)\psi_A(g)))\}$$

defines a bijection

$$\left\{ \begin{array}{l} \text{isomorphism classes} \\ \text{of central simple} \\ K[\Gamma]\text{-algebras, split} \\ \text{over } F \text{ of dimension } n^2 \end{array} \right\} \leftrightarrow H^1(\Gamma \times G(F/K); PGL_nF).$$

Here Γ acts trivially on PGL_nF, and M_nF corresponds to the base-point.

Proof. Following the proof of §2.43, it suffices to verify that the construction of the inverse, given by (2.42), does in fact yield a $K[\Gamma]$-algebra. However, if $F: \Gamma \times G(F/K) \to PGL_nF$ is a 1-cocycle, then $\Phi(\gamma) = F(\gamma, 1)$ gives a homomorphism $\Phi: \Gamma \to \text{Alg Aut}_F(M_nF)$. But, if $X \in M_nF$ satisfies $g(X) = \psi(g)^{-1}(X)\,(\psi(g) = F(1, g))$, then

$$\Phi(\gamma)(X) = \Phi(\gamma)(\psi(g)g(X))$$

$$= \psi(g)g(\Phi(\gamma))g(X)$$

$$= \psi(g)g(\Phi(\gamma)(X))$$

so that $\Phi(\gamma)$ restricts to a K-automorphism of the algebra, A, defined in (2.42). Hence the inverse construction gives a central, simple $K[\Gamma]$-algebra, A, and the proof is completed as in §2.43.   □

(2.52)   Now consider the exact sequence given by §1.26 (see §2.44):

$$\cdots \to H^1(\Gamma; GL_nF) \xrightarrow{j_*} H^1(\Gamma \times G(F/K); PGL_nF) \xrightarrow{\Delta} H^2(\Gamma \times G(F/K); F^*).$$

The image, $j_*(f)$, corresponds, in terms of central, simple algebras, to $M_n F$, with Γ acting through a homomorphism, $f: \Gamma \to GL_n F$. Hence $\mathrm{im}\,(j_*)$ corresponds to algebras of the form $\mathrm{End}_K(V)$, with V an n-dimensional $K[\Gamma]$-module.

There, by the arguments of §§2.45 and 2.47, we obtain the following:

(2.53) Theorem

Δ of §2.52 induces an isomorphism

$$\Delta: \mathrm{Br}\,(\Gamma, K) \to H^2(\Gamma \times K, \bar{K}^*).$$

In addition

$$H^2(\Gamma \times K; \bar{K}^*) \cong H^2(\Gamma; K^*) \oplus \mathrm{Br}\,(K).$$

Proof. We have only to prove the second isomorphism. However, we have a Hochschild-Serre spectral sequence

$$E_2^{s,t} = H^s(\Gamma; H^t(K; \bar{K}^*)) \Rightarrow H^{s+t}(\Gamma \times K; \bar{K}^*).$$

By §1.33, $E_2^{s,1} = 0$ so that the E_2-term looks as follows (since $H^0(K; \bar{K}^*) = (\bar{K}^*)^{\Omega_K} = K^*$):

(2.54)

It is clear that the forgetful map $H^i(\Gamma \times K; K^*) \to H^i(K; K^*)$ is split surjective, which means that d_2 and d_3 are zero on $\mathrm{Br}\,(K)$ in (2.54). Hence $H^2(\Gamma \times K; \bar{K}^*)$ has a (split) composition series

$$E_\infty^{2,0} = E_2^{2,0} \rightarrowtail H^2(\Gamma \times K; \bar{K}^*) \twoheadrightarrow E_\infty^{0,2} = E_2^{0,2},$$

as required. $\square$

(2.55) Remark

Under the isomorphism of §2.53 the second factor, $\mathrm{Br}\,(K)$, in $\mathrm{Br}\,(\Gamma, K)$ corresponds to sending a central, simple $K[\Gamma]$-algebra to its underlying K-algebra. The first factor, $H^2(\Gamma; K^*)$, corresponds to the following construction. If A is such a $K[\Gamma]$-algebra, we have the action map $\Phi: \Gamma \to \mathrm{Alg}\,\mathrm{Aut}_K(A)$. We have a central extension, if $U(A)$ are the units of A,

$$K^* \rightarrowtail U(A) \twoheadrightarrow \mathrm{Alg}\,\mathrm{Aut}\,(A).$$

This follows from the Skolem-Noether theorem for Azumaya algebras [Mi, p. 13]. *The $H^2(\Gamma; K^*)$ class of A is given by $\Delta(\Phi)$, where Δ is the coboundary in the sequence (see §1.26) for this extension.*

3. SPECIFIC GALOIS COHOMOLOGY REPRESENTATIVES

(3.1)   Let K be a field which, throughout this section, *will have char $K \neq 2$, when we are treating symmetric bilinear forms.* We will now examine the Galois representatives, in terms of the results of §§2.30–2.51, of specific bilinear forms and central simple algebras.

(3.2) Example

If $\psi: V \times V \to K$ is a symmetric, nondegenerate bilinear form of rank n, we may choose a basis $\{e_i\}$ for V such that

$$\psi(e_i, e_j) = \delta_{ij}a_i \qquad (a_i \in K^*).$$

Over the field $L = K(\sqrt{a_1}, \sqrt{a_2}, \ldots, \sqrt{a_n})$ (see §2.10), we may make the transformation of $V \otimes_K L$,

$$A(e_i) = (1/\sqrt{a_i})e_i$$

to obtain $\psi(Ae_i, Ae_j) = \delta_{ij}a_i/(\sqrt{(a_i a_j)}) = \delta_{ij}$. Hence, if we define for $a \in K^*$, a homomorphism,

(3.3)
$$\begin{cases} l(a): G(\bar{K}/K) = \Omega_K \to O_1(1) = \{\pm 1\}, \\ \text{given by } l(a)(g) = g(\sqrt{a})/(\sqrt{a}), \end{cases}$$

then, from (2.7), the Galois representative for (V, ψ) is given by the *homomorphism*

(3.4)
$$f: (g \in \Omega_K) \mapsto \begin{bmatrix} l(a_1)(g) & & & \\ & l(a_2)(g) & & \\ & & \ddots & \\ & & & l(a_n)(g) \end{bmatrix} \in O_n\bar{K}.$$

Of course, we remark that a 1-cocycle $f: \Omega_K \to O_n\bar{K}$ which lands in $O_1(\bar{K})^n$, or *more generally in $O_n(K)$*, is a homomorphism.

(3.5)   Since all (V, ψ) are diagonalizable, one might innocently form the opinion that (3.4) finishes the subject in a definitive manner. This is unfortunately not the case—as one discovers immediately one attempts to diagonalize a bilinear form!

Here is a very simple example.

(3.6) Example

Let $\langle a \rangle : K \times K \to K$ denote the form $\langle a \rangle(x, y) = axy$ $(a \in K^*)$. Suppose that $L = K(\sqrt{c})$, where $c \in K$ and $\sqrt{c} \notin K$. We have the *trace map*, $\mathrm{Tr}_{L/K}: L \to K$,

$$\mathrm{Tr}_{L/K}(a + b\sqrt{c}) = (a + b\sqrt{c}) + (a - b\sqrt{c}) = 2a \in K,$$

and the trace form of L/K is the bilinear form $\psi : L \times L \to K$ given by $\psi(x, y) = \mathrm{Tr}_{L/K}(xy)$. Explicitly,

$$\psi(a + b\sqrt{c}, \alpha + \beta\sqrt{c}) = \mathrm{Tr}_{L/K}(a\alpha + b\beta c + \sqrt{c}(\ldots))$$

$$= 2(a\alpha + b\beta c).$$

Hence, with a basis $\{1, \sqrt{c}\}$ for L/K, the trace form has the matrix

$$\begin{bmatrix} 2 & 0 \\ 0 & 2c \end{bmatrix},$$

and the trace form of L/K is $\langle 2 \rangle + \langle 2c \rangle$.

More generally, if $0 \neq x + y\sqrt{c} = z \in L$, we may form a bilinear form (a *scaled trace form*)

(3.7)
$$\begin{cases} \psi' : L \times L \to K & \text{given by} \\ \psi'(u, v) = \mathrm{Tr}_{L/K}(zuv). \end{cases}$$

This has the matrix

$$X = \begin{bmatrix} 2x & 2yc \\ 2yc & 2xc \end{bmatrix}.$$

Changing the basis by

$$Y = \begin{bmatrix} 1 & 0 \\ -(yc)/a & 1 \end{bmatrix},$$

we obtain, if $x \neq 0$,

$$YXY^t = \begin{bmatrix} 2x & 0 \\ 0 & 2(x^2c - y^2c^2)/x \end{bmatrix}.$$

Hence the scaled trace form, for L/K and $x + y\sqrt{c}$, is $\langle 2x \rangle + \langle 2(x^2c - y^2c^2)/x \rangle$, when $x \neq 0$.

If $x = 0$, using

$$Y = \begin{bmatrix} 1 & 1 \\ 0 & 1 \end{bmatrix},$$

we find that the scaled trace form is $\langle 4yc \rangle + \langle -yc \rangle$, or equivalently $\langle yc \rangle + \langle -yc \rangle$.

(3.8)   FROHLICH'S BILINEAR FORM [F,§2]

Let $G = G(L/K)$, the Galois group of a finite, Galois extension L/K. Suppose that we are given a nonsingular bilinear form $b: V \times V \to K$ (not necessarily symmetrical) and a homomorphism

$$(3.9) \qquad\qquad T: G \to O(b) \subset \operatorname{Aut}_K(V),$$

where $O(b)$ is the orthogonal group of (V, b). By means of (3.9), G acts (on the left) on V. We will define a nondegenerate bilinear form $(V^* = \operatorname{Hom}_K(V, K))$

$$(3.10) \qquad\qquad \beta^*(T, b): V^* \times V^* \to K$$

by a construction originally due to Frohlich [F.§2].

Firstly, identify V^* with $\operatorname{Hom}_K(V, L)^G$. Let $\theta \in L$ be a normal basis element [A, p. 66] so that $\{g(\theta) | g \in G\}$ is a basis for L/K. Any K-linear map, $f: V \to L$, may be written

$$(3.11) \qquad\qquad f(v) = \sum_{g \in G} f(v)_g (g^{-1}(\theta)) \qquad \text{with } f(v)_g \in K.$$

Define an isomorphism

$$(3.12) \qquad\qquad \begin{cases} \psi: \operatorname{Hom}_K(V, L)^G \to V^* \\ \text{by } \psi(f)(v) = f(v)_e. \end{cases}$$

If f is G-fixed, then $g(f(v)) = f(T(g)v)$ so that we obtain

$$(3.13) \qquad\qquad f(v) = \sum_{g \in G} \psi(f)(T(g)(v)) \cdot (g^{-1}\theta).$$

Now choose a K-basis for V, $\{v_1, \ldots, v_m\}$, and let $u_1, \ldots, u_m \in V$ satisfy $b(v_1, u_j) = \delta_{ij}$. For $f, h \in \operatorname{Hom}_K(V, L)^G$ define

$$(3.14) \qquad\qquad \beta^*(T, b)(f, h) = \sum_{j=1}^{m} f(v_j) h(u_j).$$

Henceforth we will abbreviate $\beta^*(T, b)$ to β^* for the rest of this discussion. Note that β^* is symmetric if b is. Define $\beta: V \times V \to K$ by requiring that

(3.15) $$\begin{cases} \{f \mapsto \beta^*(-, f)\}: V^* \to (V^*)^* \cong V & \text{and} \\ \{x \mapsto \beta(x, -)\}: V \to V^* & \text{be mutually inverse.} \end{cases}$$

We will now determine the manner in which $\beta = \beta(T, b)$ is represented in $H^1(G(L/K); O(b_L))$.

(3.16) Proposition

The bilinear form $\beta: V \times V \to K$ of (3.14–3.15) is represented in $H^1(G(L/K); O(b_L))$ by the homomorphism, $T: G(L/K) \to O(b) \to O(b_L)$.

Proof. Define $\hat{v}_i, \hat{u}_j \in V^*$ by the equations $\hat{v}_i(v_j) = \hat{u}_i(u_j) = \delta_{ij}$. Let $X = (b(v_i, v_j))$ be the matrix of b with respect to the $\{v_i\}$. In $\operatorname{Hom}_K(V, L)^G$, $\psi^{-1}(\hat{v}_i)$ is given by

$$\left(x \mapsto \sum_g \hat{v}_i(T(g)(x))(g^{-1}\theta) \right),$$

by (3.13). Also, if $v_j = \sum_r Y_{rj} u_r$, then

$$\begin{aligned} X_{ij} &= b(v_i, v_j) \\ &= \sum_r Y_{rj} b(v_i, u_r) \\ &= Y_{ij}. \end{aligned}$$

Hence $u_s = \sum_r (X^{-1})_{rs} v_r$. Therefore

$$\begin{aligned} \beta^*(\hat{v}_i, \hat{v}_j) &= \sum_{1 \le s \le m} (\psi^{-1}(\hat{v}_i)(v_s))(\psi^{-1}(\hat{v}_j)(u_s)) \\ &= \sum_{\substack{g, h \in G \\ 1 \le s, r \le m}} \hat{v}_i(T(g)(v_s))(X^{-1})_{rs} \hat{v}_j(T(h)(v_r)) g^{-1}(\theta) h^{-1}(\theta) \\ &= \sum_{\substack{g, h \in G \\ 1 \le s, r \le m}} T(g)_{is}(X^{-1})_{rs} T(h)_{jr} g^{-1}(\theta) h^{-1}(\theta) \\ &= \sum_{g, h \in G} (T(g)(X^{-1})^t T(h)^t)_{ij} g^{-1}(\theta) h^{-1}(\theta). \end{aligned}$$

Now set

(3.17) $$A = \sum_{g \in G} T(g) g^{-1}(\theta) \in M_m(L).$$

Therefore, if $X^* = (X^{-1})^t$, we have

$$(3.18) \quad \begin{cases} AX^*A^t = \sum_{g,h} T(g)X^*T(h)^t g^{-1}(\theta)h^{-1}(\theta) & \text{and} \\ g(A) = \sum_h T(hg^{-1})T(g)gh^{-1}(\theta) = AT(g). \end{cases}$$

Hence $A^{-1}g(A) = T(g)$ for $g \in G$. Now the matrix of β^*, with respect to $\{\hat{v}_i\}$, is $Z = AX^*A^t$ so that the first map in (3.15) sends $\hat{v}_j$ to $\sum_r Z_{rj}v_r$.

Hence the inverse map is $(v_i \mapsto \sum_r Z_{ri}^{-1}\hat{v}_r)$ so that $\beta(v_i, v_j) = \sum_r Z_{ri}^{-1}\hat{v}_r(v_j) = Z_{ji}^{-1}$. Therefore the matrix of β, with respect to the $\{v_i\}$, is $Z^* = A^*XA^{-1} = A^*X(A^*)^t$.

Hence

$$\beta(v_i, v_j) = \sum_{r,s} A_{ir}^* X_{rs} A_{sj}^{-1}$$

$$= \sum_{r,s} A_{ri}^{-1} b(v_r, v_s) A_{sj}^{-1}$$

$$= b(A^{-1}(v_i), A^{-1}(v_j)).$$

From (2.5), (2.7), and (3.18) we see that β is classified by

$$(g \mapsto A^{-1}g(A) = T(g)),$$

as required. $\qquad\qquad\qquad\qquad\qquad\qquad\qquad\qquad\qquad\qquad\qquad\square$

(3.19) THE SCHARLAU TRANSFER

Suppose that L/K is a finite, separable extension and that

$$\psi: V \times V \to L$$

is a nondegenerate, bilinear form of rank $m = \dim_L V$. We may compose ψ with the trace of L/K, $\text{Tr}_{L/K}$, to form a nondegenerate, bilinear form over K of rank md, where $[L:K] = d$. This is called the *Scharlau transfer* of ψ:

$$(3.20) \qquad\qquad \text{Tr}_{L/K}^S(\psi): V \times V \to L \to K.$$

(3.21) We are now going to determine, in terms of a representative for ψ, a Galois cohomology representative for $\text{Tr}_{L/K}^S(\psi)$.

Let us restrict to the case in which ψ is symmetric so that it determines a class

$$[\psi] \in H^1(L; O_m L),$$

where $O_m L = \{X \in GL_m(L) \mid XX^t = I_m\}$ and, of course, char $K \neq 2$.

It is always possible to represent ψ by a 1-cocycle, which is a *homomorphism* of the form

$$(3.22) \qquad\qquad F: \Omega_L \xrightarrow{\pi} G(E/L) \xrightarrow{f} O_m(K) \xrightarrow{i} O_m(L),$$

where π is the canonical surjection. For example, we may diagonalize ψ and represent it as in §3.2.

On the other hand, since $[G(E/K):G(E/L)]=d$, we may form the *induced representation*

$$(3.23) \qquad \operatorname{Ind}_{G(E/L)}^{G(E/K)}(f) \colon G(E/L) \to O_{md}(K).$$

We will frequently denote (3.23) by $\operatorname{Ind}_{L/K}(f)$, and similarly, we have

$$(3.24) \qquad \operatorname{Ind}_{L/K}(F) \colon \Omega_K \xrightarrow{\ \pi\ } G(E/K) \xrightarrow{\ \operatorname{Ind}_{L/K}(f)\ } O_{md}(K).$$

(3.25) Theorem

Suppose that L/K is a finite, separable extension, $\operatorname{char} K \neq 2$ and that $\psi \colon V \times V \to L$ is a symmetric, nondegenerate L-bilinear form represented in $H^1(L; O_m(L))$ by the homomorphism, F, in (3.22). Then $\operatorname{Tr}_{L/K}^S(\psi)$ is represented in $H^1(K; O_{md}(K))$ $(d = [L:K])$ by $\operatorname{Ind}_{L/K}(F)$ of (3.24).

(3.26)   The discussion which establishes §3.25 will occupy §§3.26–3.41.

Proof. Firstly, let us recall the structure of the separable extension L/K. Let N/K be the *normal closure* of L/K. It is a Galois extension with group $G = G(N/K)$.

As G permutes the copies of L in N, we have a homomorphism:

$$G \to \Sigma_d \qquad (d = [L:K])$$

so that if $H = \operatorname{stab}(1)$, the stabilizer of 1 as G acts on $\{1,\ldots,d\}$, then

$$[G:H] = d \quad \text{and} \quad L = N^H.$$

Choose $\{v_1,\ldots,v_d\}$ a basis for L/K, and let $\{g_1,\ldots,g_d\}$ be a set of (left) coset representatives for G/H. Choose $\hat{g}_i \in \Omega_K = G(\bar{K}/K)$ such that $\hat{g}_i$ maps to g_i under the canonical surjection, $\Omega_K \to G$. Note that if $x \in N \subset \bar{K}$, then $\hat{g}_i(x) = g_i(x)$.

The distinct K-embeddings of L into $\bar{K}$ are just $\{x \mapsto \hat{g}_i(x) = g(x)\}$. There are d of these embeddings, which is the maximum possible by [A, Thm. 13, p. 36], but if $g_i = g_j$ on L, then $H' = \langle H, g_i g_j^{-1} \rangle$ fixes L so that $N^{H'} = N^H$, which implies $H' = H$ by [A, Thm. 16, p. 47].

Now let (V, ψ) be the L-bilinear form of §3.25, and let $\{w_1,\ldots,w_m\}$ be a basis of V/L. Hence $\{w_i v_j\}$ is a K-basis for V. Let $B \in GL_m(L)$ be given by $B_{ij} = \psi(w_i, w_j)$.

We will write $md \times md$ matrices as $d \times d$ matrices of $m \times m$ blocks. The matrix for $\mathrm{Tr}^S_{L/K}(V, \psi)$, with respect to the aforementioned basis, has (i,j)th block

$$(3.27) \qquad \tilde{B}_{ij} = \sum_{u=1} g_u(v_i) g_u(v_j) g_u(B),$$

since $\mathrm{Tr}_{L/K}(x) = \sum_{u=1} g_u(x)$.

Suppose that $A \in GL_m\bar{K}$ satisfies

$$(3.28) \qquad AA^t = B.$$

We define a matrix of $m \times m$ blocks, $\tilde{A}$, by

$$(3.29) \qquad \tilde{A}_{ij} = \hat{g}_j(v_i A).$$

Now let $\langle L \rangle$ denote the matrix, with respect to the $\{v_i\}$, of the *trace form of* L/K (cf., §3.6):

$$(3.30) \qquad \begin{cases} L \times L \to K, \\ (x, y) \mapsto \mathrm{Tr}_{L/K}(xy). \end{cases}$$

Hence $\langle L \rangle \in GL_d(K)$ is the $d \times d$ matrix (nonsingular because L/K is separable)

$$(3.31) \qquad \langle L \rangle_{ij} = \sum_{u=1}^d \hat{g}_u(v_i v_j) = \sum_{u=1}^d g_u(v_i v_j).$$

From (3.31) we may form a nonsingular matrix of $m \times m$ blocks by $\langle L \rangle \otimes I_m$, the tensor product with the $(m \times m)$ identity matrix:

$$(3.32) \qquad (\langle L \rangle \otimes I_m)_{ij} = \sum_{u=1}^d g_u(v_i v_j) I_m.$$

Now define C by

$$(3.33) \qquad C_{ij} = \hat{g}_i(v_j A^{-1}),$$

then we see that

$$(\tilde{A} C)_{ij} = \sum_{u=1}^d \hat{g}_u(v_i A) \hat{g}_u(v_j A^{-1})$$

$$= \sum_{u=1}^d \hat{g}_u(v_i, v_j) I_m$$

$$= (\langle L \rangle \otimes I_m)_{ij}.$$

Hence $\tilde{A}^{-1} = C(\langle L \rangle \otimes I_m)^{-1}$. If $\langle L \rangle^{-1} = (\alpha_{ij}) \in GL_d(L)$ so that

$$(3.34) \qquad C((\alpha_{ij}) \otimes I_m)\tilde{A} = I_{mn} = I_m \otimes I_n,$$

we obtain a useful equation

(3.35)
$$\sum_{u,t=1}^{d} \hat{g}_i(v_u A^{-1})\alpha_{ut}\hat{g}_j(v_t A) = \begin{cases} I_m & \text{if } i=j, \\ 0 & \text{if } i \neq j. \end{cases}$$

The 1-cocycle associated to $\text{Tr}_{L/K}^{S}(V, \psi)$, as in (2.5)–(2.7), is

(3.36)
$$\begin{bmatrix} F: \Omega_K \to O_{md}(\bar{K}), \\ F: g \mapsto \tilde{A}^{-1}g(\tilde{A}), \end{bmatrix}$$

since $\tilde{A}\tilde{A}^t = \tilde{B}$, by (3.27) and (3.29), for

$$(\tilde{A}\tilde{A}^t)_{ij} = \sum_{u=1}^{d} \tilde{A}_{iu}\tilde{A}_{ju}$$

$$= \sum_{u=1}^{d} \hat{g}_u(v_i A)\hat{g}_u(v_j A^t)$$

$$= \sum_{u=1}^{d} \hat{g}_u(v_i)\hat{g}_u(v_j)\hat{g}_u(A A^t)$$

$$= \tilde{B}_{ij}.$$

Writing $g \in \Omega_K$ as $g = \hat{g}_s h$, $h \in G(\bar{K}/L)$, we have

(3.37)
$$(\text{Ind}_{L/K}(F))(\hat{g}_s h) = \sum_{u,t=1}^{d} \hat{g}_i(v_u A^{-1})\alpha_{ut}\{\hat{g}_s h(\hat{g}_j(v_t A))\}.$$

Temporarily, fix s, h, i, and j, then there exist elements

(3.38)
$$\begin{cases} a \in \Omega \text{ which fixes } L \text{ and} \\ \bar{g}, \text{ one of the } \{\hat{g}_u\}, \text{ such that} \\ \hat{g}_s h \hat{g}_j = \bar{g}z. \end{cases}$$

With this notation (3.37) becomes

(3.39)
$$\begin{cases} \sum_{u,t=1}^{d} \hat{g}_i(v_u A^{-1})\alpha_{ut}\bar{g}(v_t)(\bar{g}z(A)) \\ = \sum_{u,t=1}^{d} [\hat{g}_i(v_u A^{-1})\alpha_{ut}\bar{g}(v_t A)](\bar{g}(A^{-1}z(A))) \\ = \sum_{u,t=1}^{d} \delta_{i,\bar{g}}\bar{g}(F(z)), \end{cases}$$

where, by (3.35),

(3.40)
$$\delta_{i,\bar{g}} = \begin{cases} I_m & \text{if } \hat{g}_i = g, \\ 0 & \text{otherwise.} \end{cases}$$

Here $F(z) = A^{-1}z(A)$ is the representative of (V, ψ) in (3.22).

We complete the proof of Theorem (3.25) by examining the manner in which $\hat{g}h \in \Omega_K$ acts on the induced representation given by

(3.41)
$$K[G(\bar{K}/K)] \otimes_{K[G(\bar{K}/L)]} K^m,$$

where $\Omega_L = G(\bar{K}/L)$ acts on K^m via F. Now (3.41) is the direct sum of vector spaces $\hat{g}_j \otimes K^m$ and $\hat{g}_s h$, as in (3.38), acts on here by

$$\hat{g}_s h(\hat{g}_j \otimes v) = \bar{g}a \otimes v$$
$$= \bar{g} \otimes F(z)(v)$$

so that the matrix of this action is the same as that of (3.37), by (3.39–3.40) and the fact that $\bar{g}(F(z)) = F(z)$, since $F(z) \in O_m(K)$, which completes the proof of Theorem 3.25 □

(3.42) CYCLIC ALGEBRAS

Suppose that K is a field containing $1/n$ and a primitive nth root of unity, ξ. Given $a, b \in K^*$, we can form the *cyclic algebra*, $A_\xi(a, b)$ in the following manner (e.g., see [M2, p. 143]).

(3.43)
$$A_\xi(a, b) = \langle X, Y | X^n = a, Y^n = b, YX = \xi XY \rangle.$$

If L/K is a field extension containing $a^{1/n} = \alpha$ and $b^{1/n} = \beta$, we may realize $A_f(a, b) \otimes_K L$ inside $M_n L$ by the following formulae:

(3.44)
$$X = \begin{bmatrix} \alpha & 0 & & & \\ 0 & \alpha\xi & & & \\ 0 & 0 & \alpha\xi^2 & & \\ & & & \ddots & \\ & & & & \alpha\xi^{n-1} \end{bmatrix}$$

$$Y = \begin{bmatrix} 0 & \beta & 0 & & 0 \\ 0 & 0 & \beta & & 0 \\ 0 & 0 & 0 & & \vdots \\ & & & \ddots & \beta \\ \beta & & & & 0 \end{bmatrix}$$

Since YXY^{-1} is the permutation of X by the n-cycle ($j \mapsto j - 1 \pmod n$).

(3.45) Lemma

(3.44) *gives an isomorphism*

$$A_\xi(a, b) \otimes_K L \cong M_n L.$$

Proof. Explicit calculation of Lemma (3.45) is left as an exercise. Alternatively, consult [M2, p. 145].

(3.46)   On the other hand, from $a, b \in K^*$ we can construct two cohomology classes [cf., (3.3)] $l(a)$, $l(b) \in H^1(K; \mathbb{Z}/n) = \mathrm{Hom}_{ct}(\Omega_K, \mathbb{Z}/n)$:

$$\tag{3.47} \begin{aligned} g(\alpha)/\alpha &= \xi^{l(a)(g)}, \\ g(\beta)/\beta &= \xi^{l(b)(g)}. \end{aligned}$$

We will prove the following result, which represents $A_\xi(a, b)$ in the Brauer group, $H^2(K; \bar{K}^*)$.

   Firstly, we remark that the exact sequence

$$\tag{3.48} \langle \xi \rangle \cong \mathbb{Z}/n \rightarrowtail \bar{K}^* \xrightarrow{(-)^n} \bar{K}^*$$

yields a cohomology sequence

$$\cdots \to H^1(K; \bar{K}^*) \to H^2(K; \mathbb{Z}/n) \to \mathrm{Br}(K) \xrightarrow{n} \mathrm{Br}(K) \to \cdots,$$

which, by §1.34, yields an isomorphism

$$\tag{3.49} H^2(K; \mathbb{Z}/n) \xrightarrow{\cong} {}_n\mathrm{Br}(K),$$

where ${}_n A = \{a \in A \mid na = 0\}$.

(3.50) Theorem

Let K be a field containing $1/n$ and ξ, a primitive nth root of unity. By means of (3.49) the Brauer group class of $A_\xi(a, b)$ is identified with the cup-product $l(a) \cup l(b) \in H^2(K; \mathbb{Z}/n)$, where $l(a)$, $l(b)$ are defined in (3.47).

Proof. Let $g \in G(L/K)$ so that $g(\xi) = \xi$, then we must find a matrix $U \in GL_n(L)$ such that $g(X) = UXU^{-1}$ and $g(Y) = UYU^{-1}$. Suppose that $g(\alpha) = \xi^k \alpha$ and $g(\beta) = \xi^m \beta$. The equation $g(X)U = UX$ equates

$$\begin{aligned} (g(X)U)_{ij} &= g(\alpha)\xi^{i-1} u_{ij} \\ &= \alpha \xi^{k+i-1} u_{ij}, \end{aligned}$$

with

$$(UX)_{ij} = u_{ij}\alpha\xi^{j-1}.$$

(3.51) *Hence $u_{ij} \neq 0$ if and only if $i + k = j$.*

The equation $g(Y)U = UY$ equates

$$(g(Y)U)_{ij} = g(\beta)u_{i+1,j} = \beta\zeta^m u_{i+1,j},$$

with

$$(UY)_{ij} = \beta u_{i,j-1}.$$

Therefore

(3.52) $$\zeta^m u_{i+1,j} = u_{i,j-1}.$$

Setting $u_{1,k+1} = 1$ determines U completely. It has nonzero components only as follows:

(3.53) $$u_{i,k+i} = \xi^{-m(i-1)}.$$

Write $U = U(g)$, to exhibit its g-dependence. We must now evaluate $\Delta(g \mapsto U(g))$, the coboundary in §2.44. By definition of Δ (§1.22), and the classification of central simple algebras in $H^1(G(L/K); PGL_n(L))$ (§2.36), we must evaluate the representing 2-cocycle given by

$$((g, g') \xrightarrow{\Delta(U)} U(g)g(U(g'))(U(gg'))^{-1}),$$

which equals

(3.54) $$((g, g') \xrightarrow{\Delta(U)} U(g)U(g')(U(gg'))^{-1}).$$

Since we know this matrix to lie in the centre, L^*, of $GL_n(L)$, it suffices to see what it does to any vector in the standard basis $(e_1, \ldots, e_n)$. Suppose $l(a)(g') = \xi^s$ and $l(b)(g) = \zeta^m$. However,

$$U(g')(e_1) = e_{s+1}$$
$$U(g)(e_s) = (\xi^{-m})^s e_v,$$

for some v, but

$$U(gg')(e_1) = e_v.$$

Hence

$$\Delta(U)(gg') = (-l(b)(g) \cdot l(a)(g')) \in \mathbb{Z}/n,$$

which, by Chapter 1, §2.14, represents

$$-(l(b) \cup l(a)) = l(a) \cup l(b) \in H^2(G(L/K); \mathbb{Z}/n). \qquad \square$$

(3.55)   Let K be a field, as in §3.42, containing $1/n$ and ξ, a primitive nth root of unity. From the exact sequence

$$\cdots \to H^0(K; \bar{K}^*) \xrightarrow{(-)^n} K^* \to H^1(K; \mathbb{Z}/n) \to 0,$$
$$\cong \downarrow$$
$$K^*$$

we see that $H^1(K; \mathbb{Z}/n) \cong K^*/(K^*)^n$, where $(K^*)^n = \{x \in K^* \mid x = a^n \text{ for some } a \in K^*\}$. The isomorphism is given by $(a \mapsto l(a))$ of (3.47). In terms of the cup-product

$$(K^*/(K^*)^n) \otimes (K^*/(K^*)^n) \to H^2(K; \mathbb{Z}/n),$$

we have seen that $l(a) \cup l(b)$ corresponds to the cyclic algebra, $A_\xi(a, b)$.

I will close this section by showing that $A_\xi(a, 1 - a)$ ($a \neq 0$ or 1) is trivial in $_n\mathrm{Br}(K)$.

(3.56) Theorem

Let K be as in §3.55. If $a \neq 0$ or 1, then

$$l(a) \cup l(1 - a) \in {}_n\mathrm{Br}(K)$$

is zero.

Proof. Let $x^n - a \in K[x]$ split into irreducible factors $f_1(x)f_2(x)\cdots f_n(x) \in \bar{K}[x]$, where $\bar{K}$ is the separable closure of K. Let $a_i \in \bar{K}$ be the root of $f_i(x)$, and set $K_i = K(a_i)$.

Let $t_i : H^1(K_i; \mathbb{Z}/n) \to H^1(K; \mathbb{Z}/n)$, or equivalently $t_i : K_i^*/(K_i^*)^n \to K^*/(K^*)^n$, be the cohomology transfer (Chapter 1, 2.50).

(3.57) $\begin{cases} \text{For } a \in K^*, \ y \in K_i^*, \text{ we have} \\ t_i(l(a) \cup l(y)) = l(a) \cup t_i(l(y)). \end{cases}$

From Chapter 1, §2.50, $t_i(l(y))$ is the sum of the action on y by all the coset representatives. In other words,

(3.58) $$t_i(l(y)) = l\left(\prod_{g \in G(K_i/K)} g(y) \right).$$

Furthermore

$$1 - a = \prod_{i=1}^{n} f_i(1)$$

$$= \prod_{i=1}^{n} \prod_{g \in G(K_i/K)} (1 - g(a_i)).$$

Finally, $l(\alpha\beta) = l(\alpha) + l(\beta)$, where the product is in K^* and the sum in cohomology. Thus

$$l(a) \cup l(1-a) = l(a) \cup l\left(\prod_{i=1}^{n} \prod_{g \in G(K_i/K)} (1 - g(a_i)) \right)$$

$$= \sum_{i=1}^{n} t_i(l(a) \cup l(1 - a_i))$$

$$= \sum_{i=1}^{n} t_i(l(a_i^n) \cup l(1 - a_i))$$

$$= \sum_{i=1}^{n} n(t_i(l(a_i) \cup l(1 - a_i)))$$

$$\equiv 0 \pmod{n}. \qquad \square$$

Chapter Three

Characteristic Classes of Forms and Algebras

It's not my job to take out the bucket! They got a boy for taking out the bucket. I wasn't engaged to take out buckets. My job's cleaning the floor, clearing up the tables, doing a bit of washing-up, nothing to do with taking out buckets!

—HAROLD PINTER,
"The Caretaker" (1960)

In this chapter we specialize. We focus on characteristic classes, that is, on cohomology-valued invariants.

We begin by introducing a number of these invariants for bilinear forms. We start with Frohlich's spinor invariant of an orthogonal Galois representation. Since a Galois representation, as described in Chapter 2, gives rise to a bilinear form, the spinor class gives an invariant of bilinear forms. We examine several examples and also derive the formula which describes the behaviour of the spinor class under transfer (or induction). By the same token, since Stiefel-Whitney classes are also attached to orthogonal Galois representations, we bring them into the picture, but, in this chapter, primarily in the one- and two-dimensional classes.

Next we come to the characteristic classes, due to Delzant, which generalize the two-dimensional Hasse-Witt invariant and the one-dimensional discriminant. Henceforth we will refer to these as Hasse-Witt classes, since the use of Delzant's original nomenclature ("Stiefel-Whitney classes") would lead to a heyday for ambiguity.

We also introduce a cohomology class, which I usually refer to as the equivariant second Stiefel-Whitney class, associated to a family of central extensions with Galois actions upon them. This class, which is capable of resurrection from the literature (cf. Exercise 1.41), was introduced in [Sn] for the purpose of establishing the relation, in dimension two, between the classes mentioned above.

We examine the behaviour of our invariants under the relations which define

75

the Grothendieck group of orthogonal K-representations, $RO_K(G)$. The resulting formulae are very useful when performing calculations, since they permit us to simplify a bilinear form by means of these relations (cf., the Examples in §2).

In §2 we prove, using the equivariant second Stiefel-Whitney class, the formula of J.-P. Serre, which relates the Hasse-Witt class of a bilinear form to the Stiefel-Whitney class, both in dimension two, of a suitable Galois representation. We deduce Serre's result from a generalization, which it inspired, due to Frohlich.

In §3 the Clifford invariant is studied. It is an invariant of an orthogonal Galois representation, and we derive its relationship to the classes of §2.

Finally we examine the analogue of the equivariant second Stiefel-Whitney class that comes from the central extensions associated with the second algebraic K-group of a field. These are the universal central extensions given by the Steinberg group. We also derive the projective analogues of these extensions.

---

1. CLIFFORD ALGEBRAS

We will be concerned in this section with the construction and properties of some simple invariants (particularly of bilinear forms) which can be set up cohomologically.

Firstly, however, let us recall a purely algebraic construction, the Clifford algebra [O'M., p. 131].

(1.1) Let $\beta: V \times V \to K$ be a symmetric, nondegenerate bilinear form of rank $n = \dim_K V$. Assume that $\operatorname{char}(K) \neq 2$.

Define the *Clifford algebra of* (V, β), denoted by $C(V, \beta)$, by

$$(1.2) \qquad C(V, \beta) = T(V)/(x^2 - \beta(x, x)1 \,|\, x \in V),$$

where $T(V) = \bigoplus_{n \geq 0}(v^{\otimes n})$ is the tensor algebra of V.

If $\{v_1, \ldots, v_n\}$ is an orthogonal basis for V such that $\beta(v_i, v_j) = \delta_{ij}a_i$, then $C(V, \beta)$ has the following presentation:

$$(1.3) \qquad C(V, \beta) = \{v_i; 1 \leq i \leq n \,|\, v_i v_j + v_j v_i (i \neq j), v_i^2 - a_i 1\}.$$

Hence $\dim_K C(V, \beta) = 2^n$. In fact $C(V, \beta)$ is a $\mathbb{Z}/2$-graded algebra if we set $C_i(V, \beta)$ $(i = 0, 1)$ to be generated by the images of the $V^{\otimes n}$ for $n \equiv i \pmod 2$. By construction $C_i(V, \beta) \cdot C_j(V, \beta) \subset C_{i+j}(V, \beta)$, and if $(V, \beta) = (V_1, \beta_1) \perp (V_2, \beta_2)$, then as $\mathbb{Z}/2$-graded algebras,

$$(1.4) \qquad C(V, \beta) \cong C(V_1, \beta_1) \hat{\otimes} C(V_2, \beta_2),$$

where $\hat{\otimes}$ is the $\mathbb{Z}/2$-graded tensor product—that is,

$$(M \hat{\otimes} N)_i = \bigoplus_{a+b \equiv i(\mathrm{mod}\, 2)} M_a \otimes N_b.$$

(1.5) Examples

In $[A - B - S; §4]$ we find the following table of Clifford algebras over $K = \mathbb{R}$ and $K = \mathbb{C}$. Here $C'_k = C(\mathbb{R}^k, \langle -1 \rangle^n)$, $C_k = C(\mathbb{R}^k, \langle 1 \rangle^n)$, and $C_k \otimes_\mathbb{R} \mathbb{C} \cong C'_k \otimes_\mathbb{R} \mathbb{C} = C(\mathbb{C}^k, \langle 1 \rangle^n)$.
(1.6)

k	C_k	C'_k	$C_k \otimes_\mathbb{R} \mathbb{C}$
1	$\mathbb{C}$	$\mathbb{R} \oplus \mathbb{R}$	$\mathbb{C} \oplus \mathbb{C}$
2	$\mathbb{H}$	$M_2(\mathbb{R})$	$M_2(\mathbb{C})$
3	$\mathbb{H} \oplus \mathbb{H}$	$M_2(\mathbb{C})$	$M_2(\mathbb{C}) \oplus M_2(\mathbb{C})$
4	$M_2(\mathbb{H})$	$M_2(\mathbb{H})$	$M_4(\mathbb{C})$
5	$M_4(\mathbb{C})$	$M_2(\mathbb{H}) \oplus M_2(\mathbb{H})$	$M_4(\mathbb{C}) \oplus M_4(\mathbb{C})$
6	$M_8(\mathbb{R})$	$M_4(\mathbb{H})$	$M_8(\mathbb{C})$
7	$M_8(\mathbb{R}) \oplus M_8(\mathbb{R})$	$M_8(\mathbb{C})$	$M_8(\mathbb{C}) \oplus M_8(\mathbb{C})$
8	$M_{16}(\mathbb{R})$	$M_{16}(\mathbb{R})$	$M_{16}(\mathbb{C})$

For larger k, if C_k (or C'_k) is $M_n(F)^\varepsilon$ ($\varepsilon = 1$ or 2), then C_{k+8} (or C'_{k+8}) is $M_{16n}(F)^\varepsilon$.
(1.7)   There is a connection with $C(V, \beta)$-modules and exterior algebras, which we will pause to record in passing.

Let $\Lambda^m V$ denote the m-fold exterior power of V, the coinvariant quotient of $\otimes^m V$ by the signed permutation action of Σ_m. Write $w_1 \wedge w_2 \wedge \cdots \wedge w_m$ for the image of $w_1 \otimes w_2 \cdots \otimes w_m$ in $\Lambda^m V$ ($w_i \in V$). Define bilinear forms (nonsingular and symmetric)

(1.8)
$$\begin{cases} \beta_m : \Lambda^m V \times \Lambda^m V \to K \quad \text{by} \\ \beta_m(w_1 \wedge \cdots \wedge w_m, u_1 \wedge \cdots \wedge u_m) = \det \beta(w_i, u_j). \end{cases}$$

For a sequence of integers $\mathbf{i} = (1 \le i_1 \le i_2 \le \cdots i_m \le n = \dim V)$ set $v_i = v_{i_1} \wedge v_{i_2} \wedge \cdots \wedge v_{i_m} \in \Lambda^m V$. The $\{v_i\}$ form an orthogonal basis for $(\Lambda^m V, \beta_m)$. In fact $\beta_m(v_i, v_j) = 0$ unless $\mathbf{i} = \mathbf{j}$, in which case $\beta_m(v_i, v_j) = a_{i_1} a_{i_2} \cdots a_{i_m}$, where $\beta(v_i, v_j) = \delta_{ij} a_j$.
Set $\Lambda = \oplus_m \Lambda^m V$ with bilinear form given by $(\beta_1 \perp \beta_2 \perp \cdots) = \hat{\beta}$. Let $\alpha \in \text{End}(\Lambda)$, and define $\alpha^* \in \text{End}(\Lambda)$ to be the adjoint of α, with respect to $\hat{\beta}$. That is, $\hat{\beta}(\alpha(x), y) = \hat{\beta}(x, \alpha^* y)$ for $x, y \in \Lambda$. One may easily verify the following formulae:

(1.9) $(v_i \wedge \text{---})^*(v_{s_1} \wedge \cdots \wedge v_{s_t})$

$$= \begin{cases} 0 & \text{if } i \notin \{s_1, s_2, \ldots\}, \\ (-1)^{s_c + 1}(v_{s_1} \wedge \cdots \wedge (\hat{v}_{s_c}) \wedge \cdots) a_{s_1} \cdots a_{s_t} & \text{if } i = s_c. \end{cases}$$

Define $\phi : V \to \text{End}(\Lambda)$ by $\phi(v) = [(v \wedge \text{---}) - (v \wedge \text{---})^*]$, then (1.9) implies

$$(1.10) \quad \begin{cases} \phi(v_i)\phi(v_j) + \phi(v_j)\phi(v_i) = 0 & \text{if } i \neq j, \\ \phi(v_i)\phi(v_i) = -(1_\Lambda) & \text{if } 1 = a_1 = a_2 = \cdots a_n. \end{cases}$$

Hence we have shown the first half of the following result [the last part follows from (1.9)]:

(1.11) Lemma

If (V, β) has an orthonormal basis, then ϕ, defined in the preceding equation, extends to an algebra map $\phi: C(V, \beta) \to End_K(\Lambda)$. Also $C(V, \beta) \to \Lambda$, given by $(x \mapsto \phi(x)(1))$, is an isomorphism (of vector spaces).

(1.12)   THE SPINOR NORM

Let $\sigma \in O(V, \beta)$, the orthogonal group of (V, β) as in §1.1. The *reflection perpendicular to $v \in V$* is the map [O'M., p. 96]

$$(1.13) \qquad\qquad \tau_v(x) = x - \frac{2\beta(x, v)v}{\beta(v, v)}.$$

By [O'M., Thm. 43.3, p. 102] $O(V, \beta)$ is generated by the $\{\tau_v | v \in V\}$ of (1.13). If $\alpha \in O(V, \beta)$, we define the *spinor norm* of α, $\theta(\alpha) \in K^*/(K^*)^2$ $((K^*)^2 = \text{squares in } K^*)$, by [O'M., p. 137]:

$$(1.14) \quad \begin{cases} \theta(\alpha) = \prod_{i=1}^{s} \beta(u_i, u_i), & \text{if } \alpha = \tau_{u_2} \cdots \tau_{u_s}, \quad \text{and} \\ \theta: O(V, \beta) \to K^*/(K^*)^2 & \text{is a homomorphism.} \end{cases}$$

(1.15) Example

(i) If $\alpha_s(v_i) = \begin{cases} v_i & \text{if } i \neq s, \\ -v_s & \text{if } i = s, \end{cases}$

then $\alpha_s = \tau_{v_s}$, where $\{v_i\}$ is an orthogonal basis for (V, β). Hence in this case

$$\beta(\alpha_s) = \beta(v_i, v_i) = a_i \in K^*/(K^*)^2.$$

In particular, if $\beta = \langle 1 \rangle^n$ so that

$$O(V, \beta) = \{X \in GL_n K \mid XX^t = I_n\} = O_n K,$$

then θ is trivial on the subgroup $O(1)^n \cong (\mathbb{Z}/2)^n$.

(ii) (Zassenhaus [O'M, p. 137]) If $\sigma \in O(V, \beta)$ and is a *rotation* (i.e., $\det \sigma = 1$), then

$$\theta(\sigma) = \det\left[\frac{I_v + \sigma}{2}\right] \qquad \text{if } \det(I_v + \sigma) \neq 0.$$

For example, if $O(V, \beta) = O_2(K)$ and $\beta = \langle 1 \rangle^2$, then $(ad - bc = 1, a + d \neq -2)$

$$\theta\left(\begin{bmatrix} a & b \\ c & d \end{bmatrix}\right) = 2 + a + d \in K^*/(K^*)^2.$$

(1.16) Suppose now that $C(V, \beta)^*$ denotes the unit group of the Clifford algebra, $C(V, \beta)$. Let $\alpha: C(V, \beta) \to C(V, \beta)$ be the algebra involution extending $\alpha = -1_v$ on V. Define (see [ABS §3; FP, p. 98])

$$\Gamma = \{x \in C(V, \beta)^* \mid \alpha(x) V x^{-1} \subset V\}.$$

Hence we have a projection $\pi: \Gamma \to GL(V)$, which restricts to a map

(1.17) $\begin{cases} \pi: \text{Pin}(V, \beta) \to O(V, \beta) \\ \text{whose image is } \{x \in O(V, \beta) \mid \theta(x) = 1\} \\ \text{and such that } \ker(\pi) \cong \mathbb{Z}/2\langle \pm 1 \rangle. \end{cases}$

In particular,

$$\mathbb{Z}/2 \rightarrowtail \text{Pin}(V, \beta) \twoheadrightarrow \text{im}(\pi) = \ker(\theta)$$

is a central extension.

(1.18) FROHLICH'S SPINOR CLASS, $Sp[\rho]$

Let $v: V \times V \to K$ (char $K \neq 2$) be a nondegenerate, symmetric bilinear form, and let L/K be a finite Galois extension. Let $\rho: G(L/K) \to O(V, \beta)$ be an orthogonal Galois representation (i.e., a homomorphism).

Following [F, §3.1; F3], we consider the composition

$$G(L/K) \xrightarrow{\rho} O(V, \beta) \xrightarrow{\theta} K^*/(K^*)^2 \cong H^1(K; \mathbb{Z}/2).$$

Since

$$\theta\rho \in \text{Hom}(G(L/K), H^1(K; \mathbb{Z}/2)) \cong \text{Hom}(G(L/K), \mathbb{Z}/2) \otimes H^1(K; \mathbb{Z}/2)$$

$$\cong H^1(G(L/K), \mathbb{Z}/2) \otimes H^1(K; \mathbb{Z}/2),$$

we may define a class

(1.19)

$$
\begin{cases}
Sp[\rho] \in H^2(K; \mathbb{Z}/2) \\
\text{equal to the image of } \theta\rho \text{ under} \\
H^1(G(L/K); \mathbb{Z}/2) \otimes H^1(K; \mathbb{Z}/2) \to H^1(K; \mathbb{Z}/2)^{\otimes 2} \xrightarrow{(-\cup-)} H^2(K; \mathbb{Z}/2).
\end{cases}
$$

(1.20) Example

(i) Consider any diagonalized homomorphism

$$
\rho: G(L/K) \to O_1(K)^n \cong (\mathbb{Z}/2)^n \subset O_n(K).
$$

By §1.15 (i), $Sp[\rho] = 0$ in this case.

(ii) Suppose that $G(L/K) \cong D_{2^{n+1}} \cong \mathbb{Z}/2 \ltimes (\mathbb{Z}/2^n)$ the dihedral group of order 2^{n+1}. Hence

$$
H^*(G(L/K); \mathbb{Z}/2) \cong \frac{\mathbb{Z}/2[x_1, x_2, w]}{(x_2^2 + x_1 x_2)},
$$

by Chapter 1, §4.6.

Suppose that $\rho: G(L/K) \cong D_{2^{n+1}} \to O_2(K)$ is what we will call a *standard embedding*. That is, if (as in Chapter 1, §3.12) $D_{2^{n+1}} = \langle x, y \,|\, x^{2^n} = 1 = y^2, xyx = y \rangle$, then

$$
\rho(y) = \begin{bmatrix} 0 & 1 \\ 1 & 0 \end{bmatrix}
$$

and

$$
\rho(x) = \begin{bmatrix} \alpha_n & \beta_n \\ -\beta_n & \alpha_n \end{bmatrix} \qquad (\alpha_n^2 + \beta_n^2 = 1; \alpha_n, \beta_n \in K).
$$

Also

$$
\begin{bmatrix} \alpha_{j-1} & \beta_{j-1} \\ -\beta_{j-1} & \alpha_{j-1} \end{bmatrix} = \begin{bmatrix} \alpha_j & \beta_j \\ -\beta_j & \alpha_j \end{bmatrix}^2 = \begin{bmatrix} \alpha_j^2 - \beta_j^2 & 2\alpha_j\beta_j \\ -2\alpha_j\beta_j & \alpha_j^2 - \beta_j^2 \end{bmatrix}
$$

and

$$
\begin{bmatrix} \alpha_1 & \beta_1 \\ -\beta_1 & \alpha_1 \end{bmatrix} = \begin{bmatrix} -1 & 0 \\ 0 & -1 \end{bmatrix}.
$$

Since $\rho(y) = \tau_{v_1 - v_2}$ (where v_1, v_2 is the standard orthonormal basis) $\theta\rho(y) = 2$.

Also, by §1.15 (ii),

$$\theta\rho(x) = \frac{1}{4}\det\begin{bmatrix} 1 + \alpha_n & \beta_n \\ -\beta_n & 1 + \alpha_n \end{bmatrix}$$

$$= 1 + 2\alpha_n + \alpha_n^2 + \beta_n^2$$

$$= 2(1 + \alpha_n),$$

which is nonzero if $n \geq 2$. If $n = 1$, $\theta\rho(x) = 0$. For $i = 1, 2$ $x_i: D_{2^{n+1}} \to \mathbb{Z}/2$ is given by Chapter 1, (4.4). We have

$$x_i(y^\delta x^\varepsilon) \equiv \begin{cases} \varepsilon & \text{(mod 2) if } i = 2, \\ \delta & \text{(mod 2) if } i = 1. \end{cases}$$

Therefore, in this example,

$$Sp[\rho] = \begin{cases} x_1 l(2) & \text{if } n = 0, \\ x_1 l(2) & \text{if } n = 1, \\ (x_1 + x_2)l(2) + x_2 l(1 + \alpha_n) & \text{if } n \geq 2. \end{cases}$$

(1.21) Suppose, following the notation of Chapter 1, §§2.46–2.50, $i: H \subset G$ are finite groups, with $\{x_1, \ldots, x_m\}$ a set of right $H^{\backslash G}$ coset representatives. In the notation of Chapter 1, (2.49) if $g \in G$, then $x_i g = h(i, g)x_{\pi(i)}$, where $h(i, g) \in H$ and $\pi(i) \in \Sigma_m$.

Suppose that $\rho: H \to O(V, \beta)$ is an orthogonal representation, where (V, β) is a symmetric, nondegenerate bilinear form over K and $\text{char}(K) \neq 2$. Hence we have the spinor norm, $\theta\rho \in \text{Hom}(H, K^*/(K^*)^2) \cong H^1(H; K^*/(K^*)^2)$. By Chapter 1, §2.50 the formula for $\text{Tr}(\theta\rho) \in H^1(G; K^*/(K^*)^2) \cong \text{Hom}(G, K^*/(K^*)^2)$ is $(g \in G)$,

(1.22) $$\text{Tr}(\theta\rho)(g) = i_*(\theta\rho)(g) = \prod_{i=1}^m \theta(\rho(h(i, g))) \in K^*/K^{**}.$$

On the other hand, we may consider the representation-theoretic transfer of ρ. This is

(1.23) $$\begin{cases} \text{Ind}_H^G(\rho): G \to \Sigma_m \int O(V, \beta) \subset O\left(\bigoplus_1^m (V, \beta)\right) \text{ given by the} \\ \text{(left) } G\text{-action on } K[G] \otimes_{K[H]} V. \end{cases}$$

Now $K[G] \otimes_{K[H]} V$ is, as a K-vector space, the sum of subspaces $(\{x_i^{-1}\} \otimes V; 1 \leq i \leq m)$. In addition left multiplication by $g \in G$ is given by $(v \in V)$,

$$gx_i^{-1} \otimes v = x_{\pi(i)}^{-1} h(i, g^{-1}) \otimes v$$

$$= x_{\pi(i)}^{-1} \otimes \rho(h(i, g^{-1})^{-1})(v).$$

This means that $(g \cdot —) \in \mathrm{End}\,(\bigoplus_{i=1}^{m} \{x_i^{-1}\} \otimes V)$ is the composition

$$\bigoplus_1^m \{x_i^{-1}\} \otimes V \xrightarrow{\bigoplus_1^m \rho(h(i,g^{-1})^{-1})} \bigoplus_1^m \{x_i^{-1}\} \otimes V \xrightarrow{\pi \otimes 1} \bigoplus_1^m \{x_i^{-1}\} \otimes V,$$

where $\pi \otimes 1$ permutes the copies of V(i.e., $\{x_i^{-1}\} \otimes V$) according to $\pi \in \Sigma_m$. The spinor norm of $\pi \otimes 1$ is, by §1.20 (ii).

$$(1.24) \qquad\qquad \theta(\pi \otimes 1) = 2^{\mathrm{sign}(\pi) \cdot \dim V} \in K^*/(K^*)^2.$$

Hence the spinor norm of $\mathrm{Ind}_H^G(\rho)$ is

$$(1.25) \qquad\qquad \theta(\mathrm{Ind}_H^G(\rho)(g)) = 2^{\mathrm{sign}(\pi) \cdot \dim V} \prod_{i=1}^m \theta(\rho(i, g^{-1})^{-1}).$$

Notice that π depends on $g \in G$, as explained earlier. Also, since $K^*/(K^*)^2$ is a $\mathbb{Z}/2$-vector space,

$$\theta(\mathrm{Ind}_H^G(\rho)(g)) = \theta(\mathrm{Ind}_H^G(\rho)(g^{-1}))$$

and $\theta \rho(h(i, g)^{-1}) = \theta \rho(h(i, g))$.

Therefore we may combine (1.22) and (1.25) to obtain the following result: Write $\widetilde{Sp}[\rho]$ for $\theta \rho \in \mathrm{Hom}\,(H, K^*/(K^*)^2)$.

(1.26) Theorem

Let $i: H \subset G$ be finite groups and $\rho: H \to O(V, \beta)$ an orthogonal representation where (V, β) is a symmetric, nondegenerate bilinear form defined over K (char $K \neq 2$). Then, in $H^1(G; \mathbb{Z}/2) \otimes K^*/(K^*)^2 \cong \mathrm{Hom}\,(G, K^*/(K^*)^2)$, $\widetilde{Sp}[\mathrm{Ind}_H^G \rho] = i_*(\widetilde{Sp}[\rho]) + (\dim V)\omega_1 \otimes l(2)$, where ω_1 is the determinant (or first Stiefel-Whitney class; see Chapter 1, §3.16) of the permutation representation of G acting on $K[G] \otimes_{K[H]} K = \mathrm{Ind}_H^G(1)$.

(1.27) In the context of §1.18, if $K \overset{i'}{\subset} L \subset N$ are finite extensions, with $i: G(N/L) \subset G(N/K)$ being Galois groups, then $\widetilde{Sp}[\rho]$ maps, under the natural map $\Omega_L \to G(N/L)$, and cup-product to $Sp[\rho] \in H^2(L; \mathbb{Z}/2)$. Since the transfer is an $H^*(K; \mathbb{Z}/2)$-module map, we have a commutative diagram

$$(1.28)$$

$$
\begin{array}{ccc}
H^1(L; \mathbb{Z}/2) \otimes H^1(K; \mathbb{Z}/2) & \xrightarrow{\mathrm{Tr}_{L/K} \otimes 1} & H^1(K; \mathbb{Z}/2) \otimes H^1(K; \mathbb{Z}/2) \\
{\scriptstyle (1 \cup (i')^*)} \downarrow & & \downarrow {\scriptstyle (- \cup -)} \\
H^2(L; \mathbb{Z}/2) & \xrightarrow{\mathrm{Tr}_{L/K}} & H^2(K; \mathbb{Z}).
\end{array}
$$

In addition the class w_1 maps to a class

$$\mathrm{SW}_1(\mathrm{Ind}_{G(N/L)}^{G(N/K)}(1)) \in H^1(K; \mathbb{Z}/2),$$

which is called the *discriminant* of L/K and is written $d_{L/K} \in H^1(K; \mathbb{Z}/2)$.

Therefore, from Theorem 1.26, we obtain the following corollary. Write $\mathrm{Ind}_{L/K}$ for $\mathrm{Ind}_{G(N/L)}^{G(N/K)}$.

(1.29) Corollary

Let L/K be a finite separable extension. Let $\rho: G(N/L) \to O(V, \beta)$ be an orthogonal representation, as in §1.26. Then

$$Sp[\mathrm{Ind}_{L/K}[\rho]] = (\dim V) d_{L/K} \cdot l(2) + \mathrm{Tr}_{L/K}(Sp[\rho]) \qquad \text{in } H^2(K; \mathbb{Z}/2).$$

(1.30)   In addition to the spinor class, $Sp[\rho]$, an orthogonal Galois representation is entitled to Stiefel-Whitney classes, $SW_i[\rho]$, generalizing those of Chapter 1, §3.22, which is the case when $K = \mathbb{R}$.

Let N/K be a finite Galois extension with group $G(N/K)$, and let $O(V, \beta)$ be the orthogonal group of the symmetric, nondegenerate bilinear form (V, β) over K (char $K \neq 2$). We have natural inclusion maps $O(V, \beta) \to O(V_{\bar{K}}, \beta_{\bar{K}})$, where $\bar{K}$ is a separable closure of K. In addition $O(V_{\bar{K}}, \beta_{\bar{K}}) \cong O_m(\bar{K})$ ($m = \mathrm{rank}_K(V, \beta)$) so that a Galois representation, ρ, gives rise to $\rho: G(N/K) \to O(V, \beta) \to O_n(\bar{K}) \to O(\bar{K})$.

By the results of [Su; Su1], extended to orthogonal groups in [Ka],

$$H^*(O(\bar{K}); \mathbb{Z}/2) \cong \mathbb{Z}/2[w_1, w_2, \ldots]$$

where $\deg w_i = i$ and the w_i are characterized as in Chapter 1, (3.21).

Hence we may define the ith *Stiefel-Whitney* class of $\rho: G(N/K) \to O(V, \beta)$ by

(1.31) $$SW_i[\rho] = \rho^*(w_i) \in H^i(K; \mathbb{Z}/2),$$

the image of $\rho^*(w_i)$ in Galois cohomology.

(1.32) Examples

(i) $SW_1[\rho]$ is represented in $\mathrm{Hom}_{ct}(\Omega_K, \mathbb{Z}/2)$ by

$$\Omega_K \longrightarrow G(N/K) \xrightarrow{\rho} O(V, \beta) \xrightarrow{\det} \{\pm 1\}.$$

(ii) Let $\mathbb{Z}/2 \rightarrowtail \mathrm{Pin}(V_{\bar{K}}, \beta_{\bar{K}}) \xrightarrow{\pi} O(V_{\bar{K}}, \beta_{\bar{K}})$ be the central extension of (1.17) (in which π is onto because $\bar{K}$ is separable closed so that the spinor norm is trivial). We have an exact sequence, by §1.26, with G acting trivially throughout,

$$\cdots \to H^1(G; \mathrm{Pin}(V_{\bar{K}})) \to H^1(G; O(V_{\bar{K}})) \xrightarrow{\Delta} H^2(G; \mathbb{Z}/2).$$

Putting $G = G(N/K)$, the map

$$\rho: G(N/K) \to O(V) \to O(V_{\bar{K}})$$

is a 1-cocycle and $SW_2[\rho]$ is the image of $\Delta[\rho] \in H^2(G(N/K); \mathbb{Z}/2)$ in $H^2(K; \mathbb{Z}/2)$. Therefore, by Chapter 2, (1.22), the explicit formula for $SW_2[\rho]$ as a 2-cocycle is

$$(1.33) \qquad SW_2[\rho][g|g'] = h(g)h(g')h(gg')^{-1} \in \ker \pi = \{\pm 1\},$$

where $g, g' \in \Omega_K$ and $h: G(N/K) \to O(V_{\bar{R}})$ satisfies $\pi h = \rho$.

(1.34)   HASSE-WITT CLASSES [D]

The characteristic classes of bilinear forms which I am about to introduce are due in general to Delzant [D]. However, the two-dimensional invariant is older, being due to Hasse and Witt. For this reason I prefer to call these the Hasse-Witt classes, denoted by HW_i.

Suppose that $\beta: V \times V \to K$ is a symmetric, nondegenerate bilinear form of rank m over a field K (char $K \neq 2$). We may diagonalize (V, β) over K to obtain $(V, \beta) \cong \langle a_1 \rangle \oplus \langle a_2 \rangle \oplus \cdots \oplus \langle a_m \rangle$, where $\langle a_i \rangle: K \otimes K \to K$ sends (x, y) to $a_i xy$.

Define the jth Hasse-Witt class of (V, β) by

$$(1.35) \qquad HW_j(V, \beta) = \sum_{1 \le i_1 < i_2 < \cdots < i_j \le m} l(a_{i_1})l(a_{i_2}) \cdots l(a_{i_j}) \in H^j(K; \mathbb{Z}/2),$$

where $l(a) \in H^1(K; \mathbb{Z}/2)$ is defined, as in Chapter 2, (3.3), by $l(a)(g) = g(\sqrt{a})/\sqrt{a} \in \{\pm 1\}$.

Proofs that $HW_i(V, \beta)$ is well-defined may be found in [D] and in [M]. They imitate the proof of the well-definedness of HW_2 to be found in [O'M.], for example. Actually this proof proceeds by showing that any two diagonalizations of (V, β) can be connected by a chain of diagonalizations in which adjacent ones differ by a basis-change which takes place in a two-dimensional orthogonal summand. However, if we formally write $HW = 1 + HW_1 + HW_2 + \cdots$ for the total *Hasse-Witt class*, then

$$HW((V_1, \beta_1) \oplus (V_2, \beta_2)) = HW(V_1, \beta_1)HW(V_2, \beta_2).$$

If $\dim V \le 2$, $HW(V, \beta) = 1 + HW_1(V, \beta) + HW_2(V, \beta)$. Therefore one sees that $HW_i(V, \beta)$ is well-defined for all $i \ge 1$ if and only if $HW_1(V, \beta)$ and $HW_2(V, \beta)$ are well-defined.

In fact, my construction in §1.45 of $HW_2(V, \beta)$ from the equivariant second Stiefel-Whitney class, $\tilde{\omega}_2$, later in this section, is one means to show that $HW_2(V, \beta)$ is well-defined.

(1.36)   AN EQUIVARIANT COHOMOLOGY CLASS, $\tilde{\omega}_2$

Let K be a field and for each finite Galois extension, L/K, suppose we are given a *central* extension

(1.37)
$$A_L \overset{i_L}{\rightarrowtail} B_L \overset{\pi_L}{\twoheadrightarrow} C_L$$

such that

(i) $G(L/K)$ acts on (1.37), hence so does $\Omega_K = G(\bar{K}/K)$.

(ii) (1.37) is natural in L. Hence, if $N \supset L \supset K$, there exists a commutative diagram

$$
\begin{array}{ccc}
A_L & \longrightarrow B_L & \longrightarrow C_L \\
\downarrow{\scriptstyle\alpha_{N/L}} & \downarrow{\scriptstyle\beta_{N/L}} & \downarrow{\scriptstyle\gamma_{N/L}} \\
A_N & \longrightarrow B_N & \longrightarrow C_N
\end{array}
$$

of Ω_K-equivariant maps. Choose a family of maps (of sets)

$$s_L : C_L \to B_L \qquad \text{such that}$$

(1.38)
$$\pi_L \circ s_L = 1 \quad \text{and} \quad \beta_{N/L} \circ s_L = s_N \circ \gamma_{N/L}.$$

Let $(A_L)_{\Omega_K} = (A_L)/(a - g(a) \mid a \in A_L, g \in \Omega_K)$ denote the coinvariants of Ω_K acting on A_L. Also let $\Omega_K \ltimes C_L$ denote the semidirect product of Ω_K with C_L. That is, as a set $\Omega_K \ltimes C_L = \Omega_K \times C_L$ with product given by

$$(g, c)(g', c') = (gg', cg(c')),$$

according to the convention of Chapter 2, §1.3.

Define $\omega_{2,L} : (\Omega_K \ltimes C_L)^2 \to (A_L)_{\Omega_K}$ by the formula $(g, g' \in \Omega_K, c, c' \in C_L)$,

(1.39)
$$\begin{cases} \omega_{2,L}((g, c), (g', c')) \text{ is the image in } (A_L)_{\Omega_K} \text{ of} \\ s_L(c)g(s_L(c'))[s_L(cg(c'))]^{-1} \in A_L. \end{cases}$$

(1.40) Theorem

(i) *For each L/K in §1.36, $\omega_{2,L}$ is a 2-cocycle.*

(ii) *Up to 1-boundaries, $\omega_{2,L}$ is independent of the lift, s_L.*

(iii) *$\omega_{2,L}((g, c), (g', c'))$ depends only on the images of g, g' in $G(L/K)$.*

Proof. Firstly, part (iii) is trivial from (1.39) and §1.36 (i).

Now let us verify that $\omega_{2,L}$, which we will temporarily denote by ω, is a 2-cocycle. Let $\alpha = (x, A)$, $\beta = (y, B)$, $\gamma = (z, C)$ (with $x, y, z \in G(L/K)$, $A, B, C \in C_L$), then

$$(d\omega)(\alpha, \beta, \gamma) = \omega(\beta, \gamma)[\omega(\alpha\beta, \gamma)]^{-1} \omega(\alpha, \beta\gamma)[\omega(\alpha, \beta)]^{-1}.$$

Let us abbreviate s_L to s, then we have

$$\omega(\alpha\beta, \gamma)^{-1}\omega(\alpha, \beta\gamma) = \omega(\alpha, \beta\gamma)\omega(\alpha\beta, \gamma)^{-1}, \qquad \text{as } A_L \text{ is central,}$$

$$= s(A)[xs(By(C))][s(Ax(B)xy(C))]^{-1}$$

$$\cdot[s(Ax(B)xy(C))]xys(C)^{-1}[s(Ax(B))]^{-1}$$

$$= s(A)[xs(By(C))]xys(C)^{-1}[s(Ax(B))]^{-1}.$$

Therefore, since $\omega(\alpha, \beta) = s(A)xs(B)[s(Ax(B))]^{-1}$, we obtain

$$\omega(\alpha\beta, \gamma)^{-1}\omega(\alpha, \beta\gamma)\omega(\alpha, \beta)^{-1}$$

$$= s(A)\{[xs(By(C))]xys(C)^{-1}[xs(B)]^{-1}\}s(A)^{-1}$$

$$= x\{[s(By(C))]ys(C)^{-1}[s(B)]^{-1}\}, \qquad \text{since } A_L \text{ is central,}$$

$$= \omega(\beta, \gamma)^{-1} \qquad \text{in } (A_L)_{\Omega_K},$$

as required to prove part (i). If we change $s_L = s$ to $(B \mapsto s(B)t(B))$ with $t(B) \in A_L$, then we obtain a second 2-cocycle given by

$$((g, c), (g', c')) \mapsto \omega_{2,L}((g, c)(g', c'))t(c)g(t(c'))t(cg(c'))^{-1},$$

since A_L is central. However, if $T: \Omega_K \ltimes C_L \to A_L$ is given by $T(g, c) = t(c)$, then

$$(dT)((g, c)(g', c')) = t(c)t(cg(c'))t(cg(c'))^{-1}$$

so that the two 2-cocycles differ by dT. This completes the proof of part (ii) and Theorem 1.40. $\qquad\qquad\qquad\qquad\qquad\qquad\qquad\qquad\qquad\qquad\qquad\square$

(1.41) Exercise (see [Ho-S])

The kernel of $\Omega_K \ltimes B_L \xrightarrow{1 \times \pi_L} \Omega_K \ltimes C_L$ is A_L, which is central if and only if Ω_K acts trivially on A_L. In this case the exact sequence of Chapter 2, §1.28 (b) yields a homomorphism

$$\delta: \operatorname{Hom}(A_L, A_L) \cong H^1(A_L; A_L) \to H^2(\Omega_K \ltimes C_L; A_L).$$

Relate $\delta(1_{A_L})$ to $\omega_{2,L}$ in this case.

When the action is nontrivial, we have a commutative pushout diagram

$$
\begin{array}{ccc}
A_L & \rightarrowtail \Omega_K \ltimes B_L \to \Omega_K \ltimes C_L \\
\Big\downarrow{\scriptstyle\rho} \qquad \Big\downarrow \qquad\qquad \Big\downarrow{\scriptstyle=} \\
(A_L)_{\Omega_K} \to \qquad H \qquad \to \Omega_K \ltimes C_L.
\end{array}
$$

Here ρ is the natural map. The bottom line is a central extension. Construct $\omega_{2,L}$ from its associated exact sequence.

(1.42)   Let $\beta: V \times V \to K$ be a symmetric, nonsingular bilinear form over K (char $K \neq 2$) of rank m.

For each Galois extension L/K consider the central extension of (1.17),

$$(1.43) \qquad \mathbb{Z}/2 \longrightarrow \mathrm{Pin}_m(L) \xrightarrow{\pi_L} \mathrm{im}\,(\pi_L) \subset O_m(L).$$

(V, β) is represented by a class in $H^1(K; O_m\bar{K})$, or equivalently by a compatible family of continuous homomorphisms

$$f_L(V, \beta): \Omega_K \to G(L/K) \ltimes O_m(L)$$

for L large enough (see Chapter 2, §§1.10–2.21). By Chapter 2, §1.11, pulling back $\omega_{2,L} \in H^2(G(L/K) \ltimes O_m(L); \mathbb{Z}/2)$ yields a well-defined class, *depending only on the isomorphism class of* (V, β).

$$(1.44) \qquad \tilde{\omega}_2(V, \beta) = f_L^*(V, \beta)(\omega_{2,L}) \in H^2(K; \mathbb{Z}/2).$$

Note that §1.40 (iii) ensures that $f_L^*(V, \beta)(w_{2,L})$ is given by a continuous 2-cocycle.

(1.45) Theorem

In (1.44),

$$\tilde{\omega}_2(V, \beta) = HW_2(V, \beta) \in H^2(K; \mathbb{Z}/2).$$

Proof. It suffices, since each (V, β) can be diagonalized, to treat the case $(V, \beta) = \bigoplus_1^m \langle a_i \rangle$ $(a_i \in K^*)$. In this case $f_L(V, \beta)$—quite independently of L—lands in $O_1^m \cong (\mathbb{Z}/2)^m$. Therefore, by naturality of $\omega_{2,L}$ with respect to extensions, it suffices to consider a simpler central extension. For we have a commutative diagram

$$
(1.46) \qquad
\begin{array}{ccc}
\mathbb{Z}/2 \longrightarrow H & \xrightarrow{\pi} & (\mathbb{Z}/2)^m \\
= \downarrow \qquad \downarrow & & \downarrow \\
\mathbb{Z}/2 \longrightarrow \mathrm{Pin}_m(L) & \xrightarrow{\pi_L} & O_m(L)
\end{array}
$$

in which H is the group generated by the $\{v_i; 1 \le i \le m\}$ inside the Clifford algebra, $C(\langle 1 \rangle^m)^*$.

Notice that $G(L/K)$ acts trivially on the upper extension in (1.46). Therefore in the upper central extension of (1.46), $\omega_{2,L}$ is given by the same formula as the restriction to $O_1(L)^m \cong (\mathbb{Z}/2)^m$ of SW_2 given in §1.33.

Hence, if $f_L(V, \beta) = \text{diag}(l(a_1), \ldots, l(a_m)): \Omega_K \to (\mathbb{Z}/2)^m$, then

$$f_L^*(V, \beta)(\omega_{2,L}) = SW_2\left[\bigoplus_1^m l(a_i)\right]$$

$$= \sum_{1 \le i < j \le m} l(a_i)l(a_j)$$

$$= HW_2(V, \beta),$$

since $SW_n(V \oplus W) = \sum_a SW_a(V)SW_{n-a}(W)$ (cf., [M-St; H]) and $SW_1[l(a)] = l(a)$.

$\square$

(1.47)   Let me sketch how an explicit calculation may be made of $\omega_{2,L} = SW_2$ the upper extension of (1.46).

First, let v_i, the ith standard basis vector, be the chosen lifting of reflexion in the ith coordinate. See [Q, §4.2]. In general, if A is a diagonal matrix with (-1) in the i_jth places for $1 \le i_1 < i_2 < \cdots < i_u \le m$, define $S(A) = v_{i_1} v_{i_2} \cdots v_{i_u}$. Suppose B has (-1)'s at places $1 \le k_1 < k_2 < \cdots < k_v \le m$ so that $S(B) = v_{k_1} \cdots v_{k_v}$. One verifies that $Q(A, B) = S(A)S(B)S(AB)^{-1} \in \mathbb{Z}/2 = \{\pm 1\}$ is bilinear in A and B. From this and the description of Q in [Q, §4], one easily verifies §1.45 directly.

(1.48) Exercise   Calculate Q in §1.47 directly from the definition of the extension in (1.46).

(1.49)   THE GROTHENDIECK GROUP OF ORTHOGONAL K-REPRESENTATIONS

I will conclude this section with a discussion of the characteristic classes, SW_i, HW_i, and Sp, in relation to the Grothendieck group of orthogonal G-representations.

Let G be a finite group, and let K be a field of characteristic different from two. Following [Q2, §5], we may define $RO_K(G)$, *the Grothendieck group of orthogonal K-representations*, as a quotient of the free abelian group on the isomorphism classes of finite dimensional orthogonal representations, (ρ, V, β). Here $\beta: V \times V \to K$ is a symmetric, nondegenerate bilinear form, and $\rho: G \to O(V, \beta)$ is a homomorphism. We impose the following relations:

(i) *Scaling.* If $a \in K^*$ and $(a\beta)(x, y) = a\beta(x, y)$, then $O(V, \beta) = O(V, a\beta)$ and $(\rho, V, \beta) \sim (\rho, V, a\beta)$.

(ii) *Direct sum.* $(\rho, V, \beta) \sim (\rho', V', \beta') + (\rho'', V'', \beta'')$ if (ρ, V, β) is the orthogonal direct sum of (ρ', V', β') and (ρ'', V'', β'').

(iii) *Hyperbolic.* Let $\rho: G \to O(V, \beta)$ be an orthogonal representation. Let $W \subset V$ be a G-invariant, isotropic subspace (i.e., $W \subset W^0$, where $W^0 =$

$\{v \in V | \beta(W, v) = 0\}$). From β we have induced new bilinear forms

$$\beta_1 : (W^0/W) \times (W^0/W) \to K \quad \text{and}$$

$$\beta_2 : (W \oplus (V/W^0)) \times (W \oplus (V/W^0)) \to K$$

given by $\beta_1(v + W, v' + W) = \beta(v, v')$ and $\beta_2((w, v + W^0), (w', v' + W^0)) = \beta(w, v') + \beta(v, w')$. Also ρ induces $\rho_1 : G \to O(W^0/W, \beta_1)$ and $\rho_2 : G \to O(W \oplus (V/W^0), \beta_2)$.

With these conventions

$$(\rho, V, \beta) \sim (\rho_1, W^0/W, \beta_1) + (\rho_2, W \oplus (V/W^0), \beta_2).$$

Suppose that L/K is a finite Galois extension and that $G = G(L/K)$. We can consider $\rho : G(L/K) \to O(V, \beta)$ as a 1-cocycle giving a class $(\rho) \in H^1(K; O(V_{\bar{K}}, \beta_{\bar{K}}))$. This represents the bilinear form of Chapter 2, §3.16 and has Hasse-Witt classes, $HW_i(\rho)$. Also, as a representation, it is entitled to Stiefel-Whitney classes, $SW_i[\rho]$, and the spinor class, $Sp[\rho]$.

With this notation we have the following:

(1.50) Proposition

The effect of scaling by $a \in K^*$ is given by

(i) $SW_i[\rho]$ *is unchanged.*

(ii) *If $(\rho) = (\rho, V, \beta) \cong \oplus_1^m \langle a_i \rangle$, then $(\rho, V, a\beta) \cong \oplus_1^m \langle aa_i \rangle$ so that*

$$HW_i(\rho, V, a\beta) = \sum_{1 \le j_1 < j_2 < \cdots < j_i \le m} (l(a) + l(a_{j_1})) \cdots (l(a) + l(a_{j_i})).$$

(iii) $Sp[\rho, V, a\beta] = Sp[\rho, V, \beta] + l(a)SW_1[\rho, V, \beta]$.

Proof. Part (i) follows from the fact that $O(V, \beta) = O(V, a\beta)$ so that the homomorphism $G(L/K) \to O(V, \beta)$, upon which $SW_i[\rho]$ depends, is unaltered.

Part (ii) follows from the explicit formula for Frohlich's bilinear form, Chapter 2, §3.14. In the notation of Chapter 2, §3.8 scaling by $a \in K^*$ changes the $\{u_j\}$ to $\{u_j a^{-1}\}$ which in turn multiplies Chapter 2, §3.14 by $a = a^{-1} \in K^*/(K^*)^2$, as required.

Changing β to $a\beta$ changes the spinor norm of a reflexion by a factor, a, [O'M, p. 138]. Hence, if $\rho(g)$ is the product of t reflexions, then $\theta(\rho(g)) \in K^*/(K^*)^2$ changes by $a^t = a^{\det \rho(g)}$. Since $\det \rho = SW_1[\rho]$, part (iii) follows. $\qquad \square$

(1.51) Explicitly the low-dimensional cases of §1.50 (ii) yield (rank $V = m$) $HW_1(\rho, V, a\beta) = ml(a) + HW_1(\rho, V, \beta)$, and $HW_2(\rho, V, a\beta) = HW_2(\rho, V, \beta) + (m-1)l(a)HW_1(\rho, V, \beta) + (m(m-1)/2)l(a)^2$.

(1.52) Proposition

In the notation of §1.49 (ii) the behaviour of orthogonal direct sums is as follows:

(i) $SW_i[\rho, V, \beta] = \sum_{a=0}^{i} SW_a[\rho', V', \beta'] SW_{i-a}[\rho'', V'', \beta'']$.

(ii) $HW_i(\rho, V, \beta) = \sum_{a=0}^{i} HW_a(\rho', V', \beta') HW_{i-a}(\rho'', V'', \beta'')$.

(iii) $Sp[\rho, V, \beta] = Sp[\rho', V', \beta'] + Sp[p'', V'', \beta'']$.

Proof. Part (i) is a well-known property of Stiefel-Whitney classes [M-St; H]. Part (ii) is clear from the definition (1.35), and part (iii) follows from the fact that $\theta(\rho' \oplus \rho'') = \theta(\rho')\theta(\rho'')$, another fact which is obvious from the definition, §1.12.   □

(1.53)   Finally, we consider the effect of the hyperbolic relation. We retain the notation of §1.49 (iii). We will work in terms of matrices. By induction, we may find a basis for V over K of the form

$$w_i, \ldots, w_a \in W,$$

$$v_1, \ldots, v_a, u_1, \ldots, u_t \in V \quad \text{such that}$$

$$\beta(w_i, w_j) = \beta(v_i, v_j) = 0,$$

$$\beta(w_i, v_j) = \delta_{ij}, \quad \text{for all } 1 \le i, j \le a,$$

$$\beta(w_i, u_j) = 0 = \beta(v_i, u_j).$$

With respect to this basis, the matrix of (V, β) is as follows (where $\wedge$ is a $t \times t$ matrix):

(1.54)
$$\Phi = \begin{bmatrix} 0 & 0 & I_a \\ 0 & \wedge & 0 \\ I_a & 0 & 0 \end{bmatrix}.$$

Therefore $W = \langle w_1, \ldots, w_a \rangle$ and $W^0 = \langle w_1, \ldots, u_1, \ldots, u_t \rangle$.

Also with respect to this basis, if $g \in G(L/K)$, we have

(1.55)
$$\rho(g) = \begin{bmatrix} A(g) & D(g) & E(g) \\ 0 & B(g) & F(g) \\ 0 & 0 & C(g) \end{bmatrix}.$$

Calculating $\rho(g)\Phi\rho(g)^t = \Phi$, we obtain

$$(1.56) \quad \begin{cases} A(g) = (C(g)^t)^{-1} = C(g)^* \\ \land = B(g) \land B(g)^t \\ F(g) = C(g)B(g) \land D(g)^t \\ E(g) + E(g)^t + D(g) \land D(g)^t = 0 \end{cases}$$

Also $A(\text{---})$ and $B(\text{---})$ are homomorphisms from $G(L/K)$ to $GL_a(K)$ and $O(K^t, \land)$, respectively.

In the notation of §1.49 (iii), by (1.56) we obtain

$$(1.57) \quad \begin{cases} (\rho, V, \beta) = \left[\begin{bmatrix} A & D & E \\ 0 & B & F \\ 0 & 0 & A^* \end{bmatrix}, K^{2a+t}, \begin{bmatrix} 0 & 0 & I_a \\ 0 & \land & 0 \\ I_a & 0 & 0 \end{bmatrix} \right], \\ (\rho_1, W^0/W, \beta_1) = (B, K^t, \land), \text{ and} \\ (\rho_2, W \oplus (V/W^0), \beta_2) = \left[\begin{bmatrix} A & 0 \\ 0 & C \end{bmatrix}, K^{2a}, \begin{bmatrix} 0 & I_a \\ I_a & 0 \end{bmatrix} \right]. \end{cases}$$

Abbreviate the orthogonal representations of (1.57) to (ρ), (ρ_1) and (ρ_2), respectively.

(1.58) From (1.57) we have a homomorphism

$$\lambda : O(V, \beta) \to O(K^t, \land) \times O\left(K^{2a}, \begin{bmatrix} 0 & I_a \\ I_a & 0 \end{bmatrix} \right) = H.$$

The kernel of λ consists of matrices of the form

$$\begin{bmatrix} I & D & \bar{E} \\ 0 & I & F \\ 0 & 0 & I \end{bmatrix}.$$

Also

$$\begin{bmatrix} I & D & \bar{E} \\ 0 & I & F \\ 0 & 0 & I \end{bmatrix} \begin{bmatrix} I & D' & E'' \\ 0 & I & F' \\ 0 & 0 & I \end{bmatrix} = \begin{bmatrix} I & D+D' & ? \\ 0 & I & F+F' \\ 0 & 0 & I \end{bmatrix},$$

while

$$\begin{bmatrix} I & 0 & \bar{E} \\ 0 & I & 0 \\ 0 & 0 & I \end{bmatrix} \begin{bmatrix} I & 0 & \bar{E''} \\ 0 & I & 0 \\ 0 & 0 & I \end{bmatrix} = \begin{bmatrix} I & 0 & E+E'' \\ 0 & I & 0 \\ 0 & 0 & I \end{bmatrix}.$$

(1.59) Lemma

Let Ω'_K be the 2-Sylow prosubgroup of $\Omega_K = G(\bar{K}/K)$. If char $K \neq 2$, then the map induced by λ

$$(\lambda_K)_* : H^1_{ct}(\Omega_K; O(V_{\bar{K}}, \beta_{\bar{K}})) \to H^1(\Omega'_K; H_{\bar{K}})$$

is injective. Here

$$H_{\bar{K}} = O(\bar{K}^t, \wedge) \times O\left(\bar{K}^{2a}, \begin{bmatrix} 0 & I_a \\ I_a & 0 \end{bmatrix}\right).$$

Proof. From [Lan, p. 215] we know that $H^i_{ct}(\Omega'_K; \bar{K}) = \{*\}$. However, Ker $(\lambda_{\bar{K}})$ has a decomposition series, invariant under the $\Omega'_{\bar{K}}$-action with successive quotients isomorphic to $\bar{K}$ (additively). Hence the sequence of Chapter 2, §1.16 and induction shows that $H^1(\Omega'_K; \text{Ker}(\lambda_{\bar{K}})) = \{*\}$. Therefore, by Chapter 2, §1.16 again, $(\lambda_{\bar{K}})_*$ is one-one. $\square$

(1.60) Theorem

In the notation of §1.49 (iii) [and (1.57)]

(i) $HW_i((\rho_1) \oplus (\rho_2)) = HW_i(\rho),$
(ii) $SW_i([\rho_1] \oplus [\rho_2]) = SW_i[\rho],$
(iii) $Sp([\rho_1] \oplus [\rho_2]) = Sp[\rho].$

Proof. Consider the commutative diagram

$$\begin{array}{ccc} H^1(K; O_m\bar{K}) & \xrightarrow{HW_i} & H^i(K; \mathbb{Z}/2) \\ {\scriptstyle \text{res}} \downarrow & & \downarrow {\scriptstyle \text{res}} \\ H^1_{ct}(\Omega'_K; O_m\bar{K}) = H^1(F; O_m\bar{K}) & \xrightarrow[HW_i]{} & H^i(F; \mathbb{Z}/2), \end{array}$$

where $m = 2a + t$ and $F = L^{\Omega'_K}$. Since the right-hand restriction is one-one, we see that $HW_i(\rho)$ depends only on its image in $H^1(\Omega'_K; O_m\bar{K})$. However, by §1.59 and (1.57), the images of $(\rho_1) \oplus (\rho_2)$ and (ρ) are equal in $H^1(F; O_m\bar{K})$. This proves part (i).

Parts (ii) and (iii) follow from the fact that $H^*(O(V, \beta); \mathbb{Z}/2) \cong H^*(H; \mathbb{Z}/2)$ by a similar argument. $\square$

2.   SERRE'S FORMULA, FROHLICH'S FORMULA, AND OTHER EXAMPLES OF THE CHARACTERISTIC CLASSES

(2.1)   Our first objective in this section will be to prove a very attractive formula, due to Frohlich [F, §3.1], which is a generalization of an earlier formula, due to Serre [Ser 3]. These formulae relate low-dimensional Hasse-Witt classes, Stiefel-Whitney classes, and the spinor class. Higher-dimensional generalizations of this, generalizing slightly a theorem of Kahn [K], will be proved later.

We will pause to prove a very well-known fact concerning products in Galois cohomology. In the spirit of these notes I have given a proof based on group cohomology—specifically based upon our calculation of $H^*(D_8; \mathbb{Z}/2)$ in Chapter 1, Theorem 4.6.

(2.2) Proposition

Let K be a field of characteristic different from 2. Let $a \in K^*$ and $l(a) \in H^1(K; \mathbb{Z}/2)$ be as in §§1.34–1.35 or Chapter 2, §3.3.

   Then

$$l(a)^2 = l(a)l(-1) \in H^2(K; \mathbb{Z}/2).$$

Proof. If $\sqrt{a} \in K$ there is nothing to prove, for then $l(a)(g) = g(\sqrt{a})/(\sqrt{a}) = 1$ for all $g \in \Omega_K$ and $l(a) = 0$.

If $\sqrt{a} \notin K$, then $t^4 - a \in K[t]$ is separable, and its splitting field is $L = K(\xi, \mu)$, where $\xi^2 = -1$ and $\mu^4 = a$. The Galois group, $G(L/K)$, contains elements x, y defined, if $\xi \notin K$, by

(2.3)
$$\begin{cases} y(\xi) = -\xi, & y(\mu) = \mu, \\ x(\xi) = \xi, & x(\mu) = \xi\mu. \end{cases}$$

Clearly, $yxy(\xi) = \xi$, $yxy(\mu) = y(\xi\mu) = -\xi\mu$, so that $yxy = x^3 = x^{-1} \in G(L/K)$. Therefore, by dimensions, $G(L/K) \cong D_8 = \langle x, y \rangle$ if $\xi \notin K$. Furthermore

(2.4)
$$\begin{cases} 1 = x(\xi)/\xi = l(-1)(x), & -1 = y(\xi)/\xi = l(-1)(y), \\ -1 = x(\mu^2)/\mu^2 = l(a)(x), & 1 = y(\mu^2)/\mu^2 = l(a)(y), \end{cases}$$

so that $l(a) = x_2, l(-1) = x_1$ in the notation of Chapter 1, §4.6. Since $x_2^2 + x_1 x_2 = 0$ in $H^2(D_8; \mathbb{Z}/2)$, the result follows if $\xi \notin K$.

If $\xi \in K$, then $G(L/K) \cong \mathbb{Z}/4$, and $l(a)$ originates in $H^1(\mathbb{Z}/4; \mathbb{Z}/2)$ so that $l(a)^2 = 0 = l(a)l(-1)$, as required.   □

(2.5) Theorem

Let K be a field of characteristic different from 2. If $a, b \in K^*$, then $0 = l(a)l(b) \in H^2(K; \mathbb{Z}/2)$ if and only if a is a norm from $K(\sqrt{b})^*$.

Proof. If $N: K(\sqrt{b})^* \to K^*$ denotes the norm, then $N(\alpha + \beta\sqrt{b}) = \alpha^2 - b\beta^2$. If $\alpha^2 - \beta^2 b = a$, then $1 = b(\beta/\alpha)^2 + a(1/\alpha)^2$ so that, by Chapter 2, §3.56,

$$0 = l(b(\beta/\alpha)^2)l(a/\alpha^2)$$

$$= l(b)l(a).$$

If $\alpha = 0$, then $a = -b\beta^2$, and $l(a)l(b) = l(a)l(-a) = 0$, by §2.2.

Conversely, suppose that $l(a)l(b) = 0$. By Chapter 2, §3.29, this means that there is an isomorphism of central, simple K-algebras

$$A_{-1}(a, b) \cong M_2 K.$$

Recall from Chapter 2, (3.43)

$$A_{-1}(a, b) = \langle X, Y \mid X^2 = a, Y^2 = b, YX = -XY \rangle.$$

Hence there exists $Y \in M_2 K$, and choosing a suitable basis for K^2, we may write

$$Y = \begin{bmatrix} 0 & b \\ 1 & 0 \end{bmatrix}.$$

Now observe that

$$Y \begin{bmatrix} -1 & 0 \\ 0 & 1 \end{bmatrix} Y^{-1} = \begin{bmatrix} 1 & 0 \\ 0 & -1 \end{bmatrix},$$

whereas the eigenspace of $\{Z \in M_2 K \mid YZY^{-1} = -Z\}$ is spanned by X and XY. Therefore

$$\begin{bmatrix} 1 & 0 \\ 0 & -1 \end{bmatrix} = cX + dXY = X(c + dY)$$

for some $c, d \in K$. Squaring this

$$I_2 = X(c + dY)(X)(c + dY)$$

$$= X^2(c - dY)(c + dY)$$

$$= a(c^2 - bd^2)I_2$$

from which we see that $1/a$ (and hence a, too) is a norm from $K(\sqrt{b})$.   □

(2.6)   Recall from Chapter 2, §3.8 that if $b: V \times V \to K$ is a nonsingular bilinear form and $T: G(L/K) \to O(b)$ is an orthogonal representation, then there is a bilinear form, given explicitly by the formulae of Chapter 2, (3.10)–(3.15), denoted by $\beta(T,b): V \times V \to K$ and, by 3.16, represented by the 1-cocycle

$$T: G(L/K) \to O(b) \to O(b_L)$$

in $H^1(G(L/K); O(b_L))$.

The Stiefel-Whitney classes, $SW_i[T]$ (see Chapter 1, §3.16), lie in $H^1(G(L/K);$ $\mathbb{Z}/2)$. Also the spinor class, $Sp[T]$ (see §1.18), lies in $H^2(K; \mathbb{Z}/2)$.

(2.7) Theorem [F, §3]

Let $T: G(L/K) \to O(b)$ be as in §2.6, with char $K \neq 2$ and $b: V \times V \to K$ symmetric. Then, with the notation of §2.6,

(i) $HW_1(\beta(T,b)) = HW_1(b) + SW_1[T]$ *in $H^1(K; \mathbb{Z}/2)$,*
    (where $SW_1[T] = \det T: G(L/K) \to \{\pm 1\}$)   *and*

(ii) $HW_2(\beta(T,b)) = HW_2(b) + SW_2[T] + Sp[T] + HW_1(b)SW_1[T]$
   in $H^2(K; \mathbb{Z}/2)$.

(2.8) Corollary [Ser 3]

Let L/K be a finite separable field extension with $\mathrm{char}(K) \neq 2$, and let $\beta: V \times V \to L$ be a nonsingular, symmetric bilinear form which is represented in $H^1(K; O_n(\bar{K}))$ by a (continuous) homomorphism $f: \Omega_L \to O_n(K) \subset O_n(\bar{K})$. Then

(i) $HW_1(\mathrm{Tr}^S_{L/K}(\beta)) = SW_1(\mathrm{Ind}^{\Omega_K}_{\Omega_L}[f])$ *and*

(ii) $HW_2(\mathrm{Tr}^S_{L/K}(\beta)) = SW_2(\mathrm{Ind}^{\Omega_K}_{\Omega_L}[f]) + \mathrm{Tr}_{E/K}(Sp[f]) + nl(2)d_{L/K}$.

Here $\mathrm{Tr}^S_{L/E}$ is the Scharlau transfer of Chapter 2, §3.19, $\mathrm{Tr}_{L/E}$ is the cohomology transfer, Ind^G_H is the induced representation and $d_{L/K}$ is the discriminant of L/K.

Proof. In §2.7, b is the standard form, $b = n\langle 1 \rangle$, so $HW_i(b) = 0$. By Chapter 2, §3.25, $\mathrm{Ind}^{\Omega_K}_{\Omega_L}[f]$, represents $\mathrm{Tr}^S_{L/K}(\beta)$.

Therefore the formulae follow at once from those of §2.7, together with the formula $Sp(\mathrm{Ind}[f]) = nl(2)d_{L/K} + \mathrm{Tr}_{L/K}(Sp[f])$ of §1.29.   □

(2.9) Before proving Theorem 2.7, we will need some preliminary discussion. The key to the formula of §2.7 (ii) is our description of HW_2 in §1.45. In order to capitalize on that, we will need to describe the Galois cohomology representative of $\beta(T,b)$ in $H^1(K; O_n(\bar{K}))$ ($n = \mathrm{rank}_K b = \mathrm{rank}_K \beta(T,b)$).

Choose a K-basis for V so that if $x = (x_1, \ldots, x_n)$, $y = (y_1, \ldots, y_n) \in K^r \cong V$, $b(x,y) = \sum a_i x_i y_i$. Set

$$\alpha = \begin{bmatrix} \sqrt{a_1} & & & \\ & \sqrt{a_2} & & \\ & & \ddots & \\ & & & \sqrt{a_n} \end{bmatrix} \in GL_n K$$

so that $(\alpha - \alpha^{-1}): GL_n\bar{K} \to GL_n\bar{K}$ restricts to $(\alpha - \alpha^{-1}): O(b_{\bar{K}}) \xrightarrow{\cong} O_n(\bar{K})$.

If λ is the map on Clifford algebras, $\lambda: C(\bar{K}^n; b_{\bar{K}}) \to C(\bar{K}^n, n\langle 1 \rangle)$, induced by $\alpha: \bar{K}^n \to \bar{K}^n$, then the following diagram commutes:

(2.10)

$$
\begin{array}{ccc}
\text{Pin}(b_{\bar{K}}) & \xrightarrow{\ \lambda\ } & \text{Pin}_n(\bar{K}) \\
\downarrow{\scriptstyle \pi_b} & & \downarrow{\scriptstyle \pi} \\
O(b_{\bar{K}}) & \xrightarrow[(\alpha - \alpha^{-1})]{} & O_n(\bar{K})
\end{array}
$$

In addition, sections s_b and s may be chosen to make the following diagram commute also.

(2.11)

$$
\begin{array}{ccc}
\text{Pin}(b_{\bar{K}}) & \longrightarrow & \text{Pin}_n(\bar{K}) \\
\uparrow{\scriptstyle s_b} & & \uparrow{\scriptstyle s} \\
O(b_{\bar{K}}) & \xrightarrow[(\alpha - \alpha^{-1})]{} & O_n(\bar{K})
\end{array}
$$

Suppose that $T(g) = A^{-1}g(A)$, then, if (x, y) is the standard form, $n\langle 1 \rangle$, on $\bar{K}^n$, then (writing β for $\beta(T, b)$) by Chapter 2, §2.5,

$$(2.12) \qquad \beta_{\bar{K}}(x, y) = b_{\bar{K}}(A^{-1}x, A^{-1}y) = (\alpha A^{-1}x, \alpha A^{-1}y).$$

Therefore, since $\alpha A^{-1}g(\alpha A^{-1})^{-1} = \alpha T(g)g(\alpha^{-1})$,

$$(2.13) \qquad \begin{cases} \beta(T, b) & \text{is represented in } H^1(K, O_n(\bar{K})) \\ \text{by } (g \to (\alpha T(g)\alpha^{-1})\alpha g(\alpha^{-1}). \end{cases}$$

Write $X(g) = \alpha g(\alpha^{-1})$, which is the 1-cocycle representing b in $H^1(K, O_n(\bar{K}))$ since $b(x, y) = (\alpha x, \alpha y)$. Also write $h(g) = \alpha T(g)\alpha^{-1}$ so that h is a homomorphism, $h: \Omega_K \to O_n(\bar{K})$, with $SW_i[T] = SW_i[h] \in H^i(K; \mathbb{Z}/2)$.

(2.14) Lemma

With the preceding notation

$$\gamma: \Omega_K \times \Omega_K \to \Omega_K \ltimes O_n(\bar{K}),$$

given by

$$\gamma(g, g') = (g, h(g)X(g')),$$

is a homomorphism. The composition

$$\gamma \circ (\text{diag}): \Omega_K \to \Omega_K \ltimes O_n(\bar{K})$$

classifies $\beta(T, b)$ in $H^1(K; O_n(\bar{K}))$ (see Chapter 2, §§1.1–1.5).

Proof. The last statement merely recapitulates (2.13). Therefore we have only to check that γ is a homomorphism.

If $g, g', g_1, g'_1 \in \Omega_K$, then we have to show that

$$(g, h(g)X(g'))(g_1, h(g_1)X(g'_1)) = (gg_1, h(gg_1)X(g', g'_1)).$$

However,

$$(g, h(g)X(g'))(g_1, h(g_1)X(g'_1)) = (gg_1, h(g)X(g')g(h(g_1))g(X(g'_1))),$$

whereas

$$(gg_1, h(gg_1)X(g'g'_1)) = (gg_1, hg)h(g_1)X(g')g'(X(g'_1))).$$

Since X and $(g \mapsto h(g)X(g))$ are 1-cocycles,

$$h(g)X(g)g(h(g_1))g(X(g_1)) = h(g)h(g_1)X(g)g(X(g_1))$$

so that $h(g_1)X(g) = X(g)g(h(g_1))$. In addition $X(g) \in O_1(K)^n$ so that $X(g)$ is fixed by Ω_K. Therefore both expressions reduce to $(gg_1, h(g)X(g')h(g_1)X(g'_1))$, which completes the proof of §2.14. □

(2.15)   The Künneth formula [C-E; H-S; Spa], which is proved using Chapter 1, §2.26 in a manner similar to Chapter 1, (2.25)–(2.33), states that the cohomology product induces an isomorphism (with trivial coefficients, $\mathbb{Z}/2$)

$$H^n(H \times G; \mathbb{Z}/2) \cong \bigoplus_{a+b=n} H^a(H; \mathbb{Z}/2) \otimes H^b(G; \mathbb{Z}/n).$$

(2.16) Proof of Theorem 2.7 (i). Part (i) follows from (2.13) since, by definition, $HW_1(\beta)$ is the image of β under $\det_*: H^1(K; O_n(\bar{K})) \to H^1(K; \mathbb{Z}/2)$. However, this image sends $g \in \Omega_K$ to

$$(\det \alpha T(g)\alpha^{-1}) \cdot \det(\alpha g(\alpha)^{-1}), \qquad \text{by (2.13).}$$

The first factor is $\det T(g)$, representing $SW_1[T]$, while the second factor is $HW_1(b)$ since $(g \mapsto \alpha g(\alpha)^{-1})$ represents b in $H^1(K; O_n(\bar{K}))$. □

(2.17) Lemma

With the notation of §2.7, let $\tilde{\omega}_2 \in H^2(\Omega_K \ltimes O(b_{\bar{K}}); \mathbb{Z}/2)$ denote the class of §1.36 *derived from*

$$\mathbb{Z}/2 \to \mathrm{Pin}(b_{\bar{K}}) \twoheadrightarrow O(b_{\bar{K}}).$$

Then

$$(1, T)^*(\tilde{\omega}_2) = SW_2[T] + Sp[T] \in H^2(K; \mathbb{Z}/2).$$

Proof. We may factorise $(1, T)$ as

$$\Omega_K \xrightarrow{(1,T)} \Omega_K \times O(b) \xrightarrow{j} \Omega_K \ltimes O(b_{\bar{K}}).$$

Hence $j^*(\tilde{\omega}_2) \in H^2_{ct}(\Omega_K \times O(b); \mathbb{Z}/2)$ which is

$$H^2(K; \mathbb{Z}/2) \oplus (H^1(K; \mathbb{Z}/2) \otimes H^1(O(b); \mathbb{Z}/2)) \oplus H^2(O(b); \mathbb{Z}/2)$$

by §2.15. The first factor of $j^*(\omega_2)$ is zero because it is represented by the 2-cocycle

$$\{(g, g') \mapsto \tilde{\omega}_2((g, 1), (g', 1)) = 1\}.$$

To calculate the second factor, we must evaluate

$$\{(g, X) \mapsto \tilde{\omega}_2((g, 1), (1, X)) = g(s_b(X))(s_b(X))^{-1}\},$$

where s_b is the section map in §2.11. If X is written $X = \tau_{v_1} \cdots \tau_{v_r}$, then

$$s_b(X) = \left[\frac{v_1}{\sqrt{b(v_1, v_1)}} \right] \circ \cdots \circ \left[\frac{v_r}{\sqrt{b(v_r, v_r)}} \right]$$

for a suitable choice of section, s_b [O'M]. Here $(\text{---}\circ\text{---})$ denotes the Clifford algebra product. Hence $g(s_b(X)) \cdot s_b(X)^{-1}$ equals $g(\sqrt{\theta(X)})/\sqrt{\theta(X)}$, by (1.14). By definition, in §1.18, pulling this back by $(1, T)^*$ gives $Sp[T] \in H^2(K; \mathbb{Z}/2)$.

The third factor is represented by

$$\{(X, Y) \mapsto \tilde{\omega}_2((1, X)(1, Y)) = s_b(X)s_b(Y)s_b(XY)^{-1}\},$$

which is the second Stiefel-Whitney class, by (1.33), and this completes the proof. □

(2.18) Proof of Theorem 2.7 (ii). Let $\tilde{\omega}_2 \in H^2(K; O_n\bar{K})$ denote the ("continuous") class given by the family $\{\omega_{2,L}\}$ of §1.42. By §2.14 and §1.45,

$$HW_2(\beta(T, b)) = HW_2(\beta) = (\mathrm{diag})^* \gamma^*(\tilde{\omega}_2).$$

Hence we will evaluate, again using the Künneth formula of §2.15, $\gamma(\tilde{\omega}_2)$ in

$$H^2(K; \mathbb{Z}/2) \oplus (H^1(K; \mathbb{Z}/2) \otimes H^1(K; \mathbb{Z}/2)) \oplus H^2(K; \mathbb{Z}/2).$$

The first factor is represented by the 2-cocyle $\{(g, g_1) \mapsto \tilde{\omega}_2(g, h(g)),$ $(g_1, h(g_1))\}$. However, $h(g) = \alpha T(g)\alpha^{-1}$, and in (2.11), $s(\alpha T(g)\alpha^{-1}) = \lambda(s_b(T(g)))$ which implies that this term is equal to $(1, T)^*(\tilde{\omega}_2)$, as calculated in §2.17. Therefore the first factor is $SW_2[T] + Sp[T]$.

The third factor is represented by

$$\{(g', g_1') \mapsto \tilde{\omega}_2((1, X(g')), (1, X(g_1')))\},$$

which is equal to $HW_2(b) = SW_2[X]$, since $X(g')$, $X(g_1')$ are diagonal.

The second factor is represented by the 2-cocyle

$$\{(g', g_1) \mapsto \tilde{\omega}_2((1, X(g')), (g_1, h(g_1))$$

$$= s(X(g'))s(h(g_1))[s(X(g')h(g_1))]^{-1}.$$

The only one-dimensional characteristic class is the determinant. However, this represents a term in $H^1(K; \mathbb{Z}/2) \otimes H^1(K; \mathbb{Z}/2)$. Therefore it must depend on $SW_1[T]$ and $HW_1(b)$ (since X classifies b and $h(g_1) = \alpha T(g_1)\alpha^{-1}$) in such a way as to vanish if either $SW_1[T] = 0$ or $HW_1(b) = 0$. Therefore the second factor maps to $qSW_1[T] \cdot HW_1(b)$ (for some $q \in \mathbb{Z}/2$) under (diag)*.

So far we have shown that

$$HW_2(\beta) = HW_2(b) + SW_2[T] + Sp[T] + qSW_1[T]HW_1(b)$$

in $H^2(K; \mathbb{Z}/2)$. To decide between the cases $q = 0$ and $q = 1$ is simple.

Method 1. Apply the preceding formula with T, b replaced by $T \oplus (\det T)$, $b \oplus \langle 1 \rangle$, and use the fact that $\beta(T \oplus (\det T), b + \langle 1 \rangle) = \beta(T, b) \oplus \beta(\det T, \langle 1 \rangle)$. The formulae of §1.49 for characteristic classes readily show that $q = 1$.

Method 2. By changing T to $T \oplus (1)$ and b to $b \oplus \langle 1 \rangle$, we may assume that n is odd. Now we observe that, in §1.49, scaling by $a \in K^*$ alters $HW_2(\beta)$ by $(n(n-1)/2)l(a)^2$ (see §1.51). However, if $q = 0$ the other side of our equation changed by $(n(n-1)2)l(a)^2 + l(a)SW_1[T]$ so that $l(a)SW_1[T] = 0$ for all $a \in K^*$. Therefore $HW_1(b)SW_1[T] = 0$, so we might as well take $q = 1$. This completes the proof of Theorem 2.7.   □

(2.19) We will now give some applications of the formulae of §§2.7/2.8. These examples are all originally due to Serre [Ser 3, §3].

First, suppose that L/K is a finite, separable extension with char $(K) \neq 2$. Let N denote the normal closure of L/K. That is, if $\bar{K}$ is a separable closure of K

containing L, then

$$N = \langle \phi(L) | \phi: L \to \bar{K}, \text{ a } K\text{-homomorphism}\rangle$$

is the field generated by all K-embeddings of L into $\bar{K}$. The extension N/K is Galois, and $G(N/K) \subset \Sigma_n$ acts *transitively* on the set $\{1,\ldots,n\}$ (by means of the permutation action of Σ_n). Let $H = \text{stab}(1)$, then $L \cong N^H$ and the inclusion homomorphism is given by

$$(2.20) \qquad \lambda = \text{Ind}_H^{G(N/K)}(1): G(N/K) \to \Sigma_n,$$

the permutation representation given by the (left) action on the set $G(N/K)/H$.

Consequently, applied to λ in (2.20), Serre's formula of §2.8 yields

$$(2.21) \qquad HW_2\langle L/K\rangle = SW_2(\lambda) + l(2)d_{L/K}$$

in $H^2(K; \mathbb{Z}/2)$. Here $\langle L/K\rangle$ denotes the *trace form*, $L \times L \to K$, given by $(x, y) \mapsto \text{Tr}_{L/K}(xy)$, which is classified by λ (see Chapter 2, §§3.9–3.25).

(2.22) The vanishing of $SW_2[\lambda] \in H^2(K; \mathbb{Z}/2)$ is, by (1.33), the obstruction to lifting $\lambda: \Omega_K \to G(N/K)$ to the double covering $\tilde{G}(N/K)$ which is induced by the following diagram of central extensions:

(2.23)
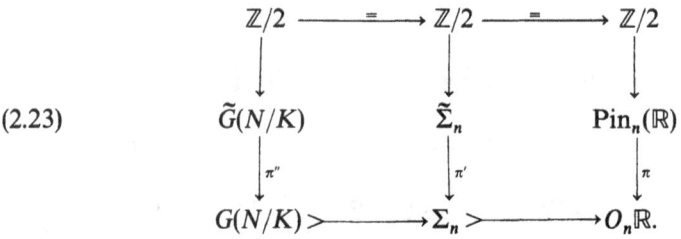

The existence of a commutative diagram

(2.24)
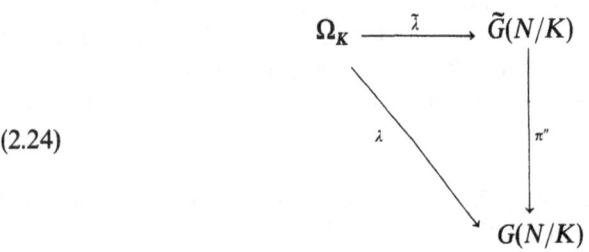

is equivalent to the following:

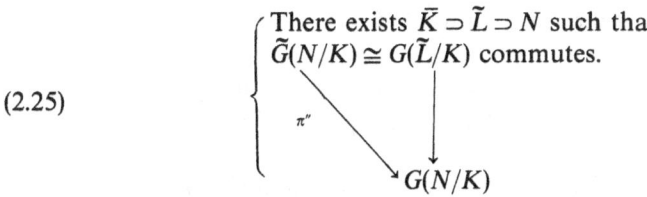

(2.25)

(2.26)   Let K be a number field (i.e., a finite extension of $\mathbb{Q}$), and let $b: V \times V \to K$ be a nonsingular symmetric bilinear form. For each infinite real place (i.e., an embedding $V: K \to \mathbb{R} = K_v$), we may form b_{K_v}, which is equivalent to $p\langle 1 \rangle + (n - p)\langle -1 \rangle$ and take its *signature at* v [Sch, p. 42]

(2.27) $\sigma_v(b) = \operatorname{sign}(b_{K_v}) = p - (n - p) = 2p - n.$

In addition, if $j \geq 3$ [M],

(2.28) $H^j(K; \mathbb{Z}/2) \cong \bigoplus_{v \text{ real}} (\mathbb{Z}/2)$

in which the vth component is $H^j(K_v; \mathbb{Z}/2)$ generated by the image of $l(-1)^j$. Since $HW_j(p\langle 1 \rangle + (n - p)\langle -1 \rangle)$ equals $HW_j((n - p)\langle -1 \rangle) = \begin{bmatrix} n-p \\ j \end{bmatrix} l(-1)^j$, we have the following lemma:

(2.29) Lemma

The v-component of $HW_j(V, b)$ in (2.28) is $\begin{bmatrix} n-p \\ j \end{bmatrix}$, *where* $n = \operatorname{rank}(b)$ *and* $2p - n = \sigma_v(b)$. *In particular, the rank and the* $\{HW_j(V, b); j \geq 1\}$ *determine all the signatures of* (V, b) *at the real places of* K.

Proof. Although (2.28) holds only for $j \geq 3$ the formula for $HW_j(b_{K_v})$ still holds for $j = 1, 2$. Hence, by choosing j to run through powers of 2, we can read off the dyadic expansion of $n - p$ (and have $2p - n = \sigma_v(b)$) from the $HW_j(b_{K_v})$'s. This is because

$$\begin{bmatrix} m \\ 2^\alpha \end{bmatrix} \equiv 1 \ (\mathrm{mod}\ 2) \qquad \text{if and only if } 2^\alpha$$

appears in the dyadic expansion of m.   □

We will conclude this section with some examples.

(2.30) Example

Let K be a number field, and let L/K be an extension of order 4 or 5. Then

(2.25) holds if and only if the trace form satisfies

$$\langle L/K \rangle = \begin{cases} 2\langle 1 \rangle + \langle 2 \rangle + \langle 2d_{L/K} \rangle & \text{if } n = 4, \\ 3\langle 1 \rangle + \langle 2 \rangle + \langle 2d_{L/K} \rangle & \text{if } n = 5. \end{cases}$$

The condition is equivalent by §2.22 to $SW_2(\lambda) = 0$ in (2.21) so that

$$HW_2(L/K) = l(2)d_{L/K}.$$

In addition $HW_1(L/K) = SW_1(\lambda) = d_{L/K}$, by §2.8 (i). On the other hand,

$$HW_1(\langle 2 \rangle + \langle 2d_{L/K} \rangle) = l(2) + l(2) + d_{L/K} = d_{L/K},$$

whereas

$$HW_2(\langle 2 \rangle + \langle 2d_{L/K} \rangle) = l(2)(l(2) + d_{L/K}) = l(2)d_{L/K}$$

by §2.5. This means that the trace form, $\langle L/K \rangle$, has the same rank and Hasse-Witt classes as its putative diagonalization.

A symmetrical, nondegenerate bilinear form over K is determined by HW_1, HW_2, rank, and the signatures, σ_v. Furthermore it can be shown that when $n = 4$, $\langle 1 \rangle$ splits from $\langle L/K \rangle$, while when $n = 5$, $2\langle 1 \rangle$ splits off [Ser 3, App. I, Prop. 4]. However, HW_1 and HW_2 determine all the real signatures of a form of rank three, by §2.29, and the claim is proved.

(2.31) Example

Let K be a number field and L/K a *triquadratic extension*. That is, $L = K(\sqrt{x}, \sqrt{y}, \sqrt{z})$ with $xyz = 1$, $N = L$, $[L:K] = 4$, and $G(L/K) \cong \mathbb{Z}/2 \times \mathbb{Z}/2$. Hence the trace form, $\langle L/K \rangle$, is represented by the regular representation, $1 + l(x) + l(y) + l(z)$. In other words,

$$\langle L/K \rangle = \langle 1 \rangle + \langle x \rangle + \langle y \rangle + \langle z \rangle.$$

Also $\tilde{G}(L/K) = \tilde{G}(N/K) = Q_8$, the quaternion group of order 8 (Chapter 1, (3.10)).

In this example (2.25) holds if and only if

$$\langle x \rangle + \langle y \rangle + \langle z \rangle = 3\langle 1 \rangle.$$

This follows easily since

$$d_{L/K} = HW_1(\langle x \rangle + \langle y \rangle + \langle z \rangle) = l(x) + l(y) + l(xy) = 0,$$

and therefore

$$HW_2(\langle x \rangle + \langle y \rangle + \langle z \rangle) = SW_2(\text{Ind}_{(1)}^{G(L/K)}(1)),$$

which vanishes if and only if (2.25) holds. As in §2.30, HW_1 and HW_2 characterize forms of rank three over a number field.

(2.32) Example

Suppose L/K is an extension of number fields with $[L:K] = 5$, $d_{L/K} = 1$ and 5 a square in K^*. Then the following conditions are equivalent:

(a) $SW_2(\lambda) = l(-1)^2$, for λ as in (2.20).
(b) $\langle L/K \rangle = 3\langle 1 \rangle + 2\langle -1 \rangle$.
(c) There exists $x \in L^*$ such that $\text{Tr}_{L/K}(x) = 0 = \text{Tr}_{L/K}(x^2)$.

Since $d_{L/K} = 1$, we have $HW_1\langle L/K \rangle = 0$ and $HW_2\langle L/K \rangle = SW_2[\lambda]$ by (2.21). Therefore $\langle L/K \rangle$ and $3\langle 1 \rangle + 2\langle -1 \rangle$ have the same rank, HW_1 and HW_2 if and only if $SW_2[\lambda] = l(-1)^2$. However, one can show that $\langle L/K \rangle = 2\langle 1 \rangle + \langle a_1 \rangle + \langle a_2 \rangle + \langle a_3 \rangle$ (see [Ser 2, App. I, Prop. 4]) so that these invariants classify $\langle L/K \rangle$, thereby showing that (a) and (b) are equivalent. Furthermore it is clear that (b) implies (c).

Conversely, if x exists in (c), then we may choose a K-basis for L containing 1 and x. Thus we can write the matrix for $\langle L/K \rangle$ as $\langle 5 \rangle + \beta$, where β is $\langle L/K \rangle$ restricted to the subspace

$$\{z \in L \,|\, \text{Tr}_{L/K}(z) = 0\}$$

and $\langle 5 \rangle$ is $\langle L/K \rangle$ restricted to $\{1\}$. Since 5 is a square $\langle L/K \rangle = \langle 1 \rangle + \beta$. However, by [Ser 2, App. I, Prop. 4] $\beta = \langle 1 \rangle + \langle c_1 \rangle + \langle c_2 \rangle + \langle c_3 \rangle$ or $\beta = \langle 1 \rangle + b$ with rank $b = 3$.

Since $b(x, x) = 0$, we readily find that $b = \langle a \rangle + \left\langle \begin{bmatrix} 0 & 1 \\ 1 & 0 \end{bmatrix} \right\rangle = \langle a \rangle + \langle 1 \rangle + \langle -1 \rangle$ so that $b = 2\langle -1 \rangle + \langle 1 \rangle$ since $d_{L/K} = 1$.

The conditions (a) and (c) are equivalent to the fact that the extension L/K is constructible by an icosahedral construction due to Klein (see [Ser 4]).

(2.33) Remark

Suppose that $f(t) = t^n + at + b \in K[t]$ is irreducible, K being a number field. If $n \geq 2$ and $L = K[t]/(f(t))$, then

$$d_{L/K} = \begin{cases} n^n b^{n-1} + (n-1)^{n-1} a^n & \text{if } n \equiv 1 \ (\text{mod } 2), \\ n^n b^{n-1} - (n-1)^{n-1} a^n & \text{if } n \equiv 0 \ (\text{mod } 2). \end{cases}$$

Also

$$c\langle L/K \rangle = \begin{cases} (d_{L/K})l(1-n) & \text{if } n \equiv 1 \ (\text{mod } 2), \\ (d_{L/K})l(-n) & \text{if } n \equiv 0 \ (\text{mod } 2), \end{cases}$$

where

$$c\langle L/K\rangle = \begin{cases} HW_2\langle L/K\rangle & \text{if } n \equiv 0,1 \ (\text{mod } 8), \\ HW_2\langle L/K\rangle + l(-1)d_{L/K} & \text{if } n \equiv 2,3 \ (\text{mod } 8), \\ HW_2\langle L/K\rangle + l(-1)^2 & \text{if } n \equiv 4,5 \ (\text{mod } 8), \\ HW_2\langle L/K\rangle + l(-1)(l(-1) + d_{L/K}) & \text{if } n \equiv 6,7 \ (\text{mod } 8). \end{cases}$$

These calculations may be found in [Ser 2, App. II] *and* [C-P] *and are summarized, along with many more examples, in* [C-Y, §10].

(2.34) Example

The following family of examples was brought to my attention by Pierre Conner, who very kindly showed me his calculations.

Let $f(t) = t^5 + at^2 + b \in \mathbb{Q}[t]$ be an irreducible quintic with $a \neq 0$. Let $F/\mathbb{Q}$ be the associated simple extension of degree five with normal closure $N/\mathbb{Q}$. Hence $F = \mathbb{Q}[t]/(f(t))$. Let $\theta \in F$ be the primitive generator whose minimal polynomial is $f(t)$. Suppose that $\theta = \theta_1, \theta_2, \theta_3, \theta_4, \theta_5$ are the roots of $f(t)$, then the trace is given by

(2.35) $$\mathrm{Tr}_{F/\mathbb{Q}}(\theta^i) = \sum_{j=1}^5 \theta_j^i = N_i(\theta_1, \ldots, \theta_5).$$

From the equation $t^{i-s}f(t) = t^i + at^{i-3} + bt^{i-5}$, we see that for $i \geq 5$

(2.36) $$0 = N_i + aN_{i-3} + bN_{i-5},$$

while $N_1 = 0$, $N_2 = 0$, $N_3 = -3a$, and $N_4 = 0$. Therefore we obtain

(2.37) $$\begin{cases} \mathrm{Tr}(1) = 5, \\ \mathrm{Tr}(\theta) = 0, \\ \mathrm{Tr}(\theta^2) = 0, \\ \mathrm{Tr}(\theta^3) = -3a, \\ \mathrm{Tr}(\theta^4) = 0, \\ \mathrm{Tr}(\theta^5) = -5b, \\ \mathrm{Tr}(\theta^6) = 3a^2, \\ \mathrm{Tr}(\theta^7) = 0, \\ \mathrm{Tr}(\theta^8) = 8ab. \end{cases}$$

Let $W = \langle\theta\rangle$ be the one-dimensional $\mathbb{Q}$-subspace of F spanned by θ. Let W' denote the annihilator of W with respect to the trace form

$$\langle F/\mathbb{Q}\rangle : F \times F \to \mathbb{Q},$$

$$(x, y) \mapsto \mathrm{Tr}_{F/\mathbb{Q}}(xy).$$

Hence $W^0 \supset W$ and $(A, B, C, D, E \in \mathbb{Q})$

$$v = A.1 + B\theta + C\theta^2 + D\theta^3 + E\theta^4 \in W^0$$

if and only if $E = (-3ac)/(5b)$ so that

(2.38) $$W^0 = \langle 1, \theta, \theta^3, 5b\theta^2 - 3a\theta^4 \rangle.$$

From Chapter 3, §1.60, if ρ_1 and ρ_2 are the orthogonal representations which appear when the hyperbolic relation (Chapter 3, §1.49 (iii)) is applied to $W^0 \supset W$, then

(2.39) $$HW_i(\langle F/\mathbb{Q} \rangle) = HW_i((\rho_1) \oplus (\rho_2)).$$

Taking $1, \theta^3$ and $5b\theta^2 - 3a\theta^4$ as a basis for W^0/W, we find that (ρ_1) has matrix

(2.40) $$R_1 = \begin{bmatrix} 5 & -3a & 0 \\ -3a & 3a^2 & -25b^2 \\ 0 & -25b^2 & -18a^3b \end{bmatrix}.$$

Taking $(\theta, 0)$ and $(0, \theta^2)$ as a basis for $W \oplus (F/W^0)$, we find that (ρ_2) has matrix

(2.41) $$R_2 = \begin{bmatrix} 0 & -3a \\ -3a & 0 \end{bmatrix}.$$

Hence $(\rho_2) = \langle 6a \rangle + \langle -6a \rangle$ and, by Chapter 3, §2.2,

(2.42) $$\begin{cases} HW_1(\rho_2) = l(-1), \\ HW_2(\rho_2) = 0. \end{cases}$$

Combining (2.42) with (2.39), we obtain

(2.43) $$HW_i(\langle F/\mathbb{Q} \rangle) = \begin{cases} HW_1(\rho_1) + l(-1) & \text{if } i = 1, \\ HW_i(\rho_1) + l(-1)HW_{i-1}(\rho_1) & \text{if } i \geq 2. \end{cases}$$

Now letting

$$X = \begin{bmatrix} 1 & 0 & 0 \\ (3a/5) & 1 & 0 \\ 0 & 0 & 1 \end{bmatrix},$$

we find that

$$XR_1X^t = \begin{bmatrix} 5 & 0 & 0 \\ 0 & (6a^2/5) & -25b^2 \\ 0 & -25b^2 & -18a^3b \end{bmatrix}.$$

Hence $(\rho_1) = \langle 5 \rangle + (\rho_3)$, where (ρ_3) has matrix

(2.44)
$$R_3 = \begin{bmatrix} 6a^2/5 & -25b^2 \\ -25b^2 & -18a^3b \end{bmatrix}.$$

Letting

$$Y = \begin{bmatrix} 1 & 0 \\ (125b^2/6a^2) & 1 \end{bmatrix},$$

we find that YR_3Y^t is diagonal and that

(2.45)
$$(\rho_3) = \langle (6a^2)/5 \rangle + \langle -(5^5b^4 + 4 \cdot 3^3 a^5 b)/(6a^2) \rangle.$$

If we set $d = 5^5 b^4 + 4 \cdot 3^3 a^5 b$, then, by (2.43) and the multiplicativity of $l(-)$,

$$HW_1(\langle F/\mathbb{Q} \rangle) = l(d) \in H^1(\mathbb{Q}; \mathbb{Z}/2).$$

Also, as $\langle x^2 y \rangle = \langle y \rangle$,

$$\begin{aligned} HW_2(\langle F/\mathbb{Q} \rangle) &= HW_2(\langle 5 \rangle + \langle 30 \rangle + \langle -6d \rangle) + l(-1)l(-d) \\ &= l(5)^2 + 2l(5)l(6) + l(5)l(-1) + l(5)l(d) \\ &\quad + (l(5) + l(6))l(-6) + l(30)l(d) \\ &= l(-6)l(d) + l(5)l(-6), \qquad \text{by Chapter 3, §2.2,} \\ &= l(-6)l(5d). \end{aligned}$$

To recapitulate, we have found

(2.46)
$$\begin{cases} HW_1(\langle F/\mathbb{Q} \rangle) = l(d), \\ HW_2(\langle F/\mathbb{Q} \rangle) = l(-6)l(5d). \end{cases}$$

Now consider the polynomial

(2.47)
$$f(t) = t^5 + 5(1/4(5n^2 - 1))^2 t^2 + 3(1/4(5n^2 - 1))^3.$$

In (2.47), $f(t)$ is irreducible if n is odd, and we find that

$$d = [5^3 3^2 n(1/4(5n^2 - 1))^6]^2.$$

Therefore we have

(2.48)
$$\begin{cases} HW_1(\langle F/\mathbb{Q} \rangle) = 0, \\ HW_2(\langle F/\mathbb{Q} \rangle) = l(-6)l(5), \\ \text{if } f(t) \text{ is as in (2.47).} \end{cases}$$

By Chapter 3, §2.5 $l(-6)l(5) \neq 0$ since $a^2 + 6b^2 = c^2 5$ has no integral solution. However, the Galois group, $G(N/\mathbb{Q})$, must be a subgroup of $A_5 \subset \Sigma_5$ which acts transitively on $\{1, 2, 3, 4, 5\}$.

Furthermore, by Chapter 3, §2.8 and (2.48),

$$(2.49) \qquad 0 \neq l(-6)l(5) = HW_2(\langle F/\mathbb{Q} \rangle) = SW_2(\mathrm{Ind}_{F/\mathbb{Q}}(1)).$$

The right side of (2.49) originates in $H^2(G(N/\mathbb{Q}))$. The only transitive subgroups of A_5 are $\mathbb{Z}/5$, D_{10}, and A_5, but $H^2(\mathbb{Z}/5; \mathbb{Z}/2) = 0$ so that (2.49) rules out this case.

If $G(N/\mathbb{Q}) \cong D_{10}$, then the right side of (2.49) is a square, since $H^*(D_{10}; \mathbb{Z}/2) \cong H^*(\mathbb{Z}/2; \mathbb{Z}/2)$ by Chapter 1, §4.6. In this case we must have

$$(2.50) \qquad \begin{cases} l(a)l(-1) = l(-6)l(5) \in H^2(\mathbb{Q}; \mathbb{Z}/2) \\ \text{for some } 1 \neq a \in \mathbb{Q}^*/(\mathbb{Q}^*)^2. \end{cases}$$

This case can occur, for example, if $n = 1$.

3. THE CLIFFORD INVARIANT AND SUNDRY OTHER CONSTRUCTIONS WITH CENTRAL EXTENSIONS

(3.1) Definition: The Clifford invariant (see [CN-T3; F])

Let K be a field of characteristic different from 2.

Let $b: V \times V \to K$ be a Γ-invariant, symmetric, nondegenerate bilinear form, where (as in Chapter 2, §2.24) Γ is a finite group.

Define the Clifford invariant of (V, b),

$$[\bar{C}(V, b)] \in \mathrm{Br}(\Gamma, K) \cong H^2(\Gamma \times K; \bar{K}^*)$$

(see Chapter 2, §2.53) to be the class in the equivariant Brauer group represented by the following central, simple $K[\Gamma]$-algebra ([Sch, p. 327; L, p. 106] or use Chapter 3, §1.7 to see this):

$$(3.2) \qquad \bar{C}(V, b) = \begin{cases} C(V, b) & \text{if rank}\, (V, b) \text{ is even,} \\ C_0(V, b) & \text{if rank}\, (V, b) \text{ is odd.} \end{cases}$$

Here $C(V, b)$ is the Clifford algebra of Chapter 3, §1.1, and $C_0(V, b)$ is the image of $\bigoplus_{n \geq 0}(V^{\otimes 2n})$ in $C(V, b)$.

If L/K is a field extension, we obtain a commutative diagram

$$(3.3) \qquad \begin{array}{ccccc} \mathbb{Z}/2 \cong \{\pm 1\} & \rightarrowtail & \mathrm{Pin}\,(V_L, b_L) & \xrightarrow{\pi} & O(V_L, b_L) \\ & {\scriptstyle c_3}\downarrow & {\scriptstyle c_2}\downarrow & & {\scriptstyle c_1}\downarrow \\ L^* & \rightarrowtail & \bar{C}(V_L, b_L)^* & \xrightarrow{\pi'} & \mathrm{Aut}\,(\bar{C}(V_L, b_L)). \end{array}$$

Here c_2 and c_3 are the standard inclusions, and π' sends x to conjugation by x (see Chapter 3, §1.16). The top row is as in Chapter 3, §1.17, $\mathrm{im}\,(\pi) = \ker\,(\theta)$, the kernel of the spinor norm. The bottom row of (3.3) is a central extension, by the Skolem-Noether Theorem [Mil, p. 138, Prop. 1.4].

The group Γ (via its action V) acts on the diagram (3.3) as does $G(L/K)$.

Since (3.2) gives a functor from the category of nonsingular, symmetric Γ-invariant bilinear forms to the category of central, simple $K[\Gamma]$-algebras, the map c_1 induces a map

$$(3.4) \qquad (c_{1_*}): H^1(\Gamma \times K; O(b_{\bar{K}})) \to H^1(\Gamma \times K; \mathrm{Aut}\,(\bar{C}(V_{\bar{K}}, b_{\bar{K}}))).$$

From the functoriality of this map, we obtain directly the following:

(3.5) Proposition

If, in the above notation, $\beta: U \times U \to K$ is represented by $f \in Z^1(\Gamma \times \Omega_K, O(b_{\bar{K}}))$, then $[\bar{C}(U, \beta)]$ is represented by $c_1 \cdot f$ in $Z^1(\Gamma \times \Omega_K, \mathrm{Aut}\,(\bar{C}(V_{\bar{K}}, b_{\bar{K}})))$.

(3.6) Corollary

Let $T: G(L/K) \to O(V, b)$ be the orthogonal Galois representation of Chapter 2, §3.8. In $H^1(K; \mathrm{Aut}\,(\bar{C}(V_{\bar{K}}, b_{\bar{K}})))$, the Clifford invariant of Frohlich's form of Chapter 2, §3.8 is represented by the 1-cocycle $(g \mapsto \bar{C}(T(g), b_{\bar{K}}))$. Here, when $g \in \Omega_K$, $\bar{C}(T(g), b_{\bar{K}})$ is the Clifford algebra automorphism which extends $T(g)$ on V.

Proof. This follows at once from §3.5 and Chapter 2, §3.16, with $\Gamma = \{1\}$. □

(3.7)   From (3.3) we have a commutative diagram of coboundaries:

$$(3.8) \qquad
\begin{array}{ccc}
H^1(\Gamma \times G(L/K); O(V, b_L)) & \xrightarrow{\;\Delta\;} & H^2(\Gamma \times G(L/K); \mathbb{Z}/2) \\
\downarrow{\scriptstyle (c_1)_*} & & \downarrow{\scriptstyle (c_3)_*} \\
H^1(\Gamma \times G(L/K); \mathrm{Aut}\,(\bar{C}(V_L, b_L))) & \xrightarrow{\;\Delta'\;} & H^2(\Gamma \times G(L/K); L^*).
\end{array}$$

We also have the analogous diagram to (3.8) in which L is replaced by $\bar{K}$, the separable closure of K.

From the exact sequence

$$\mathbb{Z}/2 \rightarrowtail \bar{K}^* \twoheadrightarrow \bar{K}^*,$$

we obtain an exact sequence (cf., Chapter 2, §2.47)

$$(3.9) \qquad
\begin{array}{ccc}
\xrightarrow{\;2\;} H^1(\Gamma; \bar{K}^*) \longrightarrow & H^2(\Gamma \times K; \mathbb{Z}/2) \longrightarrow & H^2(\Gamma \times K; \bar{K}^*) \xrightarrow{\;2\;} \\
\downarrow{\scriptstyle \cong} & & \downarrow{\scriptstyle \cong} \\
H^1(\Gamma \times K; \bar{K}^*) & & \mathrm{Br}\,(\Gamma, K)
\end{array}$$

Hence, if $f: \Gamma \times \Omega_K \to O(b_{\bar{K}})$ is a 1-cocycle, then, by Chapter 2, §§1.21–1.22,

$$\Delta'(c_1)_*[f] = (c_3)_*(\Delta[f]) \in H^2(\Gamma \times K; \bar{K}^*)$$

is represented by the 2-cocycle

(3.10) $((\gamma, g), (\gamma', g')) \mapsto h(\gamma, g)h(\gamma', g')[h(\gamma\gamma', gg')]^{-1} \in \{\pm 1\},$

where $h = sf$ and $s: O(b_{\bar{K}}) \to \text{Pin}(b_{\bar{K}})$ is a section of π in (3.3).

Notice that although $\Delta'(c_1)_*[f]$ lies in the equivariant Brauer group, it is not (in general) the class of the central simple $K[\Gamma]$-algebra, $(c_1)_*[f]$. For this is calculated by first representing $(c_1)_*[f]$ in $H^1(\Gamma \times K; PGL_n \bar{K})$ and then calculating the coboundary

$$\Delta: H^1(\Gamma \times K; PGL_n \bar{K}) \to H^2(\Gamma \times K; \bar{K}^*)$$

of it.

However, we will evaluate below $\Delta'(c_1)_*[f]$, in the case when $\Gamma = 1$. Notice that, when $\Gamma = 1$, $(c_3)_*: H^2(K; \mathbb{Z}/2) \to \text{Br}(K)$ is injective onto the 2-torsion, $_2\text{Br}(K)$, by (3.9). We will not use this fact, however.

The best way to proceed, in the long run, seems to be to take a detour into the "red light" district of 2-cocycles and to invent a cohomology class. The reader must accept my apologies for yet another 2-cocycle verification.

(3.11)   THE COHOMOLOGY CLASS S(a,b)

Let X be a central, simple K-algebra. Consider the central extension, upon which Ω_K acts continuously,

(3.12) $\bar{K}^* \rightarrowtail X_{\bar{K}}^* \xrightarrow{\pi} \text{Aut}_{\bar{K}}(X_{\bar{K}}),$

where $\pi(u)(v) = uvu^{-1}$.

Suppose that $a, b \in Z_{ct}^1(\Omega_K, \text{Aut}_{\bar{K}}(X_{\bar{K}}))$ such that $g(a(g_1))$ and $b(g)$ *commute* for all $g, g_1 \in \Omega_K$. Define a continuous 2-cochain, $S(a,b)$, by

(3.13) $\begin{cases} S(a,b): \Omega_K \times \Omega_K \to \bar{K}^*, \\ S(a,b)(g, g_1) = [g(\hat{a}(g_1)), \hat{b}(g)], \end{cases}$

where $[x, y] = xyx^{-1}y^{-1}$ is the commutator of x and y and $\hat{a}, \hat{b}: \Omega_K \to X_{\bar{K}}^*$ are continuous lifts which satisfy $\pi\hat{a} = a$, $\pi\hat{b} = b$.

(3.14) Proposition

With the notations of §3.11, $S(a,b)$ is a continuous 2-cocycle.

Proof. The continuity of $S(a,b)$ is clear. To prove the result, we must show that if $g, g_1, g_2 \in \Omega_K$, then $ABCD = 1 \in X_{\bar{K}}$, where

$$A = [gg_1\hat{a}(g_2), g\hat{b}(g_1)],$$

$$B = [\hat{b}(gg_1), gg_1\hat{a}(g_2)],$$

$$C = [g\hat{a}(g_1g_2), \hat{b}(g)], \quad \text{and}$$

$$D = [\hat{b}(g), g\hat{a}(g_1)].$$

Notice that each of $A, B, C,$ and D is central in $X_{\bar{K}}^*$, so we start by computing CD. Juxtaposing C and D and cancelling $\hat{b}(g)^{-1}\hat{b}(g)$, we obtain

$$CD = (g\hat{a}(g_1g_2))(\hat{b}(g))(g\hat{a}(g_1g_2))^{-1}(g\hat{a}(g_1)(\hat{b}(g))^{-1}(g\hat{a}(g_1))^{-1}.$$

Let $z = (gg_1\hat{a}(g_2))^{-1}(g\hat{a}(g_1))^{-1}(g\hat{a}(g_1g_2))$, and notice that $\pi(z) = g[g_1(a(g_2))^{-1}a(g_1)^{-1}a(g_1g_2)] = 1$ since a is a 1-cocycle. Therefore $z \in \bar{K}^*$ and is central. However,

$$\begin{aligned}
BCD &= B(g\hat{a}(g_1))^{-1}CD(g\hat{a}(g_1)) \\
&= \hat{b}(gg_1)(gg_1\hat{a}(g_2))(\hat{b}(gg_1))^{-1}z(\hat{b}(g))(g\hat{a}(g_1g_2))^{-1}(g\hat{a}(g_1))(\hat{b}(g))^{-1} \\
&= \hat{b}(gg_1)(gg_1\hat{a}(g_2))(\hat{b}(gg_1))^{-1}(\hat{b}(g))(gg_1\hat{a}(g_2))^{-1}(\hat{b}(g))^{-1}
\end{aligned}$$

by commuting z past $\hat{b}(g)$ and cancelling four factors. Now, inserting

$$A = (gg_1\hat{a}(g_2))(g\hat{b}(g_1))(gg_1\hat{a}(g_2))^{-1}(gb(g_1))^{-1}$$

after the penultimate factor in BCD and cancelling two factors, we obtain the following equation, in which $w = (\hat{b}(gg_1))^{-1}(\hat{b}(g))(g\hat{b}(g_1)) \in \bar{K}^*$:

$$\begin{aligned}
ABCD &= \hat{b}(gg_1)(gg_1\hat{a}(g_2))w(gg_1\hat{a}(g_2))^{-1}(g\hat{b}(g_1))^{-1}(\hat{b}(g))^{-1} \\
&= 1,
\end{aligned}$$

since w is central. $\square$

(3.15)   Now suppose that $X, Y,$ and Z are three central simple K-algebras which are $\bar{K}$-isomorphic by isomorphisms $\lambda: Z_{\bar{K}} = Z \otimes_K \bar{K} \xrightarrow{\cong} Y_{\bar{K}}$ and $\mu: Y_{\bar{K}} \xrightarrow{\cong} X_{\bar{K}}$. Consequently we have a commutative diagram of central extensions

(3.16)

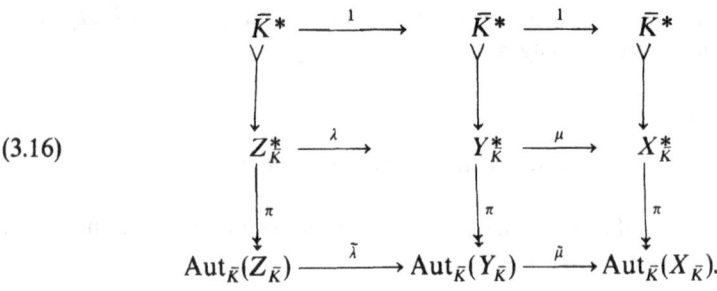

Associated to (3.16) we have coboundaries

(3.17) $\begin{cases} \Delta_X : H^1(K; \mathrm{Aut}_{\bar{K}}(X_{\bar{K}})) \to H^2(K; \bar{K}^*) \cong \mathrm{Br}\,(K), & \text{and} \\ \Delta_Y : H^1(K; \mathrm{Aut}_{\bar{K}}(Y_{\bar{K}})) \to \mathrm{Br}\,(K). \end{cases}$

Let Z be represented in $H^1(K, \mathrm{Aut}_{\bar{K}}(Y_{\bar{K}}))$ by $(g \mapsto \lambda g(\lambda^{-1}) = Z(g))$, and let Y be represented in $H^1(K; \mathrm{Aut}_{\bar{K}}(X_{\bar{K}}))$ by $(g \mapsto \mu g(\mu^{-1}) = Y(g))$. Then Z is represented in $H^1(K; \mathrm{Aut}_{\bar{K}}(X_{\bar{K}}))$ by

$$(g \mapsto \mu \lambda g(\lambda^{-1}\mu^{-1}) = \tilde{\mu} Z(g) Y(g)).$$

Let $\hat{Z}$ and $\hat{Y}$ denote lifts of Z and Y, respectively (so that $\pi\hat{Z} = Z$, $\pi\hat{Y} = Y$).

(3.18) Theorem

With the preceding notation

$$\Delta_Y[Z] = \Delta_X[Z] - \Delta_X[Y] + S(\tilde{\mu} Z, Y)$$

in $H^2(K; \bar{K}^*) \cong \mathrm{Br}\,(K)$.

Proof. Retaining the notation introduced after (3.17), let $g, g_1 \in \Omega_K$. Then $\Delta_X(Z)(g, g_1)[\Delta_X(Y)(g, g_1)]^{-1}$ may be calculated by inserting the (central) second factor after the penultimate term in the first factor. If we recall that, by Chapter 2, §1.21,

$$\Delta_X(Z)(g, g_1)^{-1} = \mu\hat{Z}(g)(\hat{Y}(g))(g\mu\hat{Z}(g_1))(g\hat{Y}(g_1))(\hat{Y}(gg_1))^{-1}(\mu\hat{Z}(gg_1))^{-1}$$

and that

$$\Delta_X(Y)(g, g_1)^{-1} = \hat{Y}(gg_1)(g\hat{Y}(g_1))^{-1}(\hat{Y}(g))^{-1},$$

we see that four terms cancel to give the following expression, in the *centre* of $X_{\bar{K}^*}$:

$$[\Delta_X(Z)\Delta_X(Y)^{-1}](g, g_1) = \mu\hat{Z}(g)(\hat{Y}(g))(g\mu\hat{Z}(g_1))(\hat{Y}(g))^{-1}(\mu\hat{Z}(gg_1))^{-1}$$

$$= \mu\hat{Z}(gg_1)^{-1}(\mu\hat{Z}(g))(\hat{Y}(g))(g\mu\hat{Z}(g_1))(\hat{Y}(g))^{-1}$$

$$= (\mu\Delta_Y[Z](g, g_1))[g\mu\hat{Z}(g_1)^{-1}, \hat{Y}(g)].$$

The first factor in this expression is $\Delta_Y[Z](g, g_1)$, since $\mu = 1$ on the centre, $\bar{K}^*$. Therefore the commutator is central and equals

$$[\hat{Y}(g), g\mu\hat{Z}(g_1)] = (S(\tilde{\mu} Z, Y)(g, g_1))^{-1},$$

as required.   □

(3.19) Remark

The formula for $\Delta_Y[Z]$ given in $[F, App.\ III.7]$ does not contain the term $S(\tilde{\mu}Z, Y)$, which is not generally zero. It may be that the slightly different conventions and formulae that we are using (e.g., $[F, §5.8]$ vs. Chapter 2, §1.21) explain this difference. In any case this difference does not affect the proofs (notably $[F, §5.12]$).

(3.20) Corollary

Let $S(a, b)$ be as in §§3.11–3.14, then the cohomology class

$$[S(a, b)] \in H^2(K; \bar{K}^*)$$

depends only on $[a], [b] \in H^1(K, \operatorname{Aut}_{\bar{K}}(X_{\bar{K}}))$, the cohomology classes of a and b.

Proof. The condition that $g(a(g_1))$ and $b(g)$ commute is precisely the condition that $(g \mapsto a(g)b(g))$ be a 1-cocycle. If this represents Z and b represents Y, then $\tilde{\mu}Z = a$, in the notation of the proof of §3.18. Hence, by Chapter 2, §1.21 and the formula of §3.18, $[S(a, b)]$ depends only on $[a]$ and $[b]$, as required.   □

(3.21) Example

Let K be a field with $\operatorname{char}(K) \neq 2$, then $M_2(K)$, the algebra of 2×2 matrices with entries in K, is generated by

$$I_2, \quad i = \begin{bmatrix} 0 & 1 \\ 1 & 0 \end{bmatrix}, \quad j = \begin{bmatrix} 0 & 1 \\ -1 & 0 \end{bmatrix}, \quad \text{and}$$

$$ij = \begin{bmatrix} -1 & 0 \\ 0 & 1 \end{bmatrix} = -ji.$$

Therefore $C(\langle 1 \rangle \oplus \langle -1 \rangle) = \bar{C}(\langle 1 \rangle \oplus \langle -1 \rangle) \cong M_2(K)$. The Clifford algebra is $\mathbb{Z}/2$-graded and, in $M_2(K)$, this becomes the "chequer-board grading"

$$M_2(K)_0 = \left\{ \begin{bmatrix} * & 0 \\ 0 & * \end{bmatrix} \right\},$$

$$M_2(K)_1 = \left\{ \begin{bmatrix} 0 & * \\ * & 0 \end{bmatrix} \right\}.$$

If A and B are two $\mathbb{Z}/2$-graded K-algebras, then $A \hat{\otimes} B$ is the $\mathbb{Z}/2$-graded algebra given by $A \otimes B$, with grading $(A \hat{\otimes} B)_\alpha = \bigoplus_{a+b \equiv \alpha (\mathrm{mod}\ 2)} A_a \otimes B_b$.

If (V_i, b_i) $(i = 1, 2)$ are symmetric bilinear forms, then

$$\Phi: V_1 \oplus V_2 \to C(b_1) \hat{\otimes} C(b_2),$$

given by $\Phi(v_1 \oplus v_2) = v_1 \otimes 1 + 1 \otimes v_2$ induces an isomorphism [L, p. 105; A-B-S]

$$\Phi: C(b_1 \oplus b_2) \xrightarrow{\cong} C(b_1) \hat{\otimes} C(b_2).$$

This shows immediately that as an ungraded K-algebra,

(3.22) $C(n(\langle 1 \rangle \oplus \langle -1 \rangle)) = \bar{C}(n(\langle 1 \rangle \oplus \langle -1 \rangle)) \cong M_{2^n}(K).$

(3.23)   It is clear from (3.8) and §3.18, for a symmetric bilinear form β, that $HW_2(\beta)$ and $[\bar{C}(\beta)]$ are related in $_2\mathrm{Br}(K) \cong H^2(K; \mathbb{Z}/2)$, the 2-torsion of the Brauer group.

Let us pursue this relation. Let β be a symmetric, nondegenerate bilinear form of rank $2n$ over K with char$(K) \neq 2$. We may represent β by a *homomorphism* $T: G(L/K) \to O(n(\langle 1 \rangle + \langle -1 \rangle))$, by diagonalization (cf., Chapter 2, §3.2).

Now consider the commutative diagram of (3.8):

$$
\begin{array}{ccc}
H^1(K; O(n(\langle 1 \rangle + \langle -1 \rangle))) & \xrightarrow{\Delta} & H^2(K; \mathbb{Z}/2) \\
{\scriptstyle (c_1)_*} \downarrow & & \downarrow {\scriptstyle (c_3)_*} \\
H^1(K, \mathrm{Aut}(M_{2^n}(K))) & \xrightarrow{\Delta'} & H^2(K; \bar{K}^*) \cong \mathrm{Br}(K).
\end{array}
$$

From Chapter 2, §1.22 (if (^) denotes a lifting),

$$\Delta(\beta)[g, g'] = \hat{T}(g)g(\hat{T}(g'))[\hat{T}(gg')]^{-1}$$
$$= (1, T)^* \tilde{\omega}_2[g, g']$$

so that §§2.17, 2.7 (ii) yield

$$\Delta(\beta) = Sp[T] + SW_2[T]$$
$$= HW_2(\beta) + HW_2(n\langle -1 \rangle) + HW_1(n\langle -1 \rangle) \cdot SW_1[T],$$

where $SW_1[T] = HW_1(\beta) + nl(-1)$.

However, $\Delta'(c_1)_*(\beta)$ is the Brauer class of $\bar{C}(\beta)$. Therefore we have shown the following:

(3.24) Proposition

Let β be a nondegenerate, symmetric bilinear form over K (char$(K) \neq 2$), then the Brauer class, $\Delta'(c_1)_*(\beta) = [\bar{C}(\beta)]$, of the Clifford invariant of β is given by

$$HW_2(\beta) + nl(-1)HW_1(\beta) + \left(\binom{n}{2} + n^2\right)l(-1)^2 \text{ in } _2\mathrm{Br}(K).$$

(3.25) Exercise

By relating $\bar{C}(\beta)$ to $\bar{C}(\beta \oplus \langle a \rangle)$ (see [L, pp. 120–121]) calculate ε, δ in the general formula

$$[\bar{C}(\beta)] = HW_2(\beta) + \varepsilon l(-1)HW_1(\beta) + \delta l(-1)^2.$$

(3.26) *THE STEINBERG CENTRAL EXTENSION*

The remainder of this section will be concerned with illustrating the equivariant cohomology class, $\tilde{\omega}_2$ of §1.36, which we obtain when the central extension of (1.37) is the Steinberg extension

$$(3.27) \qquad\qquad K_2(L, n) \rightarrowtail St_n(L) \xrightarrow{\Phi_n} E_n(L).$$

First, recall the Steinberg group [M, §5] $St_n(L)$, where L is any ring. The generators of $St_n(L)$ are

$$\{x_{ij}^\lambda; 1 \le i \ne j \le n, \lambda \in L\}$$

subject to the universal relations among elementary matrices, namely,

$$[x_{ij}^\lambda, x_{kl}^\mu] = \begin{cases} 1 & \text{if } j \ne k, i \ne l, \\ x_{ij}^{\lambda\mu} & \text{if } j = k, i \ne l, \\ x_{kj}^{-\mu\lambda} & \text{if } j \ne k, i \ne l. \end{cases}$$

Here $[a, b] = aba^{-1}b^{-1}$ as usual.

$E_n(L)$ denotes the subgroup of $GL_n(L)$ generated by the elementary matrices, e_{ij}^λ $(1 \le i \ne j \le n, \lambda \in L)$, having 1's on the diagonal and λ in the (i, j)th entry. The map, Φ, is given by $\Phi(x_{ij}^\lambda) = e_{ij}^\lambda$ in (3.27).

If L/K is a finite Galois extension, then $G(L/K)$ acts on $St_n(L)$ by the formula

$$g(x_{ij}^\lambda) = x_{ij}^{g(\lambda)} \qquad (1 \le i \ne j \le n, \lambda \in L, g \in G(L/K)).$$

Hence from §1.40 we have a compatible family of cohomology classes

$$\omega_{2,L}(n) \in H^2(G(L/K) \ltimes E_n(L); K_2(n, L)_{G(L/K)}).$$

Also, since $\det(e_{ij}^\lambda) = 1$, we see that $E_n(L) \subset SL_n(L)$. However, when n is large and L is a field (I believe $n \ge 3$ will suffice), then $SL_n(L) = E_n(L)$, and also when n is large, $E_n(L) = [GL_n(L), GL_n(L)]$, the commutator subgroup of $GL_n(L)$ [M, p. 25].

In addition $K_2(n, L)$ stabilizes very quickly with n when L is a field and $K_2(L) = \varinjlim_n K_2(L, n)$ is given by Matsumoto's theorem [M, §11.1].

(3.28) Theorem

Let L be a field. There is an isomorphism

$$K_2(L) \cong \frac{(L^* \otimes L^*)}{\{a \otimes (1-a); a \neq 0, 1\}}.$$

The image of $a \otimes b \in L^* \otimes L^*$ in $K_2(L)$ is called the *Steinberg symbol*, $\{a, b\}$ [M], which is skew-symmetric and bimultiplicative.

If $1/n \in L$ and ξ_n, a primitive nth root of unity, lies in L^*, then [Me-Su] (see also [Sou]) there is an isomorphism

(3.29) $$h: K_2(L) \otimes \mathbb{Z}/n \to H^2(L; \mathbb{Z}/n) \cong {}_n\mathrm{Br}\,(L)$$

given by $h(a \otimes b) = l(a)l(b)$. Here ${}_n\mathrm{Br}\,(L)$ is the n-torsion in the Brauer group, and $l(a) \in H^1(K; \mathbb{Z}/n)$ is given by

$$(g \mapsto g(\sqrt[n]{a})/(\sqrt[n]{a}); g \in \Omega_L).$$

For n large, therefore, we have a compatible family of cohomology classes

$$\tilde{\omega}_{2,L} \in H^2(G(L/K) \ltimes SL_n(L); K_2(L)_{G(L/K)}).$$

(3.30) Example

Let L/K be a Galois extension of local fields, of characteristic zero, so that $G(L/K)$ is solvable [Ser 2, p. 68, Cor. 5]. Suppose for a moment that L/K is cyclic. In this case "Hilbert 90 for K_2'' [Me-Su] (see also [Sou]), there is an injection induced by the norm

$$N_{L/K}: K_2(L)_{G(L/K)} \rightarrowtail K_2(K).$$

In fact this is an isomorphism. This is seen as follows: by [Me] for a local field, L,

$$K_2(L) \cong \mu(L) \times D_L,$$

where D_L is a rational vector space. Here $\mu(L)$ denotes the roots of unity in L^*. Since $K_2(K) \to K_2(L) \xrightarrow{N_{L/K}} K_2(K)$ is multiplication by $[L:K]$, we see that $\mathrm{im}\,(N_{L/K})$ contains D_K. Suppose that $\mu(K) \cong \mathbb{Z}/m$, then, by (3.29), $N_{L/K}: \mu(L) \otimes \mathbb{Z}/m \cong K_2(L) \otimes \mathbb{Z}/m \to K_2(K) \otimes \mathbb{Z}/m \cong \mathbb{Z}/m$ may be identified with the norm map $N_{L/K}: {}_m\mathrm{Br}\,(L) \to {}_m\mathrm{Br}\,(K)$. However, $\mathrm{Br}\,(L) \cong \mathbb{Q}/\mathbb{Z} \cong \mathrm{Br}\,(K)$ [Ser 2, p. 193] and $i_*: \mathrm{Br}\,(K) \to \mathrm{Br}\,(L)$ is multiplication by $[L:K]$. Since $N_{L/K} \cdot i_*(x) = [L:K]x$, we have $N_{L/K}: \mathrm{Br}\,(L) \to \mathrm{Br}\,(K)$ is an isomorphism, as required.

By induction up the composition series for $G(L/K)$ we see, in general, the following:

(3.31) Lemma

If L/K is a finite Galois extension of local fields, the norm induces an isomorphism

$$K_2(L)_{G(L/K)} \xrightarrow{\cong} K_2(K).$$

Consequently, (for $n \gg 0$),

$$\tilde{\omega}_{2,L} \in H^2(G(L/K) \ltimes SL_n(L); K_2(K))$$

in this case.

(3.32) Example

The following local field extension is taken from [Ser 5, §4, p. 413].

Let $F = \mathbb{Q}_2(i)$, where $\mathbb{Q}_2$ is the 2-adic rational field, $\mathbb{Q}_2 = \hat{\mathbb{Z}}_2[1/2]$, and $i^2 = -1$. Let $\pi = i - 1$ be the uniformizer for F, and let $(x \mapsto \bar{x})$ be the nontrivial $\mathbb{Q}_2$-automorphism of F. Hence $\bar{\pi} = i\pi = (1 + \pi)\pi$. Let $v: F^* \to \mathbb{Z}$ be the π-adic valuation (i.e., $v((x/y)\pi^s) = s$, where $\pi \nmid xy$, $x, y \in \hat{\mathbb{Z}}_2[i]$).

Let $A = \{2^n u \in F^* \mid n \in \mathbb{Z}, v(1 - u) \geq 3\}$. As $2 = \pi\bar{\pi} = (1 + \pi)\pi^2$, $v(2) = 2$, and the residue field of F is $\mathbb{F}_2$. Therefore F^*/A has order 8 because it has (from ramification theory or by inspection) a three-step filtration with decomposition factors isomorphic to $\mathbb{F}_2$. In fact,

$$\pi^2 = 2(-i) \equiv -i \qquad (\mathrm{mod}\ A)$$

so that $F^*/A \cong \mathbb{Z}/8$ generated by π.

By local class field theory F^* is dense in $(\Omega_F)_{ab}$ so that we have an epimorphism

$$\Omega_F \to (\Omega_F)_{ab} \to F^*/A \cong \mathbb{Z}/8$$

and by Galois theory we obtain a cyclic extension L/F of order 8.

In [Ser 5, §4] it is shown that $L/\mathbb{Q}_2$ is Galois with $G(L/\mathbb{Q}_2) \cong Q_{16}$, the quaternion group of order 16. In this case

$$K_2(L)_{Q_{16}} \cong K_2(\mathbb{Q}_2) \cong \mathbb{Z}/2 \oplus D_{\mathbb{Q}_2}.$$

(3.33)   The remainder of this section will be concerned with looking explicitly at the projectived version of (3.27). That is, if $PE_n(L)$ is the image of E_nL in $PSL_n(L)$ (which will be $PSL_n(L)$ if n is large enough), we will consider the homomorphism

$$\psi_n: St_n(L) \xrightarrow{\Phi_n} E_n(L) \longrightarrow PE_n(L).$$

From the outset we will assume $(n, L) \neq (3, \mathbb{F}_2)$, $(4, \mathbb{F}_2)$, or $(3, \mathbb{F}_4)$, in which case [M, p. 48] (3.27) is the universal central extension so that $\mathrm{Ker}(\Phi_n) \cong H_2(E_nL)$.

In the next series of results we will show the following:

(3.34) Theorem

Let L be a field containing ξ, a primitive nth root of unity. Suppose that $n \geq 3$ and that $(n, L) \neq (3, \mathbb{F}_2), (4, \mathbb{F}_2),$ or $(3, \mathbb{F}_4)$.

(i) *Suppose either that $n = (2s + 1)2^\beta$ with $\beta = 0$ or $\beta \geq 2$ or that $\{-1, -1\} = 0$ in $K_2(L, n)$.*
   Then there exists a universal central extension

$$(\mathbb{Z}/n) \oplus K_2(L, n) \rightarrowtail St_n(L) \twoheadrightarrow PE_n(L).$$

(ii) *If $n = 2(2s + 1)$ and $0 \neq \{-1, -1\} \in K_2(L, n)$, then there is a universal central extension of the form*

$$\frac{(\mathbb{Z}/2n) \oplus K_2(L, n)}{R_n} \rightarrowtail St_n(L) \xrightarrow{\psi_n} PE_n(L),$$

where $R_n \cong \mathbb{Z}/2$ generated by $(n, \{-1, -1\})$.

The Galois action is explained in §3.43.

(3.35) Firstly, in the Steinberg group, set [M, p. 71] $w_{ij}(\lambda) = x_{ij}^\lambda x_{ji}^{-(1/\lambda)} x_{ij}^\lambda$ and $h_{ij}(\lambda) = w_{ij}(\lambda)w_{ij}(-1)$ for $i \neq j$. One finds

$$\Phi(w_{ij}(\lambda)) = \begin{bmatrix} 0 & \lambda \\ -(1/\lambda) & 0 \end{bmatrix} \quad \text{and}$$

$$\Phi(h_{ij}(\lambda)) = \begin{bmatrix} \lambda & 0 \\ 0 & (1/\lambda) \end{bmatrix}$$

(where I have depicted only the $(i, i), (i, j), (j, i), (j, j)$ entries, the rest being diagonal 1's or off-diagonal zeros). The *Steinberg symbol*, $\{a, b\}$, is defined by the equation [M, p. 97]

$$h_{n,i}(a)h_{n,j}(b) = \{a, b\}h_{n,j}(b)h_{n,i}(a).$$

Since $\Phi_n(h_{n,1}(\xi) \cdots h_{n,n-1}(\xi)) = \xi I_n$, we have $\psi_n(h_{n,1}(\xi) \cdots) = 1$.

(3.36) Lemma

Set $s(x) = h_{n,1}(x)h_{n,2}(x) \cdots h_{n,n-1}(x) \in St_n(L)$ for $x \in L$. Then

(i) *$s(\xi)^r = \{\xi, \xi\}^\varepsilon h_{n,1}(\xi)^r \cdots h_{n,n-1}(\xi)^r$, where $\varepsilon = \binom{n-1}{2}\binom{r}{2}$.*

(ii) $h_{n,j}(\xi^r) = h_{n,j}(\xi)^r \{\xi, \xi\}^\delta$, where $\delta = \binom{r}{2}$.

(iii) $s(\xi)^r = \{\xi, \xi\}^\gamma s(\xi^r)$, where $\gamma = \frac{1}{2}\binom{r}{2}(n-1)(n-4)$.

(iv) $s(\xi)^n = \begin{cases} 1 & \text{if } n = 2^\beta(2s+1), \text{ with } \beta = 0 \text{ or } \beta \geq 2, \\ \{-1, -1\} & \text{if } n = 2(2s+1). \end{cases}$

Proof. Recall that all Steinberg symbols are central. Hence (i) and (ii) imply (iii) because $\varepsilon - (n-1)\delta = \gamma$. Furthermore, if $r = n = 2^\beta(2s+1)$, then $\gamma = 1/4(n-1)^2 n(n-4)$, which is divisible by n except when $n = 2(2s+1)$ in which case it is divisible by $(n/2)$. Since $\{a, b\}$ is bimultiplicative $\{\xi, \xi\}^n = 1$, giving the first part of (iv). Also, if $n = 2(2s+1)$ and v is a $(2s+1)$-st root of unity, then $\xi = -v$ and

$$\{\xi, \xi\} = \{-1, -1\}\{-1, v\}\{v, -1\}\{v, v\}$$

so that $\{\xi, \xi\}^{2s+1} = \{-1, -1\}^{2s+1} = \{-1, -1\}$, from which the second part of (iv) follows.

By induction on r, in (i), assume that

$$s(\xi)^r = \{\xi, \xi\}^\varepsilon h_{n,1}(\xi)^r \ldots h_{n,n-1}(\xi)^r, \quad \text{then}$$

$$s(\xi)^{r+1} = \{\xi, \xi\}^\varepsilon h_{n,1}(\xi)^r \ldots h_{n,n-1}(\xi)^r h_{n,1}(\xi) \ldots h_{n,n-1}(\xi)$$

$$= \{\xi, \xi\}^{\varepsilon + (n-2)r} h_{n,1}(\xi)^{r+1} h_{n,2}(\xi)^r$$

$$\cdots h_{n,n-1}(\xi)^r h_{n,2}(\xi) \ldots h_{n,n-1}(\xi) \quad \text{by §3.35}$$

$$= \{\xi, \xi\}^{\varepsilon + (n-2)r + (n-3)r + \cdots} h_{n,1}(\xi)^{r+1} \cdots h_{n,n-1}(\xi)^{r+1}.$$

However,

$$\varepsilon + \frac{(n-2)(n-1)r}{2} = \left\{ \left(\sum_1^{r-1} j \right) + r \right\} \binom{n-1}{2} = \binom{r+1}{2} \binom{n-1}{2},$$

which proves part (i). To prove part (ii), we use [M, §9.7] $h_{n,j}(uv) = \{u, v\} h_{n,j}(u) h_{n,j}(v)$. We obtain

$$h_{n,j}(\xi^{r+1}) = \{\xi^r, \xi\} h_{n,j}(\xi^r) h_{n,j}(\xi)$$

$$= \{\xi, \xi\}^{r+\delta} h_{n,j}(\xi)^r$$

by induction on r and, of course, $\binom{r}{2} + r = \binom{r+1}{2}$. $\qquad\square$

(3.37)   Now we show that $s(\xi) \in St_n(L)$ of §3.36 is central. From $[St, (3.8)R3]$ we have the relation

$$w_{ij}(\xi)x_{x,t}^{\lambda}w_{ij}(\xi)^{-1} = x_{\sigma_{ij}(s,t)}(\eta \xi^{-\langle \beta, \alpha \rangle}\lambda)$$

(note that $w_{ij}(\xi)^{-1} = w_{ij}(-\xi)$). Unraveling the notation of [St], or computing by hand, we obtain the following relations in our case:

$$
(3.38) \qquad w_{ij}(\xi)x_{s,t}^{\lambda}w_{ij}(\xi)^{-1} =
\begin{cases}
x_{s,t}^{\lambda} & \text{if } (s,t) \cap (i,j) = \Phi, \\
x_{j,t}^{-\lambda/\xi} & \text{if } s = i,\ t \neq j, \\
x_{x,i}^{\lambda/\xi} & \text{if } s \neq i,\ t = j, \\
x_{i,t}^{\lambda\xi} & \text{if } s = j,\ t \neq i, \\
x_{s,j}^{-\lambda\xi} & \text{if } s \neq j,\ t = i.
\end{cases}
$$

Consequently we obtain relations

$$
(3.39) \qquad
\begin{cases}
h_{ij}(\xi)x_{s,t}^{\lambda} = x_{s,t}^{\lambda}h_{ij}(\xi) & \text{if } (s,t) \cap (i,j) = \Phi, \\
h_{ij}(\xi)x_{s,i}^{\lambda} = x_{s,i}^{\lambda/\xi}h_{ij}(\xi) & \text{if } s \notin (i,j), \\
h_{ij}(\xi)x_{s,j}^{\lambda} = x_{s,j}^{\lambda\xi}h_{ij}(\xi) & \text{if } s \notin (i,j), \\
h_{ij}(\xi)x_{i,t}^{\lambda} = x_{i,t}^{\lambda\xi}h_{ij}(\xi) & \text{if } t \notin (i,j), \\
h_{ij}(\xi)x_{j,t}^{\lambda} = x_{j,t}^{\lambda/\xi}h_{ij}(\xi) & \text{if } t \notin (i,j).
\end{cases}
$$

(3.40) Lemma

In §3.36, $s(\xi)$ *is central in* $St_n(L)$.

Proof. Suppose that $i \neq j$, $i < j < n$, then

$$s(\xi)x_{i,j}^{\lambda} = h_{n,1}(\xi)\dots h_{n,j}(\xi)x_{i,j}^{\lambda}\dots$$

$$= h_{n,1}(\xi)\dots x_{i,j}^{\lambda\xi}h_{n,j}(\xi)\dots \qquad \text{by (3.39)}$$

$$= h_{n,1}(\xi)\dots h_{n,i}(\xi)x_{i,j}^{\lambda\xi}\dots$$

$$= h_{n,1}(\xi)\dots x_{i,j}^{\lambda\xi/\xi}h_{n,i}(\xi)\dots \qquad \text{by (3.39)}$$

$$= x_{ij}^{\lambda}s(\xi).$$

Similarly, we find that $s(\xi)x_{i,j}^{\lambda} = x_{i,j}^{\lambda}s(\xi)$ if $j < i < n$. To show that $s(\xi)$ commutes with $x_{i,n}^{\lambda}$ and $x_{n,j}^{\lambda}$, we proceed as follows: The preceding calculations show that $x_{i,n}^{\lambda}$, for example, commutes with

$$\hat{s}(\xi) = h_{i_0,1}(\xi)\dots \hat{h}_{i_0,i_0}(\xi)\dots h_{i_0,n}(\xi)$$

for a suitable choice of i_0 (here ($\hat{\ }$) denotes an omission). However, $\Phi_n(\hat{s}(\xi)) = \Phi_n(s(\xi))$; therefore $\hat{s}(\xi)s(\xi)^{-1}$ is central, which implies that $x_{i,n}^{\lambda}$ and $s(\xi)$ commute. $\qquad\square$

(3.41)   Since $\Phi_n(s(\xi)) = \xi I_n$, it is clear that if $n = 2^\beta(2s+1)$, with $\beta = 0$ or $\beta \geq 2$, then, by §3.36 (iv) to §3.40,

$$\mathbb{Z}/n\langle s(\xi)\rangle \oplus K_2(L, n) \to \mathrm{Ker}(\psi_n)$$

is an isomorphism in (3.34). The same is true if $\{-1, -1\} = 0$ in $K_2(L, n)$. However, if $\{-1, -1\}$ has order 2, then we have an exact sequence

$$\mathbb{Z}/2\langle\{-1, -1\}\rangle \overset{i}{\rightarrowtail} \mathbb{Z}/2n\langle s(\xi)\rangle \oplus K_2(L, n) \longrightarrow \mathrm{Ker}(\psi_n),$$

where $i\{-1, -1\} = (s(\xi)^n, \{-1, -1\})$.

(3.42) Proof of Theorem 3.34 We have only to prove that the advertised central extension is universal. This amounts (see [M]) to a calculation of $H_2(PE_n(L))$.
   However, the central extension

$$\mathbb{Z}/n \cong \mu_n(L) \rightarrowtail E_n(L) \twoheadrightarrow PE_n(L)$$

gives rise to a homology Serre spectral sequence

$$E^2_{s,t} = H_s(PE_n(L); H_t(\mathbb{Z}/n)) \Rightarrow H_{s+t}(E_n(L)).$$

In low dimensions this takes the form shown in Figure 1.

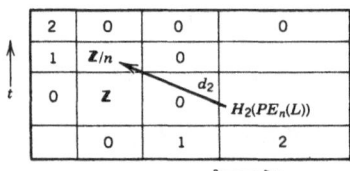

Figure 1

Since $H_1(E_n(L)) = 0$, we must have an exact sequence

$$0 \longrightarrow H_2(E_n(L)) \rightarrowtail H_2(PE_n(L)) \overset{d_2}{\twoheadrightarrow} \mathbb{Z}/n \longrightarrow 0$$
$$\cong \downarrow$$
$$K_2(L, n).$$

From this we see that $H_2(PE_n(L))/K_2(L, n)) \cong \mathbb{Z}/n$ so that, in all cases of §3.34,

$$H_2(PE_n(L)) \cong \langle K_2(L, n), s(\xi)\rangle = \mathrm{Ker}(\psi_n),$$

which proves that we have a *universal* central extension.   □

(3.43) GALOIS ACTIONS IN THE STEINBERG EXAMPLE

If $g \in G(L/K)$ acts in the Steinberg group by $g(x_{i,j}^{\lambda}) = x_{i,j}^{g(\lambda)}$, then $g\{a, b\} = \{g(a), g(b)\}$ $(a, b \in L^*)$. This gives the action on $K_2(L, n)$. The other generator of $\text{Ker}(\psi_n)$ is $s(\xi)$ which satisfies $g(s(\xi)) = s(g(\xi)) = s(\xi^r)$, where $g(\xi) = \xi^r$. Hence, from §3.36,

(3.44)
$$\begin{cases} g(s(\xi)) = \{\xi, \xi\}^{\gamma} s(\xi)^r & \text{if } g(\xi) = \xi^r, \\ \text{where } \gamma = \dfrac{1}{2}\left(\dfrac{r}{2}\right)(n-1)(n-4). \end{cases}$$

Chapter Four

Higher-Dimensional Characteristic Classes of Bilinear Forms and Galois Representations

Himmelfarb took his leave of the mistress of Xanadu. He was not in a position to dismiss her as a madwoman, as other people did, because of his involvement in the same madness. For now that the tops of the trees had caught fire, the bells of the ambulances were again ringing for him, those of the fire-engines clanging, and he shuddered to realise there could never be an end to the rescue of men from the rubble of their own ideas.

—PATRICK WHITE "Riders in the Chariot" (1961)

In this chapter we will establish higher-dimensional analogues of the formulae of Serre and Frohlich. These formulae were originally found by Bruno Kahn. I had independently found these formulae, under the restriction that the fields contain eighth roots of unity, by the use of orthogonal algebraic K-theory. The formulae relate the higher Hasse-Witt classes to the higher Stiefel-Whitney classes of a bilinear form which is induced from a one- or two-dimensional form. Our formulae are marginally more general than Khan's but our method is very similar to his.

In §1 we develop the theory of a "transfer," due to Andrej Koslowski. This theory is specifically designed to compute the Stiefel-Whitney classes of a vector bundle induced from a double covering (or, equivalently for our purposes, a representation induced from a subgroup of index 2). Kahn essentially rediscovered Koslowski's transfer in the course of his calculations—generalizing a case of it found by Deligne.

We develop Koslowski's transfer in the context of topological spaces, since the proofs are more conceptual in that setting. Actually there are three "transfers" which naturally occur when one does this, one of which originated in Steenrod's construction of mod 2 cohomology operations. Having the topological form of the transfer in hand, we explain the process whereby one passes from a representation

123

to its induced vector bundle and thereby converts the topological formulae for our algebraic use. Once again, I have tried to treat the topological aspects in such a manner that a not unreasonable act of faith will serve to deliver the nontopological reader unscathed through this Slough of Despond.

In §2 the promised formulae are derived. These formulae, and those of the previous chapter, will have interesting consequences when we construct and study the local root numbers of orthogonal Galois representations in Chapter 6.

---

1. KOSLOWSKI'S TRANSFER

In this section we make use of the mod 2 singular cohomology of spaces, $H^*(X; \mathbb{Z}/2)$. The reader familiar only with the cohomology of groups, G, should think in terms of $X = BG$ in which case $H^*(H; \mathbb{Z}/2) = H^*(BG; \mathbb{Z}/2) \cong H^*(G; \mathbb{Z}/2)$. In applications we will need only this specialized case. However, the proofs (or sketches thereof) require the more general context. I regret that these notes are not self-contained in this respect but decided that even the sketchiest development of $H^*(X; \mathbb{Z}/2)$ and its properties would take us far from our current theme—invariants with values in Galois cohomology.

The material in this section comes directly from [Kos]. The only modification I have made is to give the definition of the transfer for $BH \to BG$ (where $[G:H] = 2$) in terms of homomorphisms of groups. Specifically, I have treated the pretransfer of $[K - P]$ as a homomorphism in this case.

Throughout *this section* let $H^i(X)$ denote mod 2 cohomology of a space, X, $H^i(X; \mathbb{Z}/2)$. Thus, if G is a discrete group, $H^i(BG) = H^i(G; \mathbb{Z}/2)$.

Write $G(X)$ for the set

$$(1.1) \qquad\qquad G(X) = \prod_{i \geq 1} H^i(X).$$

We turn $G(X)$ into a group by cup-product. To be precise, let $\{x_i^\varepsilon : i \geq 1\} \in G(X)$ be identified with

$$1 + x_1^\varepsilon + x_2^\varepsilon + \cdots \in \prod_{i \geq 0} H^i(X) \qquad (x_i^\varepsilon \in H^i(X); \ \varepsilon = 0 \text{ and } 1)$$

$$(1.2) \qquad \begin{cases} \text{define } \{x_i^0\} \cdot \{x_i^1\} \in G(X) \text{ to correspond to} \\ (1 + x_1^0 + x_2^0 + \cdots) \cup (1 + x_1^1 + x_2^1 + \cdots) \in \prod_{i \geq 0} H^i(X). \end{cases}$$

Generally, we will write $\{x_i\} \in G(X)$ *as* $1 + x_1 + x_2 + \cdots$. Hence the n-dimensional component of $\{x_i^0\} \cdot \{x_i^1\}$ is $x_n^0 + x_{n-1}^0 \cup x_1^1 + \cdots + x_1^0 \cup x_{n-1}^1 + x_n^1$, where $a \cup b$ is the cohomology cup-product. Generally, write $\hat{G}(X)$ for the group

$$(1.3) \qquad\qquad \hat{G}(X) = H^0(X; \mathbb{Z}) \oplus G(X).$$

(1.4) Example

If $X = BG$ and E is a finite-dimensional orthogonal representation of G (or, more generally, for any space X, with E being a real vector bundle over X), define

$$w(E) = (\dim E, SW(E)) \in \hat{G}(X),$$

where $SW(E) = 1 + SW_1(E) + SW_2(E) + \cdots$ is the *total Stiefel-Whitney class*. If E' is a second such representation (or vector bundle), then

$$w(E) \cdot w(E') = (\dim E + \dim E', SW(E)SW(E'))$$
$$= w(E \oplus E').$$

(1.5) TRANSFERS ON G(X)

Write K_q for the Eilenberg-Maclane space $K(\mathbb{Z}/2, q)$, which is characterized up to homotopy by having only one nontrivial homotopy group, $\pi_q(K_q) \cong \mathbb{Z}/2$. It is an abelian group and can be realized by iteration of the classifying space construction, $BK_q = K_{q+1}$.

Let $M(q, n)$ denote the space of multilinear maps $\{h: K_q^n = K_q \times \cdots \times K_q \to K_{qn}\}$ such that h induces the cup-product on homotopy. The meaning of this condition is as follows. The homotopy class of maps, $[X, K_q]$, from X to K_q are naturally isomorphic $H^q(X) = H^q(X; \mathbb{Z}/2)$ [Spa]. Hence h induces a multilinear map $h_*: H^q(X) \times \cdots \times H^q(X) \to H^{qn}(X)$ sending $f_1, \ldots, f_n$ to the homotopy class of $h(f_1, \ldots, f_n) \circ \Delta$, where $\Delta: X \to X^n$ is the diagonal. We require that

$$h_*(f_1, \ldots, f_n) = f_1 \cup f_2 \cup \cdots \cup f_n.$$

Let K denote the *graded space* $K = \{K_q; q \geq 0\}$, and let $M(n)$ denote the space of *graded* multilinear maps $h': K^n = K \times \cdots \times K \to K$ which induce the cup-product on homotopy. G. B. Segal showed that $M(n)$ and $M(q, n)$ are both *contractible* spaces with a natural free Σ_n-action given by permuting the factors in K^n and K_q^n. Hence, as a Σ_n-space, each of $M(n)$ and $M(q, n)$ is a model for $E\Sigma_n$, a contractible free Σ_n-space.

Let $[f] \in G(X)$ be represented by a map $f: X \to \prod_{i \geq 1} K_q$. Set

$$E\Sigma_n \times_{\Sigma_n} X^n = (E\Sigma_n \times X^n)/\approx,$$

where $(e, x_1 \cdots x_n) \approx (\sigma(e), x_{\sigma(1)}, \ldots, X_{\sigma(n)})$, and identify this with $M(n) \times_{\Sigma_n} X^n$. Therefore, if $\alpha \in M(n)$, $\sigma \in \Sigma_n$ and $x_i \in X$

$$(\alpha, x_1, \ldots, x_n) \approx (\alpha(\sigma^{-1}-), x_{\sigma(1)}, \ldots, x_{\sigma(n)}).$$

(1.6) $\begin{cases} \text{Define } D[f] \in G(E\Sigma_n \times_{\Sigma_n} X^n) \text{ to be represented by a} \\ \text{map } D[f] \text{ given by} \\ D[f](\alpha, x_1, \ldots, x_n) = \alpha(f(x_1), \ldots, f(x_n)). \end{cases}$

This definition makes sense because

$$\sigma(\alpha)(f(x_{\sigma(1)}),\ldots) = \alpha(\sigma^{-1}(f(x_{\sigma(1)})),\ldots)$$
$$= \alpha(f(x_1),\ldots).$$

Similarly, we may define

(1.7)
$$\begin{cases} \tilde{D}: H^i(X) \to H^{ni}(E\Sigma_n \times_{\Sigma_n} X^n) \text{ to be given by} \\ \tilde{D}[f](\alpha, x_1,\ldots,x_n) = \alpha(f(x_1),\ldots,f(x_n)) \\ \text{where we interpret } E\Sigma_n \text{ as } M(i,n). \end{cases}$$

(1.8)   THE PRE-TRANSFER FOR DOUBLE-COVERINGS

Let $\pi: X \to Y$ be a double-covering of spaces. For example, if $i: H \subset G$ is a subgroup of index 2, then $Bi: BH \to BG$ is a double-covering (of connected spaces). The *pre-transfer* is a map, Φ (well-defined up to homotopy),

(1.9) $$\Phi: Y \to E\Sigma_2 \times_{\Sigma_2} X^2.$$

Several times in this section we will make use of the following fact, which is proved for the case of a double-covering of classifying spaces in §1.11: if

$$\pi: X \simeq E\Sigma_2 \times X \to B\Sigma_2 \times X$$

is the double-covering obtained by taking the product of X with $E\Sigma_2 \to B\Sigma_2$, then the associated pre-transfer

$$\Phi: B\Sigma_2 \times X \to E\Sigma_2 \times_{\Sigma_2} X^2$$

is just the "diagonal" map, Δ, given by $\Delta([e], x) = [e, x, x]$ where $[-]$ denotes the Σ_2-orbit of a point and $e \in E\Sigma_2$, $x \in X$.
When $X = BH$ and $Y = BG$, then

$$E\Sigma_2 \times_{\Sigma_2} X^2 = E\Sigma_2 \times_{\Sigma_2} (BH)^2$$

$$= (E\Sigma_2 \times (EH \times EH))\Big/ \Big(\Sigma_2 \int H \Big)$$

$$= B\Sigma_2 \int H.$$

Here $\Sigma_2 \int H$ is the wreath product generated by an involution, τ, and $H \times H$, with $H \times H \lhd \Sigma_2 \int H$ and conjugation by τ acting on $H \times H$ by interchanging the factors. $\Sigma_2 \int H$ acts freely on the contractible space, $E\Sigma_2 \times EH \times EH$, by the formulae

$$\tau(e, z_1, z_2) = (\tau(e), z_2, z_1)$$

$$(h_1, h_2)(e, z_1, z_2) = (e, h_1(z_1), h_2(z_2)) \qquad \text{for } h_i \in H, \ e \in E\Sigma_2, \ z_i \in EH.$$

In this case Φ becomes the map $\Phi: BG \to B\Sigma_2 \int H$ induced by the *homomorphism*

$$(1.10) \qquad \begin{cases} \Phi: G \to \Sigma_2 \int H \text{ given by} \\ \Phi(ht^\varepsilon) = \begin{cases} h, t^{-1}ht & \text{if } \varepsilon = 0 \\ (ht^2, t^{-1}ht)\tau & \text{if } \varepsilon = 1. \end{cases} \end{cases}$$

In (1.10) τ is the involution in $\Sigma_2 \int H$, $h \in H$, and $t \in G - H$ is a fixed choice. The homomorphism (1.10) is just a special case of the "induced homomorphism" construction associated to a subhomomorphism (in this case the subhomomorphism is $G \supset H \xrightarrow{1} H$) which we will meet again in Chapter 5, §1. However, for completeness, let us verify that Φ is a homomorphism in (1.10).

Clearly, if $h, h' \in H$, $\Phi(h)\Phi(h') = \Phi(hh')$. Also

$$\Phi(hth't) = (hth't, t^{-1}hth't^2)$$
$$= (ht^2 t^{-1}h't, t^{-1}hth't^2)$$
$$= (ht^2, t^{-1}ht)\tau(h't^2, t^{-1}ht)\tau$$
$$= \Phi(ht)\Phi(h't),$$

$$\Phi(hh't) = (hh't^2, t^{-1}hh't)\tau$$
$$= (h, t^{-1}ht)(h't^2, t^{-1}h't)\tau$$
$$= \Phi(h)\Phi(h't)$$

and

$$\Phi(hth') = ((hth't^{-1})t^2, t^{-1}hth't^{-1}t)\tau$$
$$= (ht^2, t^{-1}ht)(t^{-1}h't, h')\tau$$
$$= (ht^2, t^{-1}ht)\tau(h', t^{-1}h't)$$
$$= \Phi(ht)\Phi(h').$$

(1.11) Example (cf., [Kos, §2.4])

Suppose that $G = \Sigma_2 \times H$, then, in (1.10),

$$\Phi: \Sigma_2 \times H \to \Sigma_2 \Big\int H \quad \text{becomes}$$
$$\Phi(t^\varepsilon, h) = \tau^\varepsilon(h, h), \quad \text{where } 1 \neq t \in \Sigma_2.$$

(1.12) Definition

Let X, Y be connected spaces for the rest of this section. Let $\pi: X \to Y$ be a double covering, as in §1.8. The Koslowski transfer

$$\hat{N}: \hat{G}(X) \to \hat{G}(Y)$$

is defined by $\hat{N} = \Phi^* \hat{D}$, where $\Phi^*: \hat{G}(E\Sigma_2 \times_{\Sigma_2} X^2) \to \hat{G}(Y)$ is induced by Φ of (1.9) (or (1.10) when applicable), and

$$\hat{D}: \hat{G}(X) \to \hat{G}(E\Sigma_2 \times_{\Sigma_2} X^2)$$

is defined by the formula

$$(1.13) \begin{cases} \hat{D}(m, 1 + x_1 + x_2 + \cdots) \\ = (2m, D(1 + x_1 + x_2 + \cdots) + \sum_{i \geq 1} \tilde{D}(x_i)((1 + d)^{m-i} + 1) + (1 + d)^m + 1). \end{cases}$$

In (1.13), $d \in H^1(E\Sigma_2 \times_{\Sigma_2} X^2)$ is the image of the generator of $H^1(B\Sigma_2; \mathbb{Z}/2) \cong \mathbb{Z}/2$ under the map induced by $E\Sigma_2 \times_{\Sigma_2} X^2 \to B\Sigma_2$.

(1.14) Definition

In the situation of §1.12 we may also define

$$\tilde{N}: H^i(X) \to H^{2i}(Y)$$

by $\tilde{N} = \Phi^* \circ \tilde{D}$. In addition we may define

$$N: G(X) \to G(Y)$$

by the formula $N = \Phi^* D$, where $D: G(X) \to G(E\Sigma_2 \times_{\Sigma_2} X^2)$ is as in (1.6).

(1.15)   Let $Sq^i: H^j(W) \to H^{i+j}(W)$ denote, as usual, the (mod 2) Steenrod squaring operation [S-E]. In [Kos2] we find the following formulae:

(1.16) Lemma [Kos2]

For the double-covering $\pi: X \simeq E\Sigma_2 \times X \to B\Sigma_2 \times X$,

(i) $N(1 + x_1 + x_2 + \cdots) = 1 + \sum_{i,k} Sq^i(x_k) d^{k-i}$.

(ii) $\tilde{N}(x_i) = \sum_j Sq^j(x_i) d^{i-j}$.

(iii) $\hat{N}(m, 1 + x_1 + x_2 + \cdots) = (2m, (1 + d)^m + \sum_i \tilde{N}(x_i)(1 + d)^{m-i})$.

Here $x_i \in H^i(X)$ and $0 \neq d \in H^1(B\Sigma_2)$ (as in (1.13)).

(1.17)   Let $i: X^2 \to E\Sigma_2 \times_{\Sigma_2} X^2$ be the double-covering given by passing from $E\Sigma_2 \times X^2 \simeq X^2$ to its Σ_2-orbit space, and let $\Delta': B\Sigma_2 \times X \to E\Sigma_2 \times_{\Sigma_2} X^2$ (see §1.8) be the "diagonal" map induced by the Σ_2-map

$$\Delta': E\Sigma_2 \times X \to E\Sigma_2 \times X^2$$

given by $\Delta'(e, x) = (e, x, x)$.

From [Q2] we have the following result.

(1.18) Lemma [Q2]

For all X,

$$(\Delta^*, i^*): H^*(E\Sigma_2 \times_{\Sigma_2} X^2) \to H^*(B\Sigma_2 \times X) \oplus H^*(X \times X)$$

is injective.

(1.19) Lemma

If $f \in H^i(X)$ and $f' \in H^j(X)$, then

$$\tilde{N}(f)\tilde{N}(f') = \tilde{N}(ff') \in H^{2(i+j)}(E\Sigma_2 \times_{\Sigma_2} X^2).$$

Proof. Let $d_q: E\Sigma_2 \times_{\Sigma_2} (K_q)^2 = M(q, 2) \times_{\Sigma_2} (K_q)^2 \to K_{2q}$ be the map $d_q(\alpha, k_1, k_2) = \alpha(k_1, k_2)$ which is used in the definition of $\tilde{D}$ in (1.7).

Define maps

$$\lambda: E\Sigma_2 \times_{\Sigma_2} X^2 \to E\Sigma_2 \times_{\Sigma_2} (K_i \times K_j)^2$$

and

$$v: E\Sigma_2 \times_{\Sigma_2} (K_i \times K_j)^2 \to (E\Sigma_2 \times_{\Sigma_2} (K_i)^2) \times (E\Sigma_2 \times_{\Sigma_2} (K_j)^2)$$

by

$$\lambda(e, x_1, x_2) = (e, f(x_1), f'(x_1), f(x_2), f'(x_2))$$

and

$$v(e, k_1, k_2, k_1', k_2') = ((e, k_1, k_1'), (e, k_2, k_2'))$$

$(e \in E\Sigma_2, x_u \in X, k_u, k_u' \in K_u)$.

Let $m: K_i \times K_j \to K_{i+j}$ and $m': K_{2i} \times K_{2j} \to K_{2i+2j}$ be maps which induce the cup-product.

Consider the following diagram:

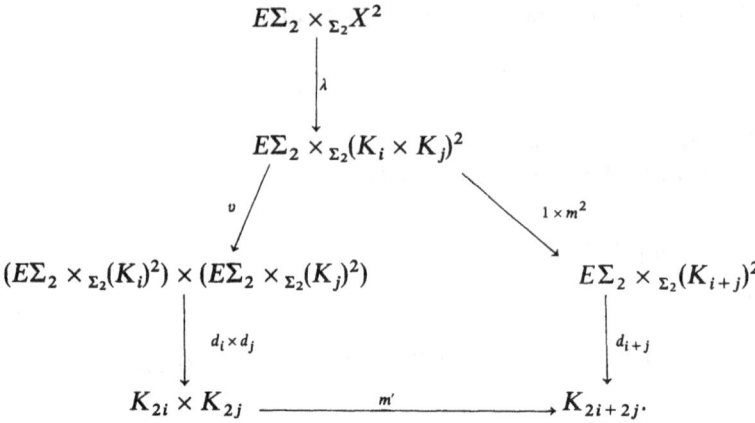

We must verify that the two routes around the diagram are homotopic. By §1.18 it suffices to show that

(1.20)
$$\begin{cases} \Delta^*(m'(d_i \times d_j)v) = \Delta^*(d_{i+j}(1 \times m^2)) & \text{and} \\ i^*(m'(d_i \times d_j)v) = i^*(d_{i+j}(1 \times m^2)). \end{cases}$$

Let $\alpha_i \in H^i(K_i)$ denote the generator. Then, in $H^*((K_i \times K_j)^2)$,

$$\begin{aligned} i^*(m'(d_i \times d_j)(v)) &= (i^*d_i^* \otimes i^*d_j^*)(m')^*(\alpha_{2i+2j}) \\ &= (i^*d_i^* \otimes i^*d_j^*)(\alpha_{2i} \otimes \alpha_{2j}) \\ &= \alpha_i \otimes \alpha_j \otimes \alpha_i \otimes \alpha_j \\ &= (m^* \otimes m^*)(i^*d_{i+j}^*(\alpha_{2i+2j})) \\ &= i^*(d_{i+j}(1 \times m^2)), \end{aligned}$$

which proves half of (1.20).

To prove the other half of (1.20), we use the fact ([Kos, §2.4] or §1.8) that

$$\Delta: B\Sigma_2 \times X \to E\Sigma_2 \times_{\Sigma_2} X^2$$

is the pre-transfer map, Φ, of (1.9) for the double-covering $X \to B\Sigma_2 \times X = Y$ (a fact which we verified in §1.11 when $X = BH$). Therefore, in $H^{2i+2j}(B\Sigma_2 \times K_i \times K_j)$,

$$\begin{aligned} \Delta^*(d_{i+j}(1 \times m^2)) &= \tilde{N}_{i+j}(\alpha_i \otimes \alpha_j) \\ &= \sum_a Sq^a(\alpha_i \otimes \alpha_j)d^{i+j-a}, \qquad \text{by §1.16 (ii),} \end{aligned}$$

$$= \sum_{u+v=a} (Sq^u(\alpha_i)d^{i-u})(Sq^v(\alpha_j)d^{j-v}), \quad \text{by the Cartan formula [S-E],}$$

$$= \tilde{N}_i(\alpha_i)\tilde{N}_j(\alpha_j), \qquad \text{by §1.16 (ii),}$$

$$= \Delta^*(v^*(d_i^* \otimes d_j^*)(\alpha_{2i} \otimes \alpha_{2j}))$$

$$= \Delta^*(m'(d_i \times d_j)v),$$

as required. In this calculation the $\tilde{N}_u$ is $\tilde{N}$ for the double cover

$$K_u \to B\Sigma_2 \times K_u. \qquad\qquad \square$$

(1.21) Lemma

For any double-covering $\pi: X \to Y$ and $a, b \in G(X)$, we have

$$N(a)N(b) = N(ab)$$

(the products being defined as in (1.2)).

Proof. This is a consequence of general facts about transfer constructions in generalized cohomology theories. Namely, if $G^*(X)$ is a cohomology theory, then the Kahn-Priddy transfer [K-P]

$$T: G^n(X) \to G^n(Y)$$

is a homomorphism. However, in [Seg], it is shown that there is such a $G^*(X)$ with $G^0(X) = G(X)$ for which (as remarked in [Kos]) $T = N$. This establishes the formula. $\qquad\qquad \square$

(1.22)   We have a double-covering

$$\pi: X^2 \simeq E\Sigma_2 \times X^2 \to E\Sigma_2 \times_{\Sigma_2} X^2,$$

from which there results an additive transfer in mod 2 cohomology,

$$\text{Tr}: H^n(X^2) \to H^n(E\Sigma_2 \times_{\Sigma_2} X^2).$$

I will not define Tr in general (see [S-E] for the general definition) since we will use only the case $X = BG$ in which case Tr is the (discrete) group cohomology transfer of Chapter 1, §2 for $G \times G \subset \Sigma_2 \int G$. I will freely use properties (e.g., the $H^*(E\Sigma_2 \times_{\Sigma_2} X^2)$-module property) which are analogous to those we have encountered in the context of group cohomology.

In particular, if $x, y \in H^n(X)$, then

$$i^*(\text{Tr}(x \otimes y)) = i^*(\text{Tr}(y \otimes x)) = x \otimes y + y \otimes x \in H^n(X \times X).$$

Also we have a commutative diagram of double-coverings

(1.23)

$$
\begin{array}{ccc}
X^2 & \xrightarrow{\ \pi\ } & E\Sigma_2 \times_{\Sigma_2} X^2 \\
\Big\uparrow{\scriptstyle e} & & \Big\uparrow{\scriptstyle \Delta} \\
X & \xrightarrow{\ \pi\ } & B\Sigma_2 \times X,
\end{array}
$$

in which e is the diagonal. By naturality of Tr in (1.23),

$$
\begin{aligned}
\Delta^* \text{Tr}(x \otimes y) &= \text{Tr}(e^*(x \otimes y)) \\
&= \text{Tr}(xy) \\
&= \text{Tr}(\pi')^*(1 \otimes xy) \\
&= 2 \otimes xy \\
&= 0.
\end{aligned}
$$

In addition, if $0 \neq d \in H^1(B\Sigma_2)$,

$$
\begin{aligned}
d\,\text{Tr}(x \otimes y) &= \text{Tr}(\pi^*(d)(x \otimes y)) \\
&= 0,
\end{aligned}
$$

by the $H^*(E\Sigma_2 \times_{\Sigma_2} X^2)$-module property, since $\pi^*(d) = 0$. This could also easily be seen by observing that $(\Delta^*, i^*)(d\,\text{Tr}(x \otimes y)) = (0, 0)$ and invoking §1.18.

(1.24) Lemma

   (i) *If, as in §1.22, $x, y \in H^n(X)$ $(n > 0)$, then $\tilde{D}(x + y) = \tilde{D}(x) + \tilde{D}(y) + \text{Tr}(x \otimes y) \in H^{2n}(E\Sigma_2 \times_{\Sigma_2} X^2)$.*

   (ii) *If $0 \neq d \in H^1(B\Sigma_2)$, $d\tilde{D}(x + y) = d\tilde{D}(x) + d\tilde{D}(y) \in H^{2n+1}(E\Sigma_2 \times_{\Sigma_2} S^2)$.*

Proof. From §1.8 (see the proof of §1.19 and also the remarks made there concerning §1.11),

$$\Delta: B\Sigma_2 \times X \to E\Sigma_2 \times_{\Sigma_2} X^2$$

is the pre-transfer of (1.9) associated to the double-covering, $X \to B\Sigma_2 \times X$. Hence

$$(\Delta^*, i^*)(\tilde{D}(x + y)) = (\tilde{N}(x + y), (x + y) \otimes (x + y))$$

$$= \left(\sum_i Sq^i(x + y)d^{n-i}, i^*(\tilde{D}(x) + \tilde{D}(y) + \mathrm{Tr}\,(x \otimes y)) \right),$$

by §1.16(ii) and §1.22,

$$= \left(\sum_i (Sq^i(x) + Sq^i(y))d^{n-i}, i^*(\tilde{D}(x) + \tilde{D}(y) + \mathrm{Tr}\,(x \otimes y) \right)$$

$$= (\tilde{N}(x) + \tilde{N}(y), i^*(\tilde{D}(x) + \tilde{D}(y) + \mathrm{Tr}\,(x \otimes y))$$

$$= (\Delta^*, i^*)(\tilde{D}(x) + \tilde{D}(y) + \mathrm{Tr}\,(x \otimes y)), \qquad \text{by §1.22.}$$

Hence part (i) follows from §1.18, and part (ii) follows from the fact, derived in §1.22, that $d\,\mathrm{Tr}\,(x \otimes y) = 0$. $\qquad\square$

(1.25) Lemma

(i) $\hat{D}: \hat{G}(X) \to \hat{G}(E\Sigma_2 \times_{\Sigma_2} X^2)$ is a homomorphism.

(ii) For any double-covering, $\pi: X \to Y$, $\hat{N}: \hat{G}(X) \to \hat{G}(Y)$ is a homomorphism.

Proof. To prove (ii), from (i) we observe that the pre-transfer $\Phi: Y \to E\Sigma_2 \times_{\Sigma_2} X^2$, induces a homomorphism

$$\Phi^*: \hat{G}(E\Sigma_2 \times_{\Sigma_2} X^2) \to \hat{G}(Y)$$

and that $\hat{N} = \Phi^*\hat{D}$, by definition.

To prove part (i), it suffices, by §1.18, to show that $(\Delta^*, i^*)\hat{D}$ is a homomorphism. However, from §1.8, Δ^* is the pre-transfer for $\pi: X \to B\Sigma_2 \times X$, as used in the proofs of §§1.19, 1.24. Hence, by §1.16 (iii),

$$\Delta^*\hat{D}(m, 1 + x_1 + x_2 + \cdots) = (2m, (1 + d)^m + \Sigma_i \tilde{N}(x_i)(1 + d)^{m-i}).$$

Therefore, for $\tilde{N}$ as in §1.16 (ii),

$$\Delta^*\hat{D}(m, 1 + x_1 + x_2 + \cdots)\Delta^*\hat{D}(n, 1 + y_1 + y_2 + \cdots)$$

$$= (2m + 2n, (1 + d)^{m+n} + \sum_{i,j} \tilde{N}(x_i)\tilde{N}(y_j)(1 + d)^{m+n-i-j})$$

$$= (2m + 2n, (1 + d)^{m+n} + \sum_a \sum_{i+j=a} \tilde{N}(x_i y_j)(1 + d)^{m+n-a}) \qquad \text{by §1.19,}$$

$$= (2m + 2n, (1 + d)^{m+n} + \sum_a \tilde{N}\left(\sum_{i+j=a} (x_i y_j)(1 + d)^{m+n-a} \right))$$

($\tilde{N}$ is additive by §1.16 (ii))

$$= \Delta^*\hat{D}((m, 1 + x_1 + x_2 + \cdots)(n, 1 + y_1 + y_2 + \cdots)) \qquad \text{by §1.16 (iii), as required.}$$

Finally, since $i^*(d) = 0$, we have from (1.13)

$$i^*\hat{D}(m, 1 + x_1 + x_2 + \cdots) = (2m, D(1 + x_1 + x_2 + \cdots)),$$

which is a homomorphism because D is a homomorphism.

The reason that D is a homomorphism is similar to that given in the proof of §1.21. Namely, D is the Segal transfer for the covering $X^2 \to E\Sigma_2 \times_{\Sigma_2} X^2$ and the cohomology theory, $G^*(X)$, where $G^0(X) = G(X)$. $\qquad\square$

(1.26)   We return now to the homomorphism, w, of §1.4. Let G be a finite group, and let $RO(G)$ denote the Grothendieck ring of finite-dimensional real representations of G. This is the free abelian group on the isomorphism classes, $[\alpha]$, of irreducible representations, $\alpha: G \to O_n(\mathbb{R})$. Alternatively, $RO(G)$ is the quotient of the free abelian group on isomorphism classes of representations modulo the relation

$$[\alpha \oplus \beta] = [\alpha] + [\beta].$$

If $SW[\alpha] = 1 + SW_1(\alpha) + SW_2(\alpha) + \cdots$ in $G(BG) = \prod_{i \geq 1} H^i(G; \mathbb{Z}/2)$, then we obtain a *homomorphism* (see §1.4)

(1.27) $\begin{cases} w: RO(G) \to \hat{G}(BG) & \text{given by} \\ w[\alpha] = (\dim(\alpha), SW[\alpha]), \end{cases}$

where $\dim: RO(G) \to \mathbb{Z}$ is the dimension function.

More generally, if X is a space, we may form $KO(X)$, the Grothendieck group of isomorphism classes of real vector bundles, E, over X modulo the relation

$$[E \oplus F] = [E] \oplus [F].$$

A real vector bundle, E, over X is a topological space together with a continuous map $p: E \to X$ such that, for each $x \in X$, there is a neighborhood U and a homeomorphism $p^{-1}(U) \cong U \times \mathbb{R}^m$ such that

$$p^{-1}(U) \xrightarrow{\;\cong\;} U \times \mathbb{R}^m$$

$$\begin{array}{ccc} & & \\ {\scriptstyle p} \searrow & & \swarrow {\scriptstyle \text{proj}} \\ & U & \end{array}$$

commutes.

Each representation, $\alpha: G \to O_m(\mathbb{R})$, gives rise to a vector bundle

$$p: EG \times_G \mathbb{R}^m \to BG$$

$$p(e, v) = [e],$$

where G acts on $\mathbb{R}^m$ via α. This construction yields a homomorphism

(1.28) $$\psi: RO(G) \to KO(BG).$$

The Stiefel-Whitney classes, $SW_i[E] \in H^i(X; \mathbb{Z}/2)$, of a vector bundle E over X yield a *homomorphism*

(1.29) $$\begin{cases} w: KO(X) \to \hat{G}(X) & \text{given by} \\ w[E] = (\dim E, SW[E]). \end{cases}$$

(1.28) and (1.29) are connected by means of a commutative diagram:

(1.30)
$$\begin{array}{ccc} RO(G) & \xrightarrow{\psi} & KO(BG) \\ {\scriptstyle w}\downarrow & & \downarrow{\scriptstyle w} \\ \prod_{i \geq 1} H^i(G; \mathbb{Z}/2) & \xrightarrow{\cong} & \hat{G}(BG). \end{array}$$

(1.31) Let $\pi: X \to Y$ be a n-fold covering, and let $p: E \to X$ be a vector bundle. The *induced bundle*, $\operatorname{Ind}_{X/Y}(E)$ or $\pi_*(E)$ over Y is the vector bundle whose fibre at $y \in Y$ is $\oplus_{i=1}^n p^{-1}(x_i)$, where $\pi^{-1}(y) = \{x_1, \ldots, x_n\}$.

This is related to induced representations in the following manner. Given groups $H \subset G$, with $[G:H]$ finite, suppose that V is a real representation of H. Then there is a natural isomorphism of bundles over BG, where π is $BH \to BG$,

(1.31a) $$EG \times_G(\operatorname{Ind}_H^G V) \cong \pi_*(EH \times_H V).$$

In other words, if π_* is the homomorphism induced by the above construction, we have a commutative diagram of the following form:

(1.31b)
$$\begin{array}{ccc} RO(H) & \xrightarrow{\psi} & KO(BH) \\ {\scriptstyle \operatorname{Ind}_H^G}\downarrow & & \downarrow{\scriptstyle \pi_*} \\ RO(G) & \xrightarrow{\psi} & KO(BG). \end{array}$$

(1.32) THE SPLITTING PRINCIPLE

Suppose that we have a natural transformation

$$\beta: KO(X) \to H^*(X; \mathbb{Z}/2) = H^*(X).$$

Given a vector bundle, $p: E \to X$, there exists a space and a map, $s: P(E) \to X$ [Spa] such that $s: H^*(X) \to H^*(P(E))$ is injective and such that the pullback vector bundle

$$s^*(E) = \{(e, q) \in E \times P(E) | p(e) = s(q)\}$$

is a sum of line bundles,

$$s^*(E) = \bigoplus_j L_j.$$

Naturality means that

$$s^*(\beta(E)) = \beta(s^*(E))$$

$$= \beta\left(\bigoplus_j L_j\right).$$

Therefore β is determined uniquely by its effect on sums of line bundles. This is the "splitting principle". We will use it, with a slight modification, to prove the next result.

(1.33) Theorem [Kos]

With the preceding notation and conventions, let $\pi: X \to Y$ be a double-covering. Then the following diagram is commutative:

$$
\begin{array}{ccc}
KO(X) & \xrightarrow{\;w\;} & \hat{G}(X) \\
{\scriptstyle \pi_*}\downarrow & & \downarrow{\scriptstyle \hat{N}} \\
KO(Y) & \xrightarrow{\;w\;} & \hat{G}(Y).
\end{array}
$$

Proof. This is a diagram of homomorphisms which are either natural for maps of spaces (w) or for maps of double-coverings (π_* and $\hat{N}$). If $p: E \to X$ is a vector bundle, the map $s: P(E) \to X$ [Spa] fits into a commutative diagram

(1.33b)
$$
\begin{array}{ccc}
P(s^*(E)) & \xrightarrow{\;s'\;} & Y \\
{\scriptstyle \pi'}\uparrow & & \uparrow{\scriptstyle \pi} \\
P(E) & \xrightarrow{\;s\;} & X
\end{array}
$$

in which π' is also a double-covering. Hence, by the splitting principle of §1.32, it suffices to prove that

(1.34) $$\hat{N}(w(E)) = w(\pi_*(E))$$

when E is a line bundle, since w, $\hat{N}$, and π^* are homomorphisms.

Let $\pi_1 : X \times X \to X$ denote the first projection, and let $\pi'' : X^2 \to E\Sigma_2 \times_{\Sigma_2} X^2$ be the natural map. The induced bundle construction may be described as

$$(1.35) \qquad \pi_*(E) = \Phi^*(\pi''_*(\pi_1^*(E))).$$

The reader is invited, in Exercise 1.37, to verify this directly in the case of a double-covering of classifying spaces.

From (1.35) we have

$$w(\pi_*(E)) = w(\Phi^*(\pi''_*(\pi_1^*(E))))$$

$$= \Phi^*(w(\pi''_*(\pi_1^*(E)))),$$

whereas

$$\hat{N}(w(E)) = \Phi^*(\hat{D}(w(E)))$$

so that it suffices to show, in

$$\hat{G}(E\Sigma_2 \times_{\Sigma_2} X^2), \quad \text{that}$$

$$(1.36) \qquad \hat{D}(w(E)) = w(\pi''_*(\pi_1^*(E)))$$

when E is a line bundle.

By §1.18 it suffices to verify that the images of both sides of (1.36) under (Δ^*, i^*) are equal.

Firstly consider $\Delta^* \hat{D}(w(E)) \in \hat{G}(B\Sigma_2 \times X)$. Since (see §1.8) Δ^* is the pre-transfer of $\tilde{\pi} : X \to B\Sigma_2 \times X$, we find from §1.16 (iii) that, if $u = SW_1[E]$,

$$\Delta^* \hat{D}(w(E)) = \hat{N}(1, 1 + u)$$

$$= (2, 1 + d + \tilde{N}(u))$$

$$= (2, 1 + d + ud + u^2).$$

On the other hand,

$$\Delta^* w(\pi''_*(\pi_1^*(E))) = w(\Delta^* \pi''_*(\pi_1(E))).$$

The diagram of double-coverings

$$
\begin{array}{ccc}
X & \xrightarrow{\;e\;} & X \times X \\
{\scriptstyle \tilde{\pi}} \downarrow & & \downarrow {\scriptstyle \pi} \\
B\Sigma_2 \times X & \xrightarrow{\;\Delta\;} & E\Sigma_2 \times_{\Sigma_2} X^2
\end{array}
$$

shows that

$$w(\Delta^*(\pi''_*(\pi^*_1(E)))) = w(\tilde{\pi}_*(e^*\pi^*_1(E)))$$
$$= w(\tilde{\pi}_*(E))$$
$$= w(E \oplus (E \otimes H)),$$

where H is the line bundle over $B\Sigma_2$ which corresponds, under ψ of §1.30, to the nontrivial one-dimensional representation of Σ_2. However,

$$w(E \oplus (E \otimes H)) = w(E)w(E \otimes H)$$
$$= (1, 1 + u)(1, 1 + u + d)$$
$$= (2, 1 + d + u^2 + ud),$$

which equals $\Delta^*\hat{D}(w(E))$, as required.

On the other hand, from the general properties of transfers in cohomology theories (applied to Segal's cohomology theory $G^*(X)$, with $G^0(X) = G(X)$; see §1.21 (proof)),

$$i^*D(1 + u) = (1 + u) \otimes (1 + u) \in G(X \times X).$$

Therefore

$$i^*(\hat{D}(w(E))) = i^*(\hat{D}(1, 1 + u))$$
$$= i^*(2, D(1 + u) + 2\tilde{D}(u) + 1 + d + 1) \qquad \text{by (1.13)},$$
$$= (2, i^*D(1 + u) + i^*(d))$$
$$= (2, 1 + u \otimes 1 + 1 \otimes u + u \otimes u) \qquad \text{as } i^*(d) = 0.$$

Finally, if $T: X \times X \to X \times X$ is the switch map,

$$i^*(w(\pi''_*(\pi^*_1(E)))) = w(i^*\pi''_*(\pi^*_1(E)))$$
$$= w(\pi^*_1 E \oplus (\pi_1 T)^*(E))$$
$$= w(\pi^*_1 E)T^*(w(\pi^*_1 E))$$
$$= (1, 1 + u \otimes 1)(1, 1 + 1 \otimes u)$$
$$= (2, 1 + u \otimes 1 + 1 \otimes u + u \otimes u)$$

in $\hat{G}(X \times X)$, which completes the proof of Theorem 1.33.   □

(1.37) Exercise

Let $H \subset G$ be discrete groups with $[G:H] = 2$. Use the group-theoretic description of the pre-transfer, $\Phi: BG \to B\Sigma_2 \int H$, given in (1.10), to verify the following: Let V be a real representative of H and $\pi_1^*(V)$ the resulting representation of $H \times H$, acting through the first factor. Then, in $KO(BG)$,

$$\psi(\Phi^* \operatorname{Ind}_{H \times H}^{\Sigma_2 \int H}(\pi_1^*(V))) = \psi(\operatorname{Ind}_H^G(V)).$$

(1.38) We will conclude this section with some sample calculations of $\hat{N}$. Suppose that $i: H \subset G$ is an inclusion of finite groups with $[G:H] = 2$. Let $0 \neq d \in H^1(B\Sigma_2)$, and, as usual, also denote by $d \in H^1(E\Sigma_2 \times_{\Sigma_2}(BH)^2) = H^1(B\Sigma_2 \int H)$ its image. As in §1.24 let

$$\operatorname{Tr}: H^*(BH \times BH) \to H^*\left(B\Sigma_2 \int H\right)$$

denote the additive transfer associated with $H \times H \subset \Sigma_2 \int H$.

We identify $H^*(BG)$ with $H^*(G; \mathbb{Z}/2)$ as in Chapter 1, §3.17. The following result concerns

$$\hat{N}: \hat{G}(BH) \to \hat{G}(BG) \quad \text{and} \quad \tilde{N}: H^m(H; \mathbb{Z}/2) \to H^{2m}(G; \mathbb{Z}/2):$$

(1.39) Lemma

In the situation of §1.38, let $\alpha \in H^2(H; \mathbb{Z}/2)$. Then, abbreviating $\Phi^*(d)$ to d,

$$\hat{N}(0, 1 + \alpha) = (0, 1 + i_*(\alpha) + \tilde{N}(\alpha)(1 + d)^{-2}).$$

Proof. By definition, $\hat{N} = \Phi^* \hat{D}$, where $\Phi: G \to \Sigma_2 \int H$ is the homomorphism of (1.10). Let us consider, from (1.13),

$$\hat{D}(0, 1 + \alpha) = (0, D(1 + \alpha) + \tilde{D}(\alpha)((1 + d)^{-2} + 1)).$$

Since $\Delta: B\Sigma_2 \times BH \to B\Sigma_2 \int H$ is the pre-transfer for $\pi: BH \to B\Sigma_2 \times BH$ (see §1.8), §1.16 implies that

$$\Delta^* \hat{D}(0, 1 + \alpha) = (0, 1 + \Delta^* \tilde{D}(\alpha)(1 + d)^{-2})$$

$$= \Delta^*(0, 1 + \tilde{D}(\alpha)(1 + d)^{-2}), \quad \text{as } 1 + \tilde{N}(\alpha) = N(1 + \alpha),$$

$$= \Delta^*(0, 1 + \operatorname{Tr}(\alpha \otimes 1) + \tilde{D}(\alpha)(1 + d)^{-2})$$

whereas by (1.13),

$$i^*(\hat{D}(0, 1 + \alpha)) = (0, i^* D(1 + \alpha)),$$

since $i^*(d) = 0$. However, as explained in §1.33 (proof),

$$i^*(D(1 + \alpha)) = 1 + \alpha \otimes 1 + 1 \otimes \alpha + \alpha \otimes \alpha$$

$$= i^*(1 + \text{Tr}(\alpha \otimes 1) + \tilde{D}(\alpha)(1 + d)^{-2}),$$

since $i^*\tilde{D}(\alpha) = \alpha \otimes \alpha$. By §1.18, these calculations establish the formula

(1.40) $\hat{D}(0, 1 + \alpha) = (0, 1 + \text{Tr}(\alpha \otimes 1) + \tilde{D}(\alpha)(1 + d)^{-2}).$

However, from (1.10), we have a commutative diagram

(1.41)

$$
\begin{array}{ccc}
H & \xrightarrow{\ i\ } & G \\
\Big\downarrow{\scriptstyle j} & & \Big\downarrow{\scriptstyle \Phi} \\
H \times H & \xrightarrow{\hspace{2cm}} & \Sigma_2 \int H,
\end{array}
$$

in which $j(h) = (h, t^{-1}ht)$. Therefore $\Phi^* \text{Tr}(\alpha \otimes 1) = i_*(j^*(\alpha \otimes 1)) = i_*(\alpha)$, and the required formula follows from (1.40) upon applying Φ^*.   □

(1.42)   Corollary

Let $t \in G - H$ in §§1.38–1.39, and let $t^*: H^i(H; \mathbb{Z}/2) \to H^i(H; \mathbb{Z}/2)$ be induced by conjugation by t, $(h \mapsto tht^{-1}; h \in H)$. Then, if $\alpha = \sum_{s=1}^{v} \alpha_s \in H^2(H; \mathbb{Z}/2)$,

$$\hat{N}(0, 1 + \alpha) = \left(0, 1 + \sum_s i_*(\alpha_s) + \sum_{s_1 < s_2} i_*(\alpha_{s_1} t^*(\alpha_{s_2})) + \sum_s \tilde{N}(\alpha_s)(1 + d)^{-2} \right).$$

Proof. By §1.24 (i),

$$\tilde{D}\left(\sum_s \alpha_s \right) = \tilde{D}(\alpha_1) + \tilde{D}\left(\sum_{s \geq 2} \alpha_s \right) + \text{Tr}\left(\alpha_1 \otimes \left(\sum_{s \geq 2} \alpha_s \right) \right)$$

so that, by induction on v,

(1.43) $\tilde{D}\left(\sum_{s=1}^{v} \alpha_s \right) = \sum_s \tilde{D}(\alpha_s) + \sum_{s_1 < s_2} \text{Tr}(\alpha_{s_1} \otimes \alpha_{s_2}).$

However, from (1.41),

$$\Phi^*(\text{Tr}(\alpha_{s_1} \otimes \alpha_{s_2})) = i_*(j^*(\alpha_{s_1} \otimes \alpha_{s_2}))$$

$$= i_*(\alpha_{s_1}(t^{-1})^*(\alpha_{s_2})).$$

Since t^{-1} and t differ by an element $h \in H$ and $h^*(\alpha_s) = \alpha_s$ (inner automorphisms act trivially on cohomology), the result follows by applying Φ^* to (1.43).   □

(1.44)   We continue with the notation of §§1.38–1.42. If $x = (t, 1)\tau \in \Sigma_2 \int G$ the following diagram commutes:

(1.45)

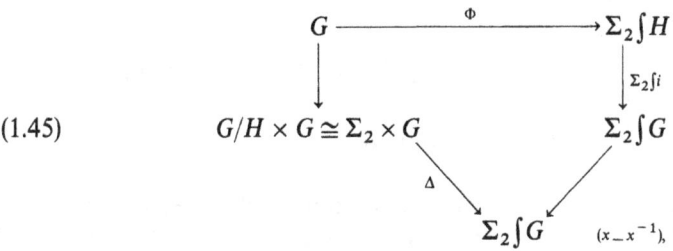

where $\lambda(g) = (gH, g)$. If $h \in H$,

$$x\left(\left(\Sigma_2 \int i\right)(\Phi(h))\right)x^{-1} = (t, 1)\tau(h, t^{-1}ht)\tau(t^{-1}, 1), \qquad \text{by (1.10)},$$

$$= (t, 1)(t^{-1}ht, h)(t^{-1}, 1)$$

$$= (h, h)$$

while

$$x\left(\left(\Sigma_2 \int i\right)(\Phi(ht))\right)x^{-1} = (t, 1)\tau(ht^2, t^{-1}ht)\tau^2(t^{-1}, 1)$$

$$= (t, 1)(t^{-1}ht, ht^2)(1, t^{-1})\tau$$

$$= (ht, ht)\tau,$$

as required.

We use (1.45) to prove the following result, which is useful in conjunction with §1.42, for example:

(1.46) Proposition

In the notation of §§1.38–1.42, *if* $a \in H^n(H; \mathbb{Z}/2)$ *is in the image of* $i^*: H^n(G; \mathbb{Z}/2) \to H^n(H; \mathbb{Z}/2)$, *say* $a = i^*(b)$, *then*

$$\tilde{N}(i_*(b)) = \sum_j Sq^j(b)d^{n-j} \in H^{2n}(G; \mathbb{Z}/2).$$

Proof. $\tilde{N}(a) = \Phi^*(\tilde{D}(a))$, and by naturality of $\tilde{D}$,

$$\tilde{D}(a) = \tilde{D}(i^*(b))$$

$$= \left(\Sigma_2 \int i\right)^* (\tilde{D}(b))$$

$$= (x \_ x^{-1})^* \left(\Sigma_2 \int i\right)^* ((\tilde{D}(b)).$$

Therefore, by (1.45), applying Φ^*, we obtain

$$\tilde{N}(a) = \lambda^*(\Delta^*(\tilde{D}(b)))$$

$$= \lambda^*\left(\sum_j Sq^j(b)d^{n-j}\right) \qquad \text{by §1.16 (ii) (cf. §§1.19, 1.24),}$$

$$= \sum_j Sq^j(b)d^{n-j}, \qquad \text{as required.} \qquad \square$$

(1.47)   I will conclude this section with a few games with dihedral representations. The object of this exercise is to prove Lemma 1.54 and Proposition 1.57 which concern $d\tilde{D}(z)$ and $d\tilde{N}(z)$, where L/K is a quadratic extension with discriminant $d \in H^1(K; \mathbb{Z}/2)$ and $z \in H^1(L; \mathbb{Z}/2)$ with char $(K) \neq 2$.

Let $D_8 = \langle x, y \,|\, x^4 = y^2 = 1, \ xyx = y \rangle$ be the dihedral group of order eight, and let $x_i: D_8 \to \mathbb{Z}/2$ be as in Chapter 1, §(4.4) for $i = 1, 2$. Consider the representation

$$(1.48) \qquad\qquad \rho: \Sigma_2 \int D_8 \xrightarrow{\ \Sigma_2 \int x_2\ } \Sigma_2 \int \Sigma_2 \subset O_2(\mathbb{R}).$$

Explicitly, this representation is given by

$$(1.49) \qquad \rho(\tau^\nu(x^{\varepsilon_1} y^{\delta_1}, x^{\varepsilon_2} y^{\delta_2})) = \begin{bmatrix} 0 & 1 \\ 1 & 0 \end{bmatrix}^\nu \begin{bmatrix} (-1)^{\varepsilon_1} & 0 \\ 0 & (-1)^{\varepsilon_2} \end{bmatrix}.$$

If we restrict ρ to $\Sigma_2 \times D_8$ by means of the "diagonal", Δ, we obtain

$$(1.50) \qquad\qquad \Delta\rho(\tau^\nu, x^\varepsilon y^\delta) = \begin{bmatrix} 0 & 1 \\ 1 & 0 \end{bmatrix}^\nu \begin{bmatrix} -1 & 0 \\ 0 & -1 \end{bmatrix}^\varepsilon.$$

Hence as a representation of $\Sigma_2 \times D_8$, we find that as a two-dimensional real representation,

$$(1.51) \qquad\qquad \Delta\rho = d \otimes x_2 + x_2,$$

where a homomorphism $f: G \to \mathbb{Z}/2$ is considered as a one-dimensional representation. However, in $H^2(\Sigma_2 \times D_8; \mathbb{Z}/2)$,

$$(1.52) \qquad \begin{cases} SW_2[\Delta\rho] = SW_1[d \otimes x_2]SW_1[x_2] \\ \qquad\quad = (d + x_2)x_2. \end{cases}$$

If we restrict ρ to $D_8 \times D_8$ by means of $i: D_8 \times D_8 \to \Sigma_2 \int D_8$, we find

$$(1.53) \qquad \begin{cases} i\rho = x_2 \otimes 1 + 1 \otimes x_2 \quad \text{and} \\ SW_2[i\rho] = x_2 \otimes x_2. \end{cases}$$

By §§1.8, 1.16 (ii), and 1.18, since $\sum Sq^i(x_2)d^{1-i} = x_2d + x_2^2$, we have shown the following result:

(1.54) Lemma

If ρ is the representation of (1.48), *then*

$$SW_2[\rho] = \tilde{D}(x_2) \in H^2\left(\Sigma_2 \int D_8; \mathbb{Z}/2\right).$$

(1.55) The final result of this section is a useful little formula, concerning $\tilde{N}$ of §1.14, *which is peculiar to Galois cohomology.* I particularly like the proof which I will give because it emphasizes this fact by its use of Serre's formula (Chapter 3, §2.8(ii)).

Let L/K be a quadratic extension, $L = K(\sqrt{c})$ with char$(K) \neq 2$. Suppose that $z \in H^1(L; \mathbb{Z}/2)$, then we have

(1.56) $$\tilde{N}(z) \in H^2(K; \mathbb{Z}/2)$$

by applying §1.14 to the covering $BG(N/L) \to BG(N/K)$, where $N \supset L \supset K$ is a Galois extension (with N "large").

(1.57) Proposition

In the situation of §1.55, $\tilde{N}(z)$ *of* (1.56) *satisfies*

$$0 = d\tilde{N}(z) \in H^2(K; \mathbb{Z}/2),$$

where $l(c) = d \in H^1(K; \mathbb{Z}/2)$ *is the discriminant of* L/K.

Proof. Suppose that $z = l(a)$, where $a \in L^*$. Clearly, we may assume $a \notin (L^*)^2$.

Consider the extension $F = L(\xi, \mu)$, where $\mu^4 = a$ and $\xi^2 = 1$. As in Chapter 3, §2.2 (proof) either $G(F/L) = D_8$ or $\mathbb{Z}/4$ according to whether $\xi \notin L$ or $\xi \in L$. In either case $H' = G(F/L)$ is a subgroup of D_8. Therefore we may form the representation of (1.48):

$$\rho: \Sigma_2 \int H' \to O_2(\mathbb{R}).$$

Now assume that $N \supset F \supset L \supset K$ is a "large" Galois extension. We are going to set $H = G(N/L)$ and $G = G(N/K)$, and then we may consider $z = l(a)$ as the composition (see the proof of Chapter 3, (2.2)

$$H \xrightarrow{\pi} H' \xrightarrow{x_2} \mathbb{Z}/2.$$

By naturality of $\tilde{D}$, §1.54 shows that

$$SW_2[\rho'] = \tilde{D}(x_2) \in H^2\left(\Sigma_2 \int H; \mathbb{Z}/2\right),$$

where ρ' is the representation

(1.58) $\qquad \rho': \Sigma_2 \int H \xrightarrow{\Sigma_2 \int \pi} \Sigma_2 \int H' \to \Sigma_2 \int G(F/L) \to \Sigma_2 \int D_8 \xrightarrow{\rho} O_2(\mathbb{R}).$

Therefore $\tilde{N}(z) = \Phi^* \tilde{D}(x_2) = \Phi^* SW_2[\rho']$, where $\Phi: G \to \Sigma_2 \int H$ is as in (1.10). From (1.10) one easily sees that

(1.59) $\qquad\qquad\qquad \text{Ind}_H^G(l(a)) = \rho'\Phi: G \to O_2(\mathbb{R}),$

where $l(a): G(N/L) \to \mathbb{Z}/2$ represents $z \in H^1(G(N/L); \mathbb{Z}/2)$; since N is "large", this makes sense.

Therefore we must show, in $H^3(K; \mathbb{Z}/2)$, that

$$0 = dSW_2[\text{Ind}_{\Omega_L}^{\Omega_K}(l(a))],$$

where $d = l(c)$ is as in §1.39, for example.

However, we may consider $l(a)$ as a symmetric, nondegenerate bilinear form of rank 1, as in Chapter 3, §§2.7, 2.8. Therefore we have

$$dSW_2[\text{Ind}_{\Omega_L}^{\Omega_K}(l(a))] = dHW_2(\text{Tr}_{L/K}^S l(a)) + dl(2)d$$

$$= dHW_2(\text{Tr}_{L/K}^S \langle a \rangle), \qquad \text{since, by Chapter 3, §§2.2, 2.5,}$$

$$d^2 l(2) = dl(-1)l(2) = d \cdot 0 = 0,$$

$$= l(c)l(2x)l(2x(x^2 c - y^2 c^2)),$$

by Example 3.6 of Chapter 2, if $a = x + y\sqrt{c}$. Now let $u = x^2 c - y^2 c^2 \in K^*$.
Therefore we have

$$d\tilde{N}(z) = l(c)(l(2) + l(x))(l(2) + l(x) + l(u))$$

$$= l(c)[l(2)^2 + l(x)^2 + l(2)l(u) + l(x)l(u)]$$

$$= l(c)[l(x)l(-1) + l(2)l(u) + l(x)l(u)],$$

by Chapter 3, §§2.2 and 2.5. Since $l(-u) = l(-1) + l(u)$ and $l(-1)l(2) = 0$, we have

$$l(x)l(-1) + l(x)l(u) = l(x)l(-u) \quad \text{and}$$

$$l(2)l(-u) = l(2)l(u) + l(2)l(-1) = l(2)l(u).$$

Our equation simplifies to

$$d\tilde{N}(z) = l(c)l(-u)[l(x) + l(2)],$$

which is zero, by Chapter 3, §2.5, since $-u = y^2c^2 - cx^2$ is clearly a norm from $L = K(\sqrt{c})$. This completes the proof of §1.57.   □

2. RELATIONS BETWEEN HIGHER HASSE-WITT CLASSES AND STIEFEL-WHITNEY CLASSES

The objective of this section is to describe the total Hasse–Witt class of the Scharlau transfer of a bilinear form in terms of Stiefel-Whitney classes and spinor classes. Firstly we will require some preliminaries before describing, in §2.6, our aim more precisely.

Let K be a field with char$(K) \neq 2$. Let $G(K)$ be the Galois cohomology analogue of $G(X)$ in (1.1). That is,

$$(2.1) \qquad G(K) = \prod_{i \geq 1} H^i(K; \mathbb{Z}/2)$$

with a product defined by considering $\{x_i^e; i \geq 1\} \in G(K)$ formally as $(1 + x_1^e + x_2^e + \cdots)$ and multiplying $\{x_i^0\}$ with $\{x_i^1\}$ by the formula of (1.2). Similarly, we define the group

$$(2.2) \qquad \hat{G}(K) = \mathbb{Z} \times G(K)$$

by analogy with (1.3). Therefore, if L/K is a quadratic extension, we have Koslowski's transfer

$$(2.3) \qquad \hat{N}_{L/K}: \hat{G}(L) \to \hat{G}(K).$$

This is induced by the topological transfer $\hat{N}: \hat{G}(BG(N/L)) \to \hat{G}(BG(N/K))$ by taking the limit over Galois extension $N \supset L \supset K$ with $\bar{K}$, a separable closure of K. Similarly, we have Galois cohomology analogues of $\tilde{N}$ and N:

$$(2.4) \qquad \begin{cases} \tilde{N}_{L/K}: H^i(L; \mathbb{Z}/2) \to H^{2i}(K; \mathbb{Z}/2) & \text{and} \\ N_{L/K}: G(L) \to G(K). \end{cases}$$

The "transfers" $\hat{N}_{L/K}$, $\tilde{N}_{L/K}$, and $N_{L/K}$ enjoy properties which are analogues of the ones in the topological case. The most important of these is the following one (cf., §1.33):

(2.5) Theorem

Let L/K be a quadratic extension with char$(K) \neq 2$. *Suppose that $N \supset L \supset K$ is* *a finite Galois extension and that*

$$\rho: G(N/L) \to O_n(K)$$

is an orthogonal representation. Then

$$\hat{N}_{L/K}(n, SW[\rho]) = (2n, SW[\text{Ind}_{G(N/L)}^{G(N/K)} \rho])$$

in either $\hat{G}(BG(N/K))$ or in $\hat{G}(K)$.
    Here $SW[\rho] = 1 + SW_1[\rho] + SW_2[\rho] + \cdots$, the total Stiefel-Whitney class.

Proof. If ρ was a representation into $O_n(\mathbb{R})$, we would immediately obtain this formula from §1.33 and (1.30). However, the Stiefel-Whitney classes of ρ are obtained from the map $\rho: BG(N/L) \to BO_n K \to BO(\bar{K})$, and not from a map to $BO(\mathbb{R})$.
    Nevertheless (in the homotopy category of 2-localized spaces, which is sufficient to recover mod 2 cohomology phenomena), the orthogonal group corollaries [Ka] of the results of [Su; Su1] yield a homology equivalence

$$BO(\mathbb{C}) \cong BO(\bar{K})^+.$$

Here $BO(\bar{K})^+$ is the result of applying Quillen's "plus-construction" [F-P] to $BO(\bar{K})$. For us, its most important feature is a canonical map, $BO(\bar{K}) \to BO(\bar{K})^+$, which induces an isomorphism in mod 2 cohomology. Finally, $BO(\mathbb{C})$ is homotopy equivalent to $BO(\mathbb{R})$.
    Therefore $B\rho: BG(N/L) \to BO(\bar{K})$ induces a (2-local) map of spaces $\hat{\rho}: BG(N/L) \to BO(\mathbb{R})$ that *preserves the total Stiefel-Whitney class* and commutes with *the forming of induced representations.* The latter remark involves us in topological technicalities which I will only sketch. Namely, there is a transfer on each of the functors of homotopy classes

$$[X, BO(\mathbb{R})], \quad [X, BO(\mathbb{C})], \quad [X, BO(\bar{K})^+]$$

for double-coverings, $\pi: X \to Y$ [K-P]. On $[X, BO(\mathbb{R})]$, which is isomorphic to the reduced K-theory of X, $\widetilde{KO}(X)$, this transfer coincides with the transfer construction for vector bundles (see §1.31). When $X = BG$ and $Y = BH$ with $[G:H] = 2$, this corresponds (see (1.33b) to taking induced representations. Also SW, on $KO(X) \cong \mathbb{Z} \times \widetilde{KO}(X)$ when X is connected, factors through $\widetilde{KO}(X)$.
    The net result of all this topological transfer technology is clearly to transform §1.33 for $\hat{\rho}$ into the formula which we require.   □

(2.6)   If $\rho: G(N/L) \to O_n(K)$ is the representation of §2.5, we may consider it as a 1-cocycle, representing a class

$$(\rho) \in H^1(L; O_n(\bar{L})),$$

which may be considered as a symmetric, nonsingular bilinear form of rank n over L, by Chapter 2, §2.10. Therefore we are entitled to a total Hasse-Witt class (Chapter 3, §1.34)

$$HW(\rho) = 1 + HW_1(\rho) + HW_2(\rho) + \cdots \in G(L).$$

On the other hand, we may form the induced representation ($[L:K] = d$)

$$\text{Ind}_{G(N/L)}^{G(N/K)}(\rho): G(N/K) \to O_{2d}(K),$$

representing the bilinear form, $\text{Tr}_{L/K}^S(\rho)$, by Chapter 2, §3.25, when L/K is separable.

In this section our objective will be to calculate, for certain types of ρ, $HW(\text{Tr}_{L/K}^S(\rho))$ in terms of $SW(\text{Ind}_{L/K}(\rho))$ and the spinor class $SP[\text{Ind}_{L/K}(\rho)]$. Here, and throughout this section, we abbreviate $\text{Ind}_{\Omega_L}^{\Omega_K}(\rho)$ to $\text{Ind}_{L/K}(\rho)$. We consider $\text{Ind}_{L/K}(\rho)$ as an orthogonal representation $G(N/L)$ for some "large," finite Galois extension, $N \supset L \supset K$.

The following result, which is a corollary of Chapter 3, §2.27, illustrates the type of formula which we will derive..

(2.7) Lemma

Let K be field with $\text{char}(K) \neq 2$, and let $\rho: G(N/K) \to O_2(K)$ be an orthogonal Galois representation. If $Sp[\rho]l(-1) = 0$, then

$$HW(\rho) = SW[\rho](1 + Sp[\rho])$$

in $G(K)$.

Proof. The left side of the equation is $1 + HW_1(\rho) + HW_2(\rho)$, and the right side is

$$1 + SW_1[\rho] + (SW_2[\rho] + Sp[\rho]) + SW_1[\rho]Sp[\rho] + SW_2[\rho]Sp[\rho].$$

By Chapter 3, §2.27, with $\beta(T, b) = \rho$, $b = \langle 1 \rangle^2$, $T = \rho$, we see that

$$HW_1(\rho) = SW_1[\rho] \quad \text{and} \quad HW_2(\rho) = SW_2[\rho] + Sp[\rho].$$

Therefore we must show that

$$0 = SW_i[\rho]Sp[\rho] \quad \text{for } i = 1, 2.$$

Let $Sq^1: H^i(\text{---}; \mathbb{Z}/2) \to H^{i+1}(\text{---}; \mathbb{Z}/2)$ denote the first Steenrod operation, which is the Bockstein or cohomology boundary map in the long exact sequence associated to the sequence of coefficients $\mathbb{Z}/2 \rightarrowtail \mathbb{Z}/4 \twoheadrightarrow \mathbb{Z}/2$ (see Chapter 1, §2.18). By the Wu formula [M-St],

$$Sq^1(SW_2[\rho]) = SW_3[\rho] + SW_1[\rho]SW_2[\rho]$$
$$= SW_1[\rho]SW_2[\rho],$$

since $\dim \rho = 2$. However, if $a, b \in K^*$, then

$$Sq^1(l(a)l(b)) = (Sq^1l(a))l(b) + l(a)(Sq^1l(b)), \qquad \text{by the Cartan formula [S-E],}$$

$$= l(a)^2l(b) + l(a)l(b)^2$$

$$= 2l(-1)l(a)l(b), \qquad \text{by Chapter 3, §2.2,}$$

$$= 0.$$

Therefore $Sq^1(HW_2(\rho)) = 0 = Sq^1(Sp[\rho])$ so that

$$0 = SW_1[\rho]SW_2[\rho]$$

$$= SW_1[\rho]Sp[\rho] + HW_1(\rho)HW_2(\rho)$$

$$= SW_1[\rho]Sp[\rho],$$

since, if $(\rho) = \langle x \rangle + \langle y \rangle$, then

$$HW_1(\rho)HW_2(\rho) = (l(x) + l(y))l(x)l(y),$$

$$= 2l(-1)l(x)l(y)$$

$$= 0.$$

Finally, since $SW_1[\rho] = l(x) + l(y)$, $0 = Sp[\rho](l(x) + l(y))$, and multiplying by $l(x)$, we obtain, by Chapter 3, §2.2,

$$0 = Sp[\rho]l(-1)l(x) + Sp[\rho]l(x)l(y)$$

$$= Sp[\rho]l(x)l(y), \qquad \text{since } Sp[\rho]l(-1) = 0,$$

$$= Sp[\rho]HW_2[\rho], \qquad \text{as required.} \qquad \square$$

(2.8) Examples

(i) Suppose that L/K is a quadratic extension and that $\Phi: G(N/L) \to \{\pm 1\} = O_1(K)$ is a homomorphism. By Chapter 3, §1.29, since $Sp[\Phi] = 0$, if $\rho = \text{Ind}_{L/K}(\Phi)$, then

$$Sp[\rho] = l(2)d_{L/K}$$

and $Sp[\rho]l(-1) = l(-1)l(2)d_{L/K} = 0$ because $l(-1)l(2) = 0$, by Chapter 3, §2.5.

(ii) More generally, let $\rho: G(N/K) \to O_2(K)$ be a homomorphism of the form

$$G(N/K) \xrightarrow{\lambda} D_{2m} \xrightarrow{j} O_2(K),$$

where D_{2m} is the dihedral group, and if $m = 2^n(2s + 1)$, j restricts to a *standard embedding* (see Chapter 3, §1.20) on $D_{2^{\alpha+1}}$. Let $H = \lambda^{-1}(D_{2^{\alpha+1}})$, which is a subgroup of odd index in $G(N/K)$. If $\pi: \Omega_K \twoheadrightarrow G(N/K)$ is the canonical surjection, then $\pi^{-1}(H) = \Omega_F$ for some field, F, with

$$N \supset F \supset K$$

and

$$H^*(K; \mathbb{Z}/2) \to H^*(F; \mathbb{Z}/2)$$

is injective. Also this injection, by naturality, maps the image of $Sp[\rho] = Sp[j\lambda]$ to that of $Sp[j\lambda|H]$. The latter is computed in Chapter 3, §1.20, and since $l(-1)l(2) = 0$, we see that $Sp[j\lambda|H]l(-1)$ is either zero (if $n = 0, 1$) or a multiple of $l(-1)l(1 + \alpha_n)$ (if $n \geq 2$). But, in the notation of Chapter 3, §1.20 there is $\beta_n \in K$, with

$$\alpha_n^2 + \beta_n^2 = 1$$

so that

$$\beta_n^2 + (1 + \alpha_n)^2 = 2(1 + \alpha_n).$$

By Chapter 3, §2.2, this means that

$$0 = l(-1)l(2(1 + \alpha_n))$$
$$= l(-1)l(2) + l(-1)l(1 + \alpha_n)$$
$$= l(-1)l(1 + \alpha_n),$$

by Chapter 3, §2.2. Hence $Sp[j\lambda|H]l(-1) = 0$, and so $Sp[\rho]l(-1) = 0$ as well.

We will call such a representation $\rho = j\lambda: G(N/K) \to O_2(K)$ a *standard dihedral representation*.

(2.9) Lemma

Let L/K be a quadratic extension with $\mathrm{char}(K) \neq 2$. Suppose that $\beta: V \times V \to L$ is a symmetric, nonsingular bilinear form of rank n. Then, in $\hat{G}(K)$,

$$\hat{N}_{L/K}(n, HW(V, \beta)) = (2n, HW(\mathrm{Tr}_{L/K}^s(V, \beta)) + nl(2)d_{L/K}),$$

where $d_{L/K} \in H^1(K; \mathbb{Z}/2)$ is the discriminant of L/K.

Proof. Suppose that (V, β) is equivalent to $\sum_{i=1}^{n} \langle a_i \rangle$ $(a_i \in L^*)$. Therefore in $H^1(L; O_n(\bar{L}))$ (V, β) is represented by the sum of one-dimensional representations $\sum_{i=1}^{n} l(a_i)$. Hence

$$\hat{N}_{L/K}(n, HW(V, \beta)) = \hat{N}_{L/K}\left(n, SW\left[\sum_i l(a_i)\right]\right)$$

$$= \left(2n, SW\left[\sum_i \mathrm{Ind}_{L/K}(l(a_i))\right]\right), \qquad \text{by §2.5,}$$

$$= \left(2n, \prod_i SW[\mathrm{Ind}_{L/K}(l(a_i))]\right),$$

$$\text{since } SW(A \oplus B) = SW(A)SW(B),$$

$$= \left(2n, \prod_i (1 + HW_1(\mathrm{Tr}^S_{L/K}\langle a_i \rangle) + HW_2(\mathrm{Tr}^S_{L/K}\langle a_i \rangle) + l(2)d_{L/K})\right),$$

$$\text{by Chapter 3, §2.8 (ii).}$$

By §2.7 and Example 2.8 (a), $Sp[\mathrm{Ind}_{L/K}l(a_i)] = l(2)d_{L/K}$ and $HW_i(\mathrm{Tr}^S_{L/K}\langle a_i \rangle)l(2)d_{L/K} = 0$ for $i = 1, 2$. In addition $l(2)^2 = l(2)l(-1) = 0$ so that, if we use the formula

$$HW(\beta \oplus \beta') = HW(\beta)HW(\beta'),$$

most of the products vanish in this expression, leaving precisely the required formula.   □

(2.10) Lemma

Suppose that L/K is a quadratic extension with $\mathrm{char}(K) \neq 2$ and that

$$\alpha = \sum_{i=1}^{k} z_i l(a_i) \in H^2(L; \mathbb{Z}/2),$$

where $z_i \in H^1(L; \mathbb{Z}/2)$ and $a_i \in K^*$. If for each $1 \leq i, j \leq t$ $l(a_i)l(a_j) = 0$, then

$$\hat{N}_{L/K}(0, 1 + \alpha) = (0, 1 + \mathrm{Tr}_{L/K}(\alpha)) \in \hat{G}(K).$$

Here $\mathrm{Tr}_{L/K}: H^n(L; \mathbb{Z}/2) \to H^n(K; \mathbb{Z}/2)$ is the (additive) transfer.

Proof. By §1.42, in which $t^*l(a_i) = l(a_i)$,

$$\hat{N}_{L/K}(0, 1 + \alpha) = \left(0, 1 + \mathrm{Tr}_{L/K}(\alpha) + \sum_{i<j} \mathrm{Tr}_{L/K}(z_i t^*(z_j)(l(a_i)l(a_j)))\right.$$

$$\left. + \sum_i \tilde{N}(z_i l(a_i))(1 + d_{L/K})^{-2}\right).$$

In the second coordinate the third term vanishes, since $l(a_i)l(a_j) = 0$. However,

$$\tilde{N}(z_i l(a_i)) = \tilde{N}(z_i)(l(a_i)^2 + l(a_i)d_{L/K}) \qquad \text{by §1.46,}$$

$$= \tilde{N}(z_i)l(a_i)d_{L/K}$$

by hypothesis, and this term is zero by §1.57.   □

(2.11) Lemma

Suppose that $N \supset L \supset K$ is a finite Galois extension and that there exists

$$\rho \colon G(N/L) \to O_2(K)$$

a standard dihedral representation, which factors through a standard embedding,
$D_{2^{n+1}(2s+1)} \to O_2(K)$. *Then*

$$0 = l(2)l(1 + \alpha_n) \in H^2(K; \mathbb{Z}/2),$$

where α_n is as in Chapter 3, §1.20.

Proof. When $n = 0, 1$, this is trivial. When $n \geq 2$, we will show that either $\sqrt{2} \in K^*$ or that $\sqrt{(-2)} \in K^*$. In the first case $l(2) = 0$, and in the second

$$l(2)l(1 + \alpha_n) = l(-1)l(1 + \alpha_n) = 0,$$

as in §2.8 (ii).   □

To see this, recall from Chapter 3, §1.20 that $\beta_{j-1} = 2\alpha_j\beta_j$ and $\alpha_{j-1} = \alpha_j^2 - \beta_j^2$. Since $\alpha_1 = -1$ and $\beta_1 = 0$, we may divide into the following cases.

Case i: $\alpha_2 = 0$
In this case $\beta_2 = \pm 1 = 2\alpha_3\beta_3$ and $0 = \alpha_3^2 - \beta_3^2$. Therefore

$$1 = \alpha_3^2 + \beta_3^2 = 2\alpha_3^2.$$

Case ii: $\beta_2 = 0$
Therefore $\alpha_2^2 = -1$ and $\alpha_2 = \alpha_3^2 - \beta_3^2$, with $\alpha_3\beta_3 = 0$. If $\beta_3 = 0$, then

$$\left[\frac{1 + \alpha_2}{\alpha_3}\right]^2 = \frac{1 + 2\alpha_2 + \alpha_2^2}{\alpha_3^2} = \frac{2\alpha_2}{\alpha_2} = 2,$$

but if $\alpha_3 = 0$,

$$\left[\frac{1 + \alpha_2}{\beta_3}\right]^2 = -\frac{2\alpha_2}{\alpha_2} = -2, \qquad \text{as required.}   □$$

(2.12) Corollary

Let $L = K_n \supset K_{n-1} \supset \cdots \supset K_1 \supset K_0 = K$ be a chain of quadratic extensions with char$(K) \neq 2$. Let N/K be a finite Galois extension containing L, and let

$$\rho: G(N/L) \to O_2(K)$$

be a standard dihedral representation, as in §2.8 (ii).
Then, in $\hat{G}(K)$,

$$\hat{N}_{K_1/K} \hat{N}_{K_2/K_1} \cdots \hat{N}_{L/K_{n-1}}(0, 1 + Sp[\rho]) = (0, 1 + Sp[\mathrm{Ind}_{L/K}(\rho)]).$$

Proof. By Chapter 3, §1.20, $Sp[\rho] = zl(2) + wl(1 + \alpha_n)$ if ρ factors through $D_{2^{n+1}(2s+1)}$. Here we have used the isomorphism of Chapter 1, §4.6 (i)

$$H^*(D_{2^{n+1}}; \mathbb{Z}/2) \cong H^*(D_{2^{n+1}(2s+1)}; \mathbb{Z}/2).$$

If $\mathrm{Tr}_{L/K_j} : H^*(L; \mathbb{Z}/2) \to H^*(K_j; \mathbb{Z}/2)$ is the (additive) cohomology transfer, then

$$\mathrm{Tr}_{L/K_j}(Sp[\rho] = (\mathrm{Tr}_{L/K_j}(z))l(2) + (\mathrm{Tr}_{L/K_j}(w))l(1 + \alpha_n)$$

$$= Sp[\mathrm{Ind}_{L/K_j}(\rho)]$$

by Chapter 3, §1.29.
Suppose, by induction on j, that

$$\hat{N}_{K_{j+1}/K_j} \cdots \hat{N}_{L/K_{n-1}}(0, 1 + Sp[\rho]) = (0, 1 + Sp[\mathrm{Ind}_{L/K_j}(\rho)]).$$

By §§2.8 (ii) and 2.11, we may set $\alpha = Sp[\mathrm{Ind}_{L/K_j}(\rho)]$ in §2.10 to obtain

$$\hat{N}_{K_j/K_{j-1}} \cdots \hat{N}_{L/K_{n-1}}(0, 1 + Sp[\rho]) = \hat{N}_{K_j/K_{j-1}}(0, 1 + Sp[\mathrm{Ind}_{L/K_j}(\rho)])$$

$$= (0, 1 + \mathrm{Tr}_{K_j/K_{j-1}}(Sp[\mathrm{Ind}_{L/K_j}(\rho)]))$$

$$= (0, 1 + Sp[\mathrm{Ind}_{K_j/K_{j-1}}(\mathrm{Ind}_{L/K_j}(\rho))])$$

$$= (0, 1 + Sp[\mathrm{Ind}_{L/K_{j-1}}(\rho)]).$$

The result now follows by induction on j. $\square$

(2.13)   Suppose that L/K is a finite separable extension with char$(K) \neq 2$. If $N \supset L \supset K$ is a finite Galois extension, let S denote the Sylow 2-subgroup of $G(N/K)$. Set $F = N^S$, then

(2.14) $\begin{cases} F \otimes_K L \cong \bigoplus_{i=1}^{t} E_i & \text{and} \\ [F:K] & \text{is odd.} \end{cases}$

For each $i = 1, \ldots, t$, there is a chain of quadratic extensions

(2.15) $$E_i = E_{i,0} \supset E_{i,1} \supset \cdots \supset E_{i,n_i} = F.$$

In addition the double-coset formula for representations yields

(2.16) $$\mathrm{res}_{F/K}(\mathrm{Ind}_{L/K}(\rho)) = \sum_{i=1}^{t} \mathrm{Ind}_{E_i/F}(\mathrm{res}_{E_i/L}(\rho))$$

if $\rho: G(N/L) \to O_n(K)$ is an orthogonal Galois representation. Here "res" is induced by restriction. For example, $\mathrm{res}_{F/K}$ is induced by $G(N/F) \to G(N/K)$.

(2.17) Theorem

Let L/K be a separable extension with $char(K) \neq 2$. Suppose that

$$\rho: \Omega_L \to O_2(K)$$

is a continuous, standard dihedral representation, as in §2.8 (ii). Then, in $G(K)$.

$$HW(\mathrm{Tr}_{L/K}^S(\rho)) = SW[\mathrm{Ind}_{L/K}(\rho)](1 + Sp[\mathrm{Ind}_{L/K}(\rho)]),$$

where $\mathrm{Tr}_{L/K}^S(\rho)$ is the Scharlau transfer of the bilinear form represented by ρ in $H^1(L; O_n(\bar{K}))$.

Proof. Suppose that N/K is a finite Galois representation through which ρ factors. Choosing F/K as in §2.13, we note that (2.14) implies that

$$\mathrm{res}_{F/K}^*: G(K) \to G(F)$$

is injective. Hence it suffices to evaluate

$$\mathrm{res}_{F/K}^*(HW(\mathrm{Tr}_{L/K}^S(\rho))) = \prod_{i=1}^{t} HW(\mathrm{Tr}_{E_i/F}^S(\mathrm{res}_{E_i/L}^*(\rho))), \qquad \text{by (2.16)}.$$

We may assume that $N \supset E_i$ for each i. Set ρ_i equal to ρ restricted to $G(N/E_i)$. By §2.7,

$$HW(\rho_i) = SW[\rho_i](1 + Sp[\rho_i]).$$

Now write $\hat{N}$ temporarily for the composition of the $\hat{N}_{E_{i,j}/E_{i,j+1}}$. Therefore, applying $\hat{N}$ to this equation, in $\hat{G}(F)$ we obtain

$$(2^{n_i}, HW(\mathrm{Tr}^S_{E_i/F}(\rho))) = \hat{N}(2, HW(\rho_i)), \qquad \text{by §2.9 and induction,}$$

$$= \hat{N}((2, SW[\rho_i])(0, 1 + Sp[\rho_i]))$$

$$= \hat{N}(2, SW[\rho_i])\hat{N}(0, 1 + Sp[\rho_i]), \qquad \text{by §1.25,}$$

$$= (2^{n_i}, SW[\mathrm{Ind}_{E_i/F}(\rho_i)](0, 1 + Sp[\mathrm{Ind}_{E_i/F}(\rho_i)]))$$

by §§2.5 and 2.12 and by induction. However,

$$\prod_{i=1}^{t} SW[\mathrm{Ind}_{E_i/F}(\rho_i)] = SW\left[\sum_{i=1}^{t} \mathrm{Ind}_{E_i/F}(\rho_i)\right]$$

$$= SW[\mathrm{res}_{F/K}(\mathrm{Ind}_{L/K}(\rho))], \qquad \text{by (2.16)}$$

$$= \mathrm{res}^*_{F/K}(SW[\mathrm{Ind}_{L/K}(\rho)]).$$

Also

$$\prod_{i=1}^{t}(1 + Sp[\mathrm{Ind}_{E_i/F}(\rho_i)]) = 1 + \sum_{i=1}^{t} Sp[\mathrm{Ind}_{E_i/F}(\rho_i)] + z$$

$$= 1 + Sp\left[\sum_{i=1}^{t} \mathrm{Ind}_{E_i/F}(\rho_i)\right] + z$$

$$= 1 + Sp[\mathrm{res}_{F/K}(\mathrm{Ind}_{L/K}(\rho))] + z$$

$$= \mathrm{res}^*_{F/K}(1 + Sp[\mathrm{Ind}_{L/K}(\rho)]) + z,$$

where z is a sum of products of the form $Sp[\mathrm{Ind}_{E_i/F}(\rho_i)]Sp[\mathrm{Ind}_{E_j/F}(\rho_j)]$. If we can show that $z = 0$, our proof will be complete. However, as observed in the proof of §2.12, each $Sp[\mathrm{Ind}_{E_i/F}(\rho_i)] = z_i l(2) + w_i l(1 + \alpha_n)$ so that $z = 0$ because, by §§2.8 (ii) and 2.11,

$$l(2)^2 = l(1 + \alpha_n)^2 = l(2)l(1 + \alpha_n) = 0. \qquad \square$$

(2.18) Corollary [K]

Let L/K be a finite separable extension with $\mathrm{char}\,(K) \neq 2$. If $a \in L^*$ and $\langle a \rangle: L \times L \to L$ is given by $\langle a \rangle(x, y) = axy$, then in $G(K)$,

$$HW(\mathrm{Tr}^S_{L/K}\langle a \rangle) = SW[\mathrm{Ind}_{L/K}l(a)](1 + l(2)d_{L/K}),$$

where $d_{L/K} \in H^1(K; \mathbb{Z}/2)$ is the discriminant of L/K.

Proof. Let N, F, E_i be as in §2.13, and $E_{i,s}$ as in (2.15). Let χ_i denote the restriction of $l(a)$ to $G(N/E_i)$. By §2.8 (i), $\mathrm{Ind}_{E_i/E_{i,1}}(\chi_i): G(N/E_{i,1}) \to O_2(K)$ satisfies §2.7, although it is not special dihedral, and has spinor class $l(2)d_{E_i/E_{i,1}}$ by Chapter 3,

§1.29. Hence, by the proof of §2.17,

$$HW(\mathrm{Tr}^S_{E_i/F}(\chi_i)) = SW[\mathrm{Ind}_{E_i/F}(\chi_i)](1 + l(2)d_{E_i/F}).$$

The proof is now concluded by calculating $\mathrm{res}^*_{F/K}(HW(\mathrm{Tr}^S_{L/K}\langle a\rangle))$ in the manner of §2.17. $\qquad\qquad\square$

Chapter Five

Stable Homotopy and Induced Representations

For if þou redes hit by ryȝt & hit to resoun brynges,
Fyrst telle me þe tyxte of þe tede lettres,
& sy en þe mater of þe mode mene me þer-after,
& I schal halde þe þe hest þat I þe hyȝt haue:
Apyke þe in porpre cloþe, palle alþer-fynest,
& þe byȝe of bryȝt golde abowte þyn nekke;
& þe þryd þryuenest þat prynges me after,
Þou schal be baroun vpon benche, bede I þe no lasse.
—*ANONYMOUS "Cleanness" 1633–1640 (14th century)**

*The free translation, which I rather like, is the following:

The clerk who can tell the king,
Explaining in speech what is spelled by those letters,
And satisfy my spirit with the sense expressed,
Making its meaning manifest to me,
Shall be gowned most gaily in garments of purple,
And a collar of clear gold shall be clasped at his
throat.
["The Owl and the Nightingale/Cleanness/St. Erkenwald"—Penguin Classics (1971).

The literal translation (Belshazzar to Daniel) is, however, this:

For if you read it correctly and make sense of it
First tell me the text of the fated letters,
And then explain the substance of the inner meaning to
me afterwards
And I will keep the promise that I have made to you:
To adorn you in purple cloth, of the very finest
material,
And a necklace of bright gold,
And you shall be a baron upon the bench,
The third most noble man in my following.

("Cleanness" and the delightful tale "Sir Gawain and the Green Knight" are thought to be the work of the same, anonymous author.)

Paradoxically, this chapter is both a colossal nonsequitur and the fulcrum for the rest of the book, as I will explain presently. In this chapter we introduce the groups of monomial homomorphisms, $R_+(G, \pi)$. We will be interested in them in the case where G is a finite group and π is a compact Lie group. We will show how to describe the completion of $R_+(G, \pi)$ with respect to the Burnside ring topology. This completion is isomorphic to the group of stable homotopy classes of maps from the classifying space of G to that of π. When π is finite, this identification is known and is a corollary of the proof of the Segal conjecture, as will be explained shortly.

Actually, although this identification of the completion of $R_+(G, \pi)$ is generally valid [Sn-Z; M-Sn-Z], for our purposes—namely, the finding of a certain formula whose identity is revealed in the chapter—the case of a torus would suffice. However, the proof of that reduction would get us deeper into technicalities. To avoid that I have given here a selection of examples for π—namely, T^n (the n-dimensional torus), $\Sigma_2 \int S^1$, and $O_2(\mathbb{R})$—because I am mindful of the demands that this chapter makes upon nontopologists.

With these readers in mind I have gone exhaustively through the two-dimensional examples of the double-coset formula in stable homotopy theory, which is the subject of §3. I remind the reader to compare this double-coset formula with the one we encountered in group cohomology in Chapter 1, (2.43).

Now let me explain precisely what is the purpose of this chapter in the context of what we will do later in the book.

For suitable choices of π, the group, $R_+(G, \pi)$, may be thought of as the group of monomial representations (ones induced from a character of a subgroup) together with the knowledge of its character of origin, up to inner automorphism. By the time that such a monomial representation is considered simply as an element of the representation ring of G, one has irrevocably lost control of the manner in which it was induced. To redress this very awkward circumstance, we would like to find a canonical form for a representation in terms of monomial representations. This canonical form will be called Explicit Brauer Induction and amounts to a map from representations to the group of monomial homomorphisms. Eventually, we will derive this in several forms.

However, at this moment we are commencing a chapter about stable homotopy! How so? The point is that it is the first step in a process which will discover the canonical form for us. We are after a function, in the unitary case, which gives us an element of $R_+(G, \Sigma_n \int S^1)$ in return for an n-dimensional unitary representation of G. On the other hand, if we take a representation, $v: G \to U(n)$, and from the induced map between classifying spaces

$$Bv: BG \to BU(n),$$

then we receive a stable homotopy class of maps in the group

$$\{BG_+, BU(n)_+\}$$

of stable maps between these classifying spaces. If we then compose this element with the stable map, called the transfer,

$$\tau: BU(n)_+ \to B\left(\Sigma_n \int S^1\right)_+,$$

we have constructed an element of the following groups, which are identified in the manner which I described earlier:

$$\tau_G(v) \in \left\{ BG_+, B\left(\Sigma_n \int S^1\right)_+ \right\} \cong R_+\left(G, \Sigma_n \int S^1\right)^{\widehat{}}_{IA(G)}.$$

As we shall see in the next chapter, the double-coset formula in stable homotopy, together with some homotopy theory, enables us to evaluate $\tau_G(v)$. In addition, since the transfer is a split injection, one excepts that $\tau_G(v)$ will map again to v in the representation ring (completed), thereby rendering up the canonical form which we are seeking.

Thus the purpose of this chapter is to develop a topological method to find a formula for

$$\tau_G(v) \in R_+\left(G, \Sigma_n \int S^1\right)$$

in the uncompleted group. Once the formula is found, after completion, by this method, we will establish all its properties, in the uncompleted group, by more elementary topological techniques and eventually use them to give a new construction of local root numbers of Galois representations.

1. THE GROTHENDIECK GROUP OF MONOMIAL HOMOMORPHISMS

Let G, π be finite groups. In [LMM], as a corollary of the Segal conjecture, subsequently proved by Carlsson [Ca], the authors show how to compute the stable homotopy classes of maps $\Sigma^n(BG_+) \to B\pi_+$. Here X_+ is X disjoint union with a base point. The description of this stable homotopy group, $\{\Sigma^n(BG_+), B\pi_+\}$, is in terms of the Grothendieck group of finite sets upon which $G \times \pi$ acts, π acting *freely*. This Grothendieck group will be denoted by $A(G, \pi)$. $A(G, \pi)$ is the free abelian group on the isomorphism classes of irreducible, π-free $(G \times \pi)$-sets or, equivalently, the free group on the isomorphism classes of π-free, $(G \times \pi)$-sets modulo the relation that $[X \sqcup Y] = [X] + [Y]$, where $X \sqcup Y$ is the disjoint union of X and Y. When $\pi = 1$, $A(G, 1) = A(G)$ is the Burnside ring of G. If X is a $(G \times \pi)$-set and W is a G-set, then $W \times X$, with the diagonal action, is a $(G \times \pi)$-set on which π acts freely. Hence $A(G, \pi)$ is an $A(G)$-module. More generally, the cartesian product makes $A(G, \pi)$ into a ring.

The irreducible π-free, $(G \times \pi)$-sets are all of the following form: Let $\rho: H \to \pi$ be a *subhomomorphism* (i.e., a homomorphism defined on $H \leq G$). Form the $(G \times \pi)$-set,

$$(1.1) \qquad G \times_\rho \pi = (G \times \pi)/((g, z) \approx (gh, z\rho(h)), g \in G, z \in \pi, h \in H),$$

with the *left* translation action. This is π-free, since if $z'(g, z) = (g, z'z) \approx (g, z)$, then $(g, z'z) = (gh, z\rho(h))$ so that $h = 1$. Conversely, any irreducible $(G \times \pi)$-set is of the form $(G \times \pi)/\wedge$ for some $\wedge \subset G \times \pi$. However, if $(h, z) \wedge = (h, z') \wedge$, then $z = z'$, for if not, $(1, z'(z^{-1}))(h, z) \wedge = (h, z') \wedge$. Therefore, if $H = \{h \in G | (h, z) \in \wedge$ for some $z \in \pi\}$, then $\rho: H \to \pi$, given by $\rho(h) = z$ (for $(h, z) \in \wedge$) is well-defined and is clearly a homomorphism, since (h, z), $(h', z') \in \wedge$ implies $(hh', zz') \in \wedge$.

We will need a second description of $A(G, \pi)$ in terms of *monomial homomorphisms*. Let Σ_n denote the symmetric group which permutes $\{1, \ldots, n\}$. The wreath product, $\Sigma_n \int \pi$, is given by the semidirect product of π^n with Σ_n. Hence $\pi^n \lhd \Sigma_n \int \pi$ and $\Sigma_n \subset \Sigma_n \int \pi$ acts on π^n by

$$(1.2) \qquad \begin{cases} \sigma(\pi_1, \pi_2, \ldots, \pi_n)\sigma^{-1} = (\pi_{\sigma(1)}, \pi_{\sigma(2)}, \ldots, \pi_{\sigma(n)}) \\ (\sigma \in \Sigma_n, \pi_i \in \pi). \end{cases}$$

We have an extension of groups $\pi^n \overset{i}{>\!\!-\!\!-} \Sigma_n \int \pi \overset{j}{-\!\!\twoheadrightarrow} \Sigma_n$.

By a monomial homomorphism we will mean a homomorphism $\lambda: G \to \Sigma_n \int \pi$. To such a λ, such that G acts transitively upon the set $\{1, 2, \ldots, n\}$ via $j\lambda: G \to \Sigma_n$, we can assign the subhomomorphism $\rho: H \to \pi$ given by

$$(1.3) \qquad \begin{cases} H = \{g \in G | j\lambda(g) \in \Sigma_n \text{ fixes } 1\} \text{ and} \\ \rho(g) \text{ satisfies } \lambda(g) = \sigma(\rho(g), \pi_2, \pi_3, \ldots, \pi_n)) \text{ for some } \sigma \in \Sigma_n. \end{cases}$$

Since, if $g, g' \in H$, then

$$\lambda(gg') = \sigma(\rho(gg'), \ldots)$$

$$= \sigma'(\rho(g), \ldots)\sigma''(\rho(g'), \ldots);$$

we see that $\rho(gg') = \rho(g)\rho(g')$, since σ, σ', and σ'' all fix $1 \in \{1, \ldots, n\}$.

Conversely, if $\rho: H \to \pi$ is a subhomomorphism, let $n = [G: H]$; let x_i ($1 \leq i \leq n$) be a set of left coset representatives for G/H. Given $g \in G$, define $\sigma(g) \in \Sigma_n$ by

$$(1.4) \qquad \begin{cases} gx_i = x_{\sigma(g)(i)} h(i, g) \\ (h(i, g) \in H). \end{cases}$$

Therefore the equation

$$g_1 g_2 x_i = g_1 x_{\sigma(g_2)(i)} h(i, g_2)$$

$$= x_{\sigma(g_1)\sigma(g_2)(i)} h(\sigma(g_2)(i), g_1) h(i, g_2)$$

implies that for $g_1, g_2 \in G$,

$$(1.5) \qquad \begin{cases} \sigma(g_1 g_2) = \sigma(g_1)\sigma(g_2) \\ h(i, g_1 g_2) = h(\sigma(g_2)(i), g_1)h(i, g_2). \end{cases}$$

To $g \in G$, assign $\lambda(g) \in \Sigma_n \int \pi$ by

$$(1.6) \qquad \lambda(g) = \sigma(g)(\rho(h(1, g)), \ldots, \rho(h(n, g))).$$

(1.7) Lemma

In (1.6) $\lambda: G \to \Sigma_n \int \pi$ is a homomorphism such that G, via $j\lambda: G \to \Sigma_n$, acts transitively upon $\{1, 2, \ldots, n\}$.

Proof. We calculate, using (1.5),

$$\begin{aligned} \lambda(g_1)\lambda(g_2) &= \sigma(g_1)(\rho(h(1, g_1)), \ldots)\sigma(g_2)(\rho(h(1, g_2)), \ldots) \\ &= \sigma(g_1)\sigma(g_2)\sigma(g_2)^{-1}(\rho(h(1, g_1)), \ldots)\sigma(g_2)\cdots \\ &= \sigma(g_1 g_2)(\rho(h(\sigma(g_2)(1), g_2))\rho(h(1, g_2)), \ldots) \\ &= \sigma(g_1 g_2)(\rho(h(1, g_1 g_2)), \ldots), \qquad \text{as required} \qquad \square \end{aligned}$$

The association $(\rho: H \to \pi) \to (\lambda: G \to \Sigma_n \int \pi)$ clearly gives a bijection between subhomomorphisms defined on H and a subset of the monomial homomorphisms, λ, such that $n = [G:H]$, G acts transitively as in Lemma 1.7, and $\text{stab}(j\lambda)(1) = H$. Furthermore, if we choose different coset representatives $\{x_i\}$, this has the following effect. Suppose we change x_1 to $x_1 z$ ($z \in H$); then

$$gx_1 z = x_{\sigma(g)(1)}h(1, g)z$$

so that if $\tau(g) = \sigma(g)^{-1}$,

$$gx_{\tau(g)(1)} = x_1 h(\tau(g)(1), g) = (x, z)z^{-1}h(\tau(g)(1), g).$$

Suppose that $\tau(g)(1) = 2$, then, in $\lambda(g)$, $\sigma(g)(\rho(h(1, g)), \rho(h(2, g)), \ldots)$ is altered to

$$\sigma(g)(\rho(h(1, g)z), \rho(z^{-1}h(2, g)), \ldots).$$

This amounts to conjugation $w = w^{-1}$ by $w = (\rho(z)^{-1}, 1, 1, \ldots)$ in $\Sigma_n \int \pi$, for

$$\begin{aligned} (\rho(z)^{-1}, 1, \ldots)\sigma(g)(\rho(h(1, g)), \ldots)(\rho(z), 1, \ldots) \\ = \sigma(g)(1, \rho(z)^{-1}, 1, \ldots)(\rho(h(1, g)), \rho(h(2, g)), \ldots)(\rho(z), 1, \ldots) \\ = \sigma(g)(\rho(h(1, g)z), \rho(z^{-1}h(2, g)), \ldots). \end{aligned}$$

Therefore we have shown that *changing the choice of coset representatives for* G/H *changes* $(\rho \mapsto \lambda)$ *by an inner automorphism of* $\Sigma_n \int \pi$.

Similarly, when one considers a π-free, irreducible $(G \times \pi)$-set up to isomorphism, then the subgroup H is defined only up to conjugacy in G and $\rho: H \to \pi$ only up to inner automorphisms of π and up to conjugation on H by its normalizer, $N_G(H)$. These changes also alter $\lambda: G \to \Sigma_n \int \pi$ by conjugation.

(1.8) Remark

If $\lambda_i: G \to \Sigma_{n_i} \int \pi$ $(i = 1, 2)$ *are monomial homomorphisms, then* $\lambda_1 + \lambda_2$ *denotes the monomial homomorphism associated to the maps* $\alpha: \Sigma_n \times \Sigma_m \to \Sigma_{n+m}$, $\beta: \pi^n \times \pi^m \to \pi^{n+m}$ *given by*

$$\alpha(\delta, \tau)(j) = \begin{cases} \delta(j) & \text{if } 1 \leq j \leq n \\ \tau(j-n) + n & \text{if } n+1 \leq j \leq n+m, \end{cases}$$

$\beta((h_1, \ldots, h_n), (h'_1, \ldots, h'_m)) = (h_1, h_2, \ldots, h_n, h'_1, h'_2, \ldots, h'_m)$.

If $\lambda_2: G \to \Sigma_{n_2}$ *is a homomorphism, we denote* $\lambda_1 \lambda_2: G \to \Sigma_{n_1 n_2} \int \pi$ *by the tensor product,* $\lambda_1 \otimes \lambda_2$. *That is, if* $\lambda_1(g) = \sigma(h_1, \ldots)$ *and* $\lambda_2(g) = \tau$,

$$\lambda_1 \lambda_2(g) = (\sigma \otimes \tau)(h_1, h_2, \ldots, h_{n_1}, h_1, h_2, \ldots, h_1, h_2, \ldots),$$

where $\sigma \otimes \tau$ *is the permutation induced by* $\sigma \times \tau$ *on the product* $\{1, \ldots, n_1\} \times \{1, \ldots, n_2\}$ *(endowed with the lexicographical order* $(i, j) < (i', j')$ *if and only if* $i < i'$ *or* $i = i', j < j'$).

Warning: *Guido Mislin has pointed out to me that, in general, not all monomial homomorphisms*

$$G \to \Sigma_n \int \pi$$

may be sums of (irreducible) monomial homomorphisms of the type constructed in (1.6) *and* (1.7). *I will have a little more to say on this point when we come to the* ρ-*construction in Chapter* 6, §3.28 *(see Chapter* 6, §3.41).

(1.9) Definition

Let $R_+(G, \pi)$ *be the Grothendieck group of isomorphism classes (up to inner automorphism in* $\Sigma_n \int \pi$) *of monomial homomorphisms, which are sums of ones of the type constructed in* (1.6) *and* (1.7), $\{\lambda: G \to \Sigma_n \int \pi, n \geq 1\}$, *with addition defined as in* §1.8.

We have shown in the preceding discussion the following:

(1.10) Theorem

As groups, $A(G; \pi) \cong R_+(G, \pi)$. In addition the $A(G) \cong R_+(G, 1)$-module structure on $A(G, \pi)$ corresponds to the product, $R_+(G, \pi) \times R_+(G, 1) \to R_+(G, \pi)$, defined in §1.8.

Theorem 1.10 will afford us two ways of looking at the groups $R_+(G, \pi)$.

(1.11) CONNECTIONS WITH STABLE HOMOTOPY THEORY

If G is a finite group and $\rho: H \to \pi$ is a subhomomorphism, we may form a stable map $\Phi(\rho): BG_+ \to B\pi_+$ by the composition

$$(1.12) \qquad \Phi(\rho): BG_+ \xrightarrow{t} BH_+ \xrightarrow{B\rho_+} B\pi_+,$$

where t is the transfer [B-G; K-P] associated to the covering $BH \to BG$.

In terms of monomial homomorphisms this construction becomes the following, as can be seen from the construction of t in [K-P]: First, form $B\lambda: BG \to B\Sigma_n \int \pi$, for λ a monomial homomorphism, and compose with the inclusion $\Sigma_n \int \pi \subset \Sigma_\infty \int \pi$ to obtain $B\lambda': BG \to B\Sigma_\infty \int \pi$. As explained in [H-Se, §3, p. 23], we may apply Quillen's plus construction to $B\Sigma_\infty \int \pi$ to obtain $(B\Sigma_\infty \int \pi)^+$, which, by the Barratt-Priddy-Quillen theorem, is homotopy equivalent to

$$(1.13) \qquad Q(B\pi_+) = \varinjlim_n \Omega^n \Sigma^n (B\pi_+).$$

Composing $B\lambda'$ with the canonical map $B\Sigma_\infty \int \pi \to (B\Sigma_\infty \int \pi)^+ \simeq Q(B\pi_+)$, we obtain a base-pointed map

$$(1.14) \qquad \Phi(\lambda): BG_+ \to Q(B\pi_+),$$

whose adjoint is the stable map of (1.12), if λ corresponds in §1.10 with the subhomomorphism, ρ.

It is part of the main result of [LMM] that *when π and G are finite*,

$$(1.15) \qquad \begin{cases} (\rho: H \to \pi) \mapsto \Phi(\lambda) & \text{induces an isomorphism} \\ \Phi: R_+(G, \pi)^\wedge_{IA(G)} \xrightarrow{\cong} \{BG_+, B\pi_+\}. \end{cases}$$

Here $(-)^\wedge_{IA(G)}$ denotes completion of the $A(G)$-module in the $IA(G)$-adic topology, where $IA(G) = \ker(\text{card}: A(G) \to \mathbb{Z})$, the kernel of the function which sends a π-free, $(G \times \pi)$-set to its cardinality.

Equivalently, *when π and G are finite,*

(1.16)
$$\begin{cases} (\lambda: G \to \Sigma_n \int \pi) \mapsto \Phi(\lambda) & \text{induces an isomorphism} \\ \Phi: R_+(G, \pi)^{\wedge}_{IR_+(G)} \xrightarrow{\cong} \{BG_+, B\pi_+\}. \end{cases}$$

We are going to work toward proving the following result:

(1.17) Theorem

Let $T^n = S^1 \times S^1 \times \cdots \times S^1$ be the n-dimensional torus, and let G be a finite group. There is an isomorphism

$$\Phi_G: R_+(G, T^n)^{\wedge}_{IA(G)} \xrightarrow{\cong} \{BG_+, BT^n_+\}$$

induced by $(\lambda \mapsto \Phi(\lambda))$ as in (1.12)–(1.16).

(1.18) The first reduction is standard in dealing with natural transformations of this type. Namely, one applies Dress's theory of Green-Mackey functors, (induction theory) to determine a sufficient set of finite groups, G, on which to test §1.17. This was done, in the context of (1.15), in [MM], and that argument applies to §1.17 to yield the following proposition:

(1.19) Proposition

To prove Theorem 1.17, it suffices to verify the cases where p is a prime and G is a finite p-group. In addition, in this case, the $IA(G)$-adic topology is the p-adic topology.

Proof. As mentioned earlier, I refer the reader to [MM] for a proof. To see that $(-)^{\wedge}_{IA(G)} \cong (-)^{\wedge}_p$, just observe that if $|G:H| = n$. then $([G/H] - n)^2 = (G/H - n)|G:H|$ in $A(G)$, provided that $H \lhd G$. If H is not normal in G, then

$$(G/H - n)^2 = \sum_{\substack{H_i > H \\ \neq}} (G/H_i - n_i) + ps(G/H - n),$$

where s is an integer and $n_i = [G:H_i]$. By induction on H we see that some power of $(G/H - n)$ is p-divisible. Therefore, since $IA(G)$ has a finite set of generators of this form, we see that $IA(G)^k \subset pIA(G)$ for some $k \geq 1$. $\qquad\square$

(1.20) THE TRANSFER

In [B-Sch., §4] we find a description of Boardman's transfer construction, a type of fibrewise S-duality construction, which I will now recall.

Suppose $H \subset G$ are compact Lie groups. Let G act freely and smoothly on a closed manifold, M. Denote the adjoint representations of H and G by $\mathrm{Ad}(H)$ and $\mathrm{Ad}(G)$, respectively. Under these circumstances the transfer is an S-map

$$(1.21) \qquad t: T(M \times_G \mathrm{Ad}(G)) \to T(M \times_H \mathrm{Ad}(H)).$$

Here TV denotes the Thom space, $D(V)/S(V)$ of a vector bundle, $V \to M$.

Let $p: M/H \to M/G$ be the projection, and choose a fibrewise embedding (for some $s \geq 0$)

$$(1.22) \qquad \begin{cases} \hat{p}: M/H \rightarrowtail M/G\, \mathbb{R}^s \\ \text{such that } \hat{p} \simeq p, \text{ fibrewise.} \end{cases}$$

Let ω denote the normal bundle of $\hat{p}$ so that, if τ_X denotes the tangent bundle of X,

$$(1.23) \qquad (\tau_{M/H}) \oplus \omega \cong p^*(\tau_{M/G}) \oplus \underline{\mathbb{R}}^s.$$

Here $\underline{\mathbb{R}}^s = (M/G) \times \mathbb{R}^s$, the trivial bundle of dimension s. Let $\xi_H = M \times_H \mathrm{Ad}(H) \to M/H$, and define ξ_G similarly. Add $\xi_H \oplus p^*(\xi_G)$ to each side of (1.23) to obtain

$$(1.24) \qquad (\tau_{M/H}) \oplus \xi_H \oplus \omega \oplus p^*(\xi_G) \cong p^*((\tau_{M/G}) \oplus \xi_G \oplus \underline{\mathbb{R}}^s) \oplus \xi_H.$$

Now observe that

$$(1.25) \qquad \begin{cases} (\tau_{M/H}) \oplus \xi_H \cong (\tau_M)/H \quad \text{and} \\ p^*((\tau_{M/G})) \cong (\tau_M)/H \end{cases}$$

so that

$$(1.26) \qquad \begin{cases} ((\tau_M)_{/H}) \oplus p^*(\xi_G) \oplus \omega \cong p^*(((\tau_M)_{/G}) \oplus \underline{\mathbb{R}}^s) \oplus \xi_H \\ \qquad\qquad\qquad \cong ((\tau_M)_{/H}) \oplus \underline{\mathbb{R}}^s \oplus \xi_H. \end{cases}$$

Adding to each side of (1.26) a complementary bundle to $(\tau_M)/_H$, we obtain

$$(1.27) \qquad p^*(\xi_G) \oplus \omega \cong \underline{\mathbb{R}}^s \oplus \xi_H \qquad \text{for some } s \gg 0.$$

Now performing the fibrewise Pontrjagin-Thom construction on $\hat{p}$ gives a map

$$(1.28) \qquad \lambda: \Sigma^s((M/G)_+) = T(M/G \times \mathbb{R}^s) \to T(\omega).$$

Define a map, by the *fibrewise* smash product

$$(1.29) \qquad \lambda \wedge 1: \Sigma^s((M/G)_+) \wedge_{(M/G)} T(\xi_G) \to T(\omega) \wedge T(p^*\xi_G)$$

given by

$$(\lambda \wedge 1)(a \wedge b) = \lambda(a) \wedge [p(\lambda(a)), b].$$

The domain of $\lambda \wedge 1$ is $\Sigma^s T(\xi_G)$, and the range is, by (1.27), $\Sigma^s T(\xi_H)$.
  The map, $(\lambda \wedge 1)$, of (1.29) represents an S-map, t, of (1.21).

(1.30) Example

Let p be a prime, and let $\mathbb{Z}/p^k$ denote the cyclic subgroup of $S^1 = U(1)$, of order
p^k ($k \geq 1$). Set $H = (\mathbb{Z}/p^k)^n$ and $G = (S^1)^n = T^n$, the n-torus.
  In this case G is abelian so that $\text{Ad}(G)$ is the n-dimensional trivial
representation, $\xi_G \cong \mathbb{R}^n$, and $\text{Ad}(H) = O$. Set $M = (S^{2T+1})^n$, where S^1 acts on
$S^{2T+1} \subset \mathbb{C}^{T+1}$ by complex multiplication and

$$(z_1, \ldots, z_n)(x_1, \ldots, x_n) = (z_1 x_1, z_2 x_2, \ldots, z_n x_n) \qquad (z_i \in S^1, x_i \in S^{2T+1}).$$

  Accordingly, t takes the form

$$t_n : \Sigma^n((S^{2T+1}/S^1)^n_+) \to (S^{2T+1}/(\mathbb{Z}/p^k))^n_+.$$

Note that $(S^{2T+1}/S^1) = \mathbb{C}P^T$, complex projective space. Also $(\mathbb{C}P^T)^n_+ \cong
\wedge^n_1((\mathbb{C}P^T)_+)$, and $(S^{2T+1}/(\mathbb{Z}/p^k))^n_+ \cong \wedge^n_1((S^{2T+1}/(\mathbb{Z}/p^k))_+)$ by means of which we
may make the identification

$$(1.31) \qquad\qquad t_n = t_1 \wedge t_1 \wedge \cdots \wedge t_1 \qquad (n \text{ times}).$$

(1.31) follows from well-known properties of the transfer (cf., [B-G]), for
example).
  From (1.31) and the Künneth isomorphism, $H_*(X \times Y; \mathbb{Z}/p) \cong H_*(X; \mathbb{Z}/p) \otimes
H_*(Y; \mathbb{Z}/p)$; to calculate $(t_n)_*$ in mod p homology, it suffices to compute $(t_1)_*$.

(1.32) Lemma

$$(t_1)_* : \tilde{H}_j(\Sigma \mathbb{C}P^T; \mathbb{Z}/p) \to \tilde{H}_j((S^{2T+1}/(\mathbb{Z}/p^k)); \mathbb{Z}/p)$$

is an isomorphism when j is odd (and, of course, zero when j is even).

Proof. Let H_* denote $H_*(-; \mathbb{Z}/p)$. Since $\mathbb{C}P^T$ and $(S^{2T+1})/(\mathbb{Z}/p^n)$ are orientable
manifolds for $H_*(-; \mathbb{Z}/p)$, then the properties of Poincaré duality with respect
to S-duality yield the following commutative diagram:

$$(1.33)$$

$$
\begin{array}{ccc}
\tilde{H}_j(\Sigma \mathbb{C}P^T) & \xrightarrow{\ (t_1)_*\ } & \tilde{H}_j(S^{2T+1}/(\mathbb{Z}/p^k)) \\
\downarrow{\scriptstyle \cong} & & \\
\tilde{H}_{j-1}(\mathbb{C}P^T) & & \Big\downarrow{\scriptstyle \cong} \\
\downarrow{\scriptstyle \cong} & & \\
\tilde{H}^{2T-j+1}(\mathbb{C}P^T) & \xrightarrow{\ p^*\ } & H^{2T+1-j}(S^{2T+1}/(\mathbb{Z}/p^k)).
\end{array}
$$

The behaviour of the natural map, p^*, is well-known. It is an isomorphism in all even degrees (> 0), since it is nonzero on the two-dimensional generator, x, and $H^*(\mathbb{C}P^T; \mathbb{Z}/p) \cong \mathbb{Z}/p[x]/(x^{T+1})$.

(1.34) By transitivity of the transfer, we have homotopy commutative diagrams (for $1 \leq T \leq \infty$):

$$\Sigma(\mathbb{C}P_+^T) \xrightarrow{\ t_1\ } (S^{2T+1})/(\mathbb{Z}/p^k)$$

$$\xrightarrow{t_1} \quad \downarrow$$

$$(S^{2T+1}/(\mathbb{Z}/p^{k-1})),$$

in which t coincides with the Becker-Gottlieb or Kahn-Priddy transfer for finite coverings (see [B-G; B-Sch., App.]).

Hence on homology we obtain a map of (graded) pro-groups,

(1.35) $\{(t_{1_*}): \tilde{H}_*(\Sigma \mathbb{C}P_+^\infty; \mathbb{Z}/p) \to \tilde{H}_*((B\mathbb{Z}/p^k)_+; \mathbb{Z}/p); k \geq 1\}.$

(1.36) Lemma

(1.35) *is an isomorphism of progroups.*

Proof. In this simple case it is sufficient to check that

$$\tilde{H}_j(\Sigma(\mathbb{C}P_+^\infty); \mathbb{Z}/p) \cong \begin{cases} \mathbb{Z}/p & j \geq 1, j \text{ odd} \\ 0 & j \text{ even or } j \leq 0 \end{cases}$$

is mapped isomorphically by $(t_1)_*$ to $\varprojlim_k \tilde{H}_j(B(\mathbb{Z}/p^k)_+; \mathbb{Z}/p)$. This follows from §1.32.   □

(1.37) Let $QX = \varprojlim_n \Omega^n \Sigma^n X$, where $\Omega^n Y$ is the nth loopspace of Y. We may form the adjoint of t_n, $\tau_n = \Sigma^n(BT_+^n) \to Q((B\mathbb{Z}/p^k)_+)$ which extends, uniquely up to homotopy, to a map of infinite loopspaces, also τ_n,

$$\tau_n \colon Q(\Sigma^n(BT_+^n)) \to Q((B(\mathbb{Z}/p^k)^n)_+).$$

Incidentally, I should admit here that I have been very cavalier about taking the limits of transfers as $T \to \infty$, on (S^{2T+1}/H). However, in this example all the $\varprojlim^1$ obstructions can be shown to vanish.

(1.38) Theorem

τ_n in (1.37) induces an isomorphism of graded pro-groups

$$\{(\tau_n)_* : H_*(Q(\Sigma(BT^n_+))); \mathbb{Z}/p) \to H_*(Q((B(\mathbb{Z}/p^k)^n)_+); \mathbb{Z}/p); k \geq 1\}.$$

Proof. In §1.36, we showed this for t_n, the adjoint of τ_n. However, $H_*(Q(X_+); \mathbb{Z}/p)$, when X is connected, is obtained in a functorial manner from $H_*(X; \mathbb{Z}/p)$. Namely, it is freely generated over the Dyer-Lashof operation algebra by $H_*(X_+; \mathbb{Z}/p)$ (see [CLM]). Since the $\{\tau_n\}$ are maps of infinite loopspaces, $(\tau_n)_*$ commutes with Dyer-Lashof operations and is a Hopf algebra map. By §1.36 it is a pro-isomorphism on the quotient of indecomposables (over the Dyer-Lashof algebra). The result follows by an easy induction on the grading dimension.   □

(1.39) We may form the $H_*(-; \mathbb{Z}/p)$-localization of $Q(X_+)$, following [BK]. We denote this by $Q(X_+)^{\hat{}}_p$. We may also form the homotopy inverse limit $[BK]$ of the tower of maps

$$\cdots \to Q((B(\mathbb{Z}/p^k)^n)_+) \xrightarrow{\iota} Q((B(\mathbb{Z}/p^{k-1})^n)_+) \xrightarrow{\iota} \cdots,$$

denoted by $\underleftarrow{\mathrm{holim}}_k Q((B(\mathbb{Z}/p^k)^n)_+)$.

The following result (when $n = 1$) was told to me by Stewart Priddy:

(1.40) Theorem

Let p be a prime, then τ_n of (1.38) induces a homotopy equivalence:

$$\tau_n : Q(\Sigma^n(BT^n_+))^{\hat{}}_p \xrightarrow{\cong} \underleftarrow{\mathrm{holim}_k} Q((B(\mathbb{Z}/p^k)^n_+)).$$

Proof. In [BK] we find that the criterion for the preceding map to be a homotopy equivalence is precisely the result of Theorem 1.35.   □

(1.41)   Returning now to the situation of §§1.17–1.19, let us suppose that G is a p-group. In this case we wish to calculate

$$\{BG_+, BT^n_+\} \cong [BG_+, Q(BT^n_+)]$$

$$\cong [BG_+, Q(BT^n_+)^{\hat{}}_p], \qquad \text{as } G \text{ is a } p\text{-group},$$

$$\cong [\Sigma^n(BG_+), Q(\Sigma^n(BT^n_+))^{\hat{}}_p]$$

$$\cong \underleftarrow{\lim_k} [\Sigma^n(BG_+), Q((B(\mathbb{Z}/p^k)^n_+))], \qquad \text{by §1.40},$$

$$\cong \underleftarrow{\lim_k} \{\Sigma^n(BG_+), B(\mathbb{Z}/p^k)^n_+\}.$$

The inverse limit is taken over compositions with the transfer S-maps,

$$t: B(\mathbb{Z}/p^k)^n_+ \to B(\mathbb{Z}/p^{k-1})^n_+ .$$

In order to evaluate the inverse limit, we will need the following result:

(1.42) Lemma

Let $H \subset G$ and $\pi' \subset \pi$ be finite groups, and let $\rho: H \to \pi$ be a subhomomorphism. Then the transfer map,

$$t: \{BG_+, B\pi_+\} \to \{BG_+, B\pi'_+\},$$

satisfies, in the notation of (1.15),

$$t(\Phi_G(\rho)) = \Phi_G(\rho: \rho^{-1}(\pi') \to \pi').$$

Proof. Suppose that $\lambda: G \to \Sigma_n \int \pi$ $(n = [G:H])$ is the monomial homomorphism corresponding to ρ, as in §1.10. The transfer of $\Phi_G(\lambda)$ is represented by the composition of λ with the map

$$\Sigma_n \int \lambda': \Sigma_n \int \pi \to \Sigma_n \int \Sigma_d \int \pi' \qquad (d = [\pi:\pi']),$$

where λ' is associated to the subhomomorphism, $(\pi \supset \pi' \xrightarrow{1} \pi')$. However, $j(\Sigma_n \int \lambda')\lambda: G \to \Sigma_n \int \Sigma_d$ is easily seen to yield a G-action on $\{1, \ldots, nd\}$, with stab $(1) = \rho^{-1}(\pi')$, since this is the stabilizer of $1, \pi'$ in the (left) π-action on π/π'. Clearly, $((\Sigma_n \int \lambda')\lambda)(g)$ equals $\sigma(g)(\rho(g), \ldots)$ when $g \in \rho^{-1}(\pi')$ so that, from the constructions in §1.11, $t(\Phi_G(\rho)) = \Phi_G(G \supset \rho^{-1}(\pi') \xrightarrow{\rho} \pi')$, as required. $\qquad \square$

(1.43) Corollary

If, in §1.42, we have a commutative diagram

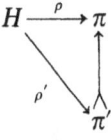

$$H \xrightarrow{\rho} \pi$$

then

$$t(\Phi_G(\rho)) = \Phi_G(\rho') \in \{BG_+, B\pi'_+\}.$$

Proof. In this case $\rho^{-1}(\pi') = (\rho')^{-1}(\pi') = H$. $\qquad \square$

(1.44) Next we must recall the evaluation of $\{BG_+, B\pi_+\}$ which is derived in

[LMM]. Let me first unravel the terminology of [LMM]. Let $(G \supset H \xrightarrow{\rho} \pi)$ be a subhomomorphism.

Define a bounty of groups according to the following list:

(i) $\Delta_\rho = \{(h, \rho(h)) \in G \times \pi \,|\, h \in H\}$.

(ii) $N_\rho = (\text{the normalizer } N_{G \times \pi}(\Delta_\rho))$
$$= \{(g, \sigma) \in N_G(H) \times \pi \,|\, \sigma\rho(-)\sigma^{-1} = \rho(g - g^{-1})\}.$$

(iii) $W_\rho = (\text{the } Weyl \text{ group of } \rho) = (N_\rho)/(\Delta_\rho)$.

(iv) $M_\rho = \{g \in G \,|\, \text{there exists } (g, \sigma) \in N_\rho\} \subset N_G H$.

(v) $V_\rho = (M_\rho)/H$.

(vi) $\pi^\rho = \{\sigma \in \pi \,|\, \sigma\rho(-)\sigma^{-1} = \rho(-)\}$.

(vii) We have a commutative diagram of group extensions:

$$
\begin{array}{ccc}
H & \!\!=\!\!\!=\!\!\!= & H \\
\downarrow{\scriptstyle \Delta} & & \downarrow \\
\pi^\rho \rightarrowtail N_\rho & \longtwoheadrightarrow & M_\rho \\
\downarrow{\scriptstyle 1} & \downarrow & \downarrow \\
\pi^\rho \rightarrowtail W_\rho & \longtwoheadrightarrow & V_\rho
\end{array}
$$

(viii) $WH = (N_G H)/H$.

From [LMM, p. 169, Prop. 5 (proof)], $[EWH \times B\pi^\rho]/V_\rho = BW_\rho$. Now M_ρ acts on $EWH \times B\pi^\rho$ through projection to V_ρ so that we have B_ρ of [LMM] given by [LMM, pp. 170–171]:

$$B_\rho = EG \times_G B_{\bar\rho}$$
$$= EG \times_G (G \times_{M_\rho} (EWH \times B\pi^\rho))$$
$$= EG \times_{M_\rho} (EWH \times B\pi^\rho)$$
$$= ((EG)/H) \times_{V_\rho} (EWH \times B\pi^\rho).$$

Therefore we have a fibration

$$((EG)/H) = BH \to B_\rho \to EWH \times_{V_\rho} B\pi^\rho = BW_\rho.$$

From the discussion of [LMM, p. 170, lines 1–6], one finds that $B_\rho = BN_\rho$, the classifying space of N_ρ, and that the preceding fibration is identifiable with that induced by (vii).

Define inclusion, with index equal to $[G:H]$ as follows:

(ix) $\qquad\qquad \begin{cases} i: N_\rho \rightarrowtail G \times W_\rho \\ \text{given by } i(g, \sigma) = (g, (g, \sigma)\Delta_\rho). \end{cases}$

(1.45) Theorem [LMM]

Let G be a p-group and π finite. The transfer maps associated to §1.43 (ix),

$$\{t:(BG_+) \wedge (BW_\rho)_+ = (BG \times W_\rho)_+ \to BN_{\rho+}\},$$

and the projections $\{N_\rho \to \pi\}$ induce an isomorphism (for each m)

$$\bigoplus_\rho \pi_m^s((BW_\rho)_+)_p^\wedge \xrightarrow{\cong} \{\Sigma^m(BG_+), B\pi_+\}.$$

The sum runs over $A(G; \pi)$-equivalence classes of subhomomorphisms, ρ.

(1.46) When $m = 0$, $\pi_0^s((BW_\rho)_+) \cong \mathbb{Z}$ so that the left side of the isomorphism of §1.44 is $\bigoplus_\rho \hat{\mathbb{Z}}_p \cong A(G; \pi)_p^\wedge$ and, in fact, the map of §1.45 may be identified with the isomorphism mentioned in (1.15) and (1.16).

(1.47)   PROOF OF THEOREM 1.17

We may assume G is a p-group. From §§1.41–1.45 we have to evaluate

$$\varprojlim_k \bigoplus_{\rho_k} \pi_n^s((BW_{\rho_k})_+)_p^\wedge,$$

where ρ_k runs through $A(G, (\mathbb{Z}/p^k)^n)$-equivalence classes of subhomomorphisms. The inverse limit is taken over transfer maps.

Firstly, choose a k_0 such that for all $H \subset G$ and $\rho: H \to T^n$, $\text{im}(\rho) \subset (\mathbb{Z}/p^{k_0})^n$. Therefore, by §1.43, the inverse limit (which may as well be taken over $k \geq k_0$) decomposes into a sum of inverse limits, one for each $R_+(G, T^n)$-equivalence class of subhomomorphisms $(G \supset H \xrightarrow{\tilde\rho} T^n)$. For each such $\tilde\rho$ we may consider it as giving a sequence of subhomomorphisms $(G \supset H \xrightarrow{\tilde\rho_k} (\mathbb{Z}/p^k)^n)$, where, for $k \geq k_0$, $\tilde\rho_k$ is merely $\tilde\rho$ considered as landing in $(\mathbb{Z}/p^k)^n$.

For $k \geq k_0$ write $W_k = W_{\tilde\rho_k}$, $T_k = (\mathbb{Z}/p^k)^n$, $N_{\tilde\rho_k} = M_{\tilde\rho_k} \times T_k = M \times T_k$ (since $M_{\tilde\rho_k} = M_\rho$ is constant). Given $f \in \pi_n^s((BW_{k+1})_+)$, it corresponds, by §§1.44–1.45, to the S-map

$$(1.48) \quad S^n \wedge (BG_+) \xrightarrow{f \wedge 1} B(W_{k+1} \times G)_+ \xrightarrow{t} B(M \times T_{k+1})_+ \xrightarrow{\pi} (BT_{k+1})_+,$$

where π is induced by projection onto T_{k+1}.

By §1.42,

$$t\pi t: B(W_{k+1} \times G)_+ \to (BT_k)_+$$

corresponds to the subhomomorphism

$$(W_{k+1} \times G \supset W_k \times G \supset N_k = M \times T_k \xrightarrow{\pi} T_k)$$

which, under $\Phi_{W_{k+1} \times G}$ of §1.11, corresponds to the S-map

$$B(W_{k+1} \times G)_+ \xrightarrow{t'} B(W_k \times G)_+ \xrightarrow{t} B(M \times T_k)_+ \xrightarrow{\pi} (BT_k)_+.$$

Hence the image of (1.48) under the transfer, $t : (BT_{k+1})_+ \to (BT_k)_+$, is

$$t\pi t(f \wedge 1) = \pi t t'(f \wedge 1).$$

This means, by §§1.44–1.45 again, that in the inverse limit $(k \geq k_0)$ $f \in \pi_n^s((BW_{k+1})_+)$ is mapped to $t'f \in \pi_n^s((BW_k)_+)$. Therefore we have shown that

$$\varprojlim_k \bigoplus_{\rho_k} \pi_n^s((BW_{\rho_k})_+)_p^{\hat{}} \cong \bigoplus_{\tilde{\rho}} \varprojlim_k \pi_n^s((BW_{\tilde{\rho}_k})_+)_p^{\hat{}},$$

where $\tilde{\rho}$ runs through $R_+(G, T^n)$-equivalence classes of subhomomorphisms.

Finally, we will show that each factor $\varprojlim_k \pi_n^S((BW_{\tilde{\rho}_k})_+)_p^{\hat{}} \cong \mathbb{Z}_p^{\hat{}}$. This will mean that $\{BG_+, BT_+^n\} \cong R_+(G, T^n)_p^{\hat{}}$ abstractly. From this, and the remark of §1.46, it is straightforward to show that this isomorphism coincides with the map of (1.15) and (1.16).

Fix $\tilde{\rho}$ as just shown. Since T_k is abelian, $T_k^{\tilde{\rho}} = T_k$ and $W_{\tilde{\rho}_k} = V_{\tilde{\rho}_k} \times T_k = V \times T_k$, say, where for $k \geq k_0$, V is independent of k. Hence $BW_{\tilde{\rho}_k} = BW \times B\mathbb{Z}/p^k \times \cdots \times B\mathbb{Z}/p^k$ and $(BW_{\tilde{\rho}_k})_+ = A \wedge B_k \wedge \cdots \wedge B_k$, where $A = BV_+$, $B_k = (B\mathbb{Z}/p^k)_+$. Also the transfer becomes

$$1 \wedge t \wedge t \wedge \cdots \wedge t : A \wedge \left(\overset{n}{\underset{1}{\wedge}} B_{k+1} \right) \to A \wedge \left(\overset{n}{\underset{1}{\wedge}} B_k \right),$$

where $t : B_{k+1} \to B_k$ is the transfer associated with $i : \mathbb{Z}/p^k \rightarrowtail \mathbb{Z}/p^{k+1}$. Inside each of A or B_{k+1}, we have a copy of $S^0 = (pt)_+$. Now $ti_* : \pi_*^s(B_k) \to \pi_*^s(B_{k+1}) \to \pi_*^s(B_k)$ is multiplication by a stable cohomotopy element of $\pi_S^0(B_k)$ which has the form [B-G; K-P]

(1.49) $p \cdot$(a unit of the stable cohomotopy group, π_S^0, of the fibre)

For any sequence $1 \leq i_1 < i_2 < \cdots < i_j \leq n+1$, consider $X_{i,k} < A \wedge (\wedge_1^n B_k)$, the subspace consisting of S^0 in i_sth coordinate for $s = 1, \ldots, j$. The S-map, $1 \wedge t \wedge \cdots \wedge t$, maps $X_{i,k+1}$ to $X_{i,k}$, and (1.49) implies that

$$\varprojlim_k \pi_n^s(X_{i,k}) = 0$$

unless $\underline{i}$ is the empty sequence. Since

$$\pi_n^s\left(A \wedge \left(\overset{n}{\underset{1}{\wedge}} B_k\right)\right) \cong \bigoplus_{\underline{i}} \pi_n^s(X_{\underline{i},k}),$$

our inverse limit is reduced to

$$(1.50) \qquad \varprojlim_k \pi_n^s((BV \wedge B\mathbb{Z}/p^k \wedge B\mathbb{Z}/p^k \wedge \cdots \wedge B\mathbb{Z}/p^k)_+)_p^{\wedge}.$$

By the Hurewicz theorem [Spa] applied to $BV \wedge B\mathbb{Z}/p^k \wedge \cdots \wedge B\mathbb{Z}/p^k$, and the universal coefficient theorem for homology,

$$\pi_n^s(BV \wedge \cdots) \cong \overset{n+1}{\underset{j=1}{\bigoplus}} U_{j,k},$$

where $U_{j,k}$ is equal to $V_{ab} \otimes \mathbb{Z}/p^k \cdots \otimes \mathbb{Z}/p^k$, with the jth factor omitted. Here $V_{ab} = H_1(V)$ and $\mathbb{Z}/p^k = H_1(\mathbb{Z}/p^k)$. By the analogue of (1.49) in homotopy [B-G; K-P] each $\varprojlim_k U_{j,k} = 0$, except for the case $j = 1$. Since $t_*: H_1(\mathbb{Z}/p^{k+1}) \to H_1(\mathbb{Z}/p^k)$ is onto $\varprojlim_k U_{1,k} = \varprojlim_k \mathbb{Z}/p^k = \hat{\mathbb{Z}}_p$, as required. $\square$

2. SOME SIMPLE NONABELIAN EXAMPLES OF STABLE HOMOTOPY THEORY

(2.1)   In §1.17 we showed that there is an isomorphism ($n \geq 1$)

$$\Phi_G: R_+(G, T^n)_{IA(G)}^{\wedge} \xrightarrow{\cong} \{BG_+, BT_+^n\}$$

for all finite groups, G.

In this section we will study the groups, $\{\Sigma^n BG_+, B\pi_+\}$ for some simple nonabelian, compact Lie groups, π. We are particularly interested in the dimension, $n = 0$, and the analogue of §1.17 for π.

We will begin with the example of the wreath product $\pi = \Sigma_2 \int S^1$ (see (1.2)). Hence we have an exact sequence

$$(2.2) \qquad T^2 \overset{i}{\rightarrowtail} \Sigma_2 \int S^1 \xrightarrow{j} \mathbb{Z}/2$$

and the resulting fibration of classifying spaces

$$(2.3) \qquad BT^2 \overset{i}{\rightarrowtail} B\Sigma_2 \int S^1 \longrightarrow \mathbb{R}P^\infty = B\mathbb{Z}/2.$$

The Serre spectral sequence for (2.3), *when p is an odd prime,*

(2.4) $$E_2^{s,t} = H^s(\mathbb{Z}/2; H^t(BT^2; \mathbb{Z}/p)) \Rightarrow H^{s+t}\left(B\Sigma_2 \int S^1; \mathbb{Z}/p\right),$$

has $E_2^{s,t} = 0$ if $s > 0$. Hence, from (2.4),

$$H^s\left(B\Sigma_2 \int S^1; \mathbb{Z}/p\right) \cong H^s(\mathbb{C}P^\infty \times \mathbb{C}P^\infty; \mathbb{Z}/p)^{\mathbb{Z}/2}.$$

From this we see that the natural map

(2.5) $$H^*(BU(2); \mathbb{Z}/p) \to H^*\left(B\Sigma_2 \int S^1; \mathbb{Z}/p\right)$$

is an isomorphism when $p \neq 2$ (see [M-St], for example, for $H^*(BU(2); \mathbb{Z})$). This means that *if* G *is a p-group and* $p \neq 2$, then

$$\left\{BG_+, B\Sigma_2 \int S_+^1\right\} \to \{BG_+, BU(2)_+\}$$

is an isomorphism. For this reason we will assume for the rest of this section, unless otherwise stated, that

(2.6) G is a finite   2-group.

(2.7)   Let $M = S^{2n+1} \times (S^{2n+1} \times S^{2n+1})$ on which $\pi = \Sigma_2 \int S^1$ acts by

(2.8) $$\begin{cases} \sigma \begin{bmatrix} z_1 & 0 \\ 0 & z_2 \end{bmatrix}(v, w_1, w_2) \\ \quad = ((-1)^{\text{sign}(\sigma)} v, z_{\sigma(1)} w_{\sigma(1)}, z_{\sigma(2)} w_{\sigma(2)}). \end{cases}$$

Write $H_k = \Sigma_2 \int(\mathbb{Z}/2^k) \subset \pi$; to the maps $\{p_k: M/H_k \to M/\pi\}$, we will apply the transfer of §1.20.

Firstly we observe that $\text{Ad}(\pi)$ is isomorphic to $\mathbb{R} \oplus \mathbb{R}$, on which $T^2 \subset \pi$ acts trivially and $\Sigma_2 \cong \pi/T^2$ acts nontrivially by switching the factors. Hence, if $\mathbb{R}_r$ is $\mathbb{R}$ with the action $(u \in \mathbb{R}, z_i \in S^1, \sigma \in \Sigma_2)$,

(2.9) $$\mathbb{R}_r: \sigma \begin{bmatrix} z_1 & 0 \\ 0 & z_2 \end{bmatrix}(u) = (-1)^{\text{sign}(\sigma)} u,$$

then

(2.10) $\text{Ad}(\pi) \cong \mathbb{R} \oplus \mathbb{R}_r,$

where $\mathbb{R}$ is the trivial one-dimensional representation. Form the vector bundle

$$(2.11) \qquad M \times_{\pi} \mathrm{Ad}\,(G) \cong (M \times_{\pi} \mathbb{R}_{\tau}) \times \mathbb{R},$$

then the transfer is an S-map of the form, if $\underline{\mathbb{R}}_{\tau} = M \times_{\pi} \mathbb{R}_{\tau}$,

$$(2.12) \qquad t_k : \Sigma T(\underline{\mathbb{R}}_{\tau}) \to (M/H_k)_{+}.$$

Passing to the limit as $n \to \infty$, this becomes the S-map

$$(2.13) \qquad t_k : \Sigma T(E\pi \times_{\pi} \mathbb{R}_{\tau}) \to (BH_k)_{+}.$$

For (2.13) we have a result similar to §1.40.

(2.14) Theorem

With the preceding notation, (2.13) induces a homotopy equivalence

$$\tau : Q(\Sigma T(E\pi \times_{\pi} \mathbb{R}_{\tau}))\hat{\,}_{2}^{\,} \xrightarrow{\cong} \underleftarrow{\mathrm{holim}}_{k}\, Q((BH_k)_{+}).$$

Proof. By the arguments used in the proofs of §§1.38–1.40 we quickly reduce to proving that the S-maps of (2.13) induce an isomorphism

$$\underleftarrow{\lim}_{k}\,(t_k)_{*} : H_{*}(\Sigma T(E\pi \times_{\pi} \mathbb{R}_{\tau}); \mathbb{Z}/2) \xrightarrow{\cong} \underleftarrow{\lim}_{k}\, H_{*}(H_k; \mathbb{Z}/2).$$

However, as in (1.33), we have a commutative diagram, by S-duality, in which $X = \Sigma T(\underline{\mathbb{R}}_{\tau})$ and $Y_k = M/H_k$:

$$(2.15)$$

$$
\begin{array}{ccc}
H_m(X; \mathbb{Z}/2) & \xrightarrow{\;(t_k)_{*}\;} & H_m(Y_k; \mathbb{Z}/2) \\[2pt]
\cong\,\downarrow & & \Big\downarrow\,\cong \\[2pt]
H_{m-2}(M/\pi; \mathbb{Z}/2) & & \\[2pt]
\cong\,\downarrow & & \\[2pt]
H^{6n+3-m}(M/\pi; \mathbb{Z}/2) & \xrightarrow{\;P_k^{*}\;} & H^{6n+3-m}(Y_k; \mathbb{Z}/2).
\end{array}
$$

Therefore we must show that

$$(2.16) \qquad \underrightarrow{\lim}_{k}\,(P_k)^{*} : H^{*}(B\pi; \mathbb{Z}/2) \to \underrightarrow{\lim}_{k}\, H^{*}(H_k; \mathbb{Z}/2)$$

is an isomorphism. However, this follows from the natural isomorphisms, proved by means of Serre spectral sequences:

$$(2.17) \quad \begin{cases} H^*(B\pi; \mathbb{Z}/2) \cong H^*(\Sigma_2; H^*(\mathbb{C}P^\infty; \mathbb{Z}/2)^{\otimes 2}) \\ H^*(BH_k; \mathbb{Z}/2) \cong H^*(\Sigma_2; H^*(B\mathbb{Z}/2^k; \mathbb{Z}/2)^{\otimes 2}). \end{cases}$$

Hence (2.16) is an isomorphism by (2.17) and by §§1.32 and 1.36. This completes the proof of Theorem 2.14. $\qquad\qquad\Box$

(2.18)   Now consider the natural double covering map of (2.3)

$$BT^2 \xrightarrow{i} B\pi = B\Sigma_2 \int S^1.$$

This map is the projection map of the sphere bundle (with fibres S^0) of the vector bundle $E\pi \times_\pi \mathbb{R}_\tau \to B_\pi$. This means that the cofibre of i is $T(E\pi \times_\pi \mathbb{R}_\tau)$, the Thom space of the line bundle introduced earlier. Hence we have a Puppe sequence for this cofibration (suspended up once):

$$(2.19) \qquad \Sigma(BT_+^2) \xrightarrow{i} \Sigma(B\pi_+) \xrightarrow{j} \Sigma T(E\pi \times_\pi \mathbb{R}_\tau) \xrightarrow{\partial} \Sigma^2(BT^2)_+.$$

However, by [Kn, Lemma 2.12], as an S-map the coboundary, ∂ in (2.19), is given by

$$(2.20) \qquad\qquad \partial = \pm (\text{transfer of } BT^2 \xrightarrow{i} B\pi).$$

Hence, changing the sign of ∂ if necessary, by transitivity of the transfer, we have a commutative diagram of S-maps:

$$(2.21)$$

$$\begin{array}{ccc} \Sigma(B\pi)_+ \longrightarrow \Sigma T(E\pi \times_\pi \mathbb{R}_\tau) \xrightarrow{(\pm\partial)} \Sigma^2(BT_+^2) \\ \Big\downarrow t \qquad\qquad\qquad \Big\downarrow t \\ (BH_k)_+ \xrightarrow{\qquad t \qquad} B(\mathbb{Z}/2^k)_+^2. \end{array}$$

Applying $Q(-)$ turns cofibrations into fibrations and $(-)_2^\wedge$ preserves fibrations so that (2.21), §§1.40 and 2.14 imply the following result:

(2.22) Theorem

There is a fibration

$$Q\Sigma\left(B\left(\Sigma_2 \int S^1\right)_+\right)_2^\wedge \to \underset{k}{\text{holim}}\, Q((BH_k)_+) \xrightarrow{t} \underset{k}{\text{holim}}\, Q((B(\mathbb{Z}/s^k)_+^2),$$

where $H_k = \Sigma_2 \int (\mathbb{Z}/2^k)$ and t is induced by the transfers associated with $(\mathbb{Z}/2^k) \subset H_k$.

(2.23)   We now begin collecting the necessary facts to enable us to calculate $\{BG_+, B\Sigma_2 \int S^1_+\}$ by means of Theorem 2.22, in a manner analogous to that used to prove Theorem 1.7 (see §1.47).

Recall that G is a 2-group, by the assumption of (2.6). Hence $IR_+(G, \pi)^{\wedge}_{IA(G)} \cong IR_+(G, \pi) \otimes \hat{Z}_2$, by §1.19.

(2.24) Lemma

Let G be a finite 2-group, then

$$\mathrm{Ker}\left(R_+(G; T^2)^{\wedge}_2 \rightarrow R_+\left(G, \Sigma_2 \int S^1\right)^{\wedge}_2 \right)$$

is the free $\hat{Z}_2$-module on generators

$$\{(H: \xrightarrow{\rho} T^2) - (H \xrightarrow{\tau \rho \tau} T^2)\},$$

where ρ runs through $R_+(G; T^2)$-equivalence classes of subhomomorphisms and

$$\tau = \begin{bmatrix} 0 & 1 \\ 1 & 0 \end{bmatrix} \in \Sigma_2 \int S^1.$$

Proof. This is clear because the equivalence classes of subhomomorphisms, $\rho: H \rightarrow \pi$, are defined by the equivalence relations of conjugation by $N_G(H)$ on H and by inner automorphisms of π. The only extra inner automorphism on $\rho: H \rightarrow T^2$ that is available if we consider ρ as landing in $\Sigma_2 \int S^1$ is $\tau(-)\tau$.   □

(2.25)   Now let us examine subhomomorphisms of the type $G \supset H \xrightarrow{\rho} \Sigma_2 \int S^1$, where G is a finite 2-group. For any such ρ there exists a k_0 such that

$$\mathrm{im}(\rho) \subset H_{k_0},$$

where $H_k = \Sigma_2 \int (\mathbb{Z}/2^k)$. Write $B_k = \mathbb{Z}/2^k \times \mathbb{Z}/2^k$, the 2^k-torsion part of the diagonal torus, T^2, in $\Sigma_2 \int S^1$:

(2.26) $\begin{cases} \text{For } k \geq k_0, \text{ let } \rho_k: H \rightarrow H_k \text{ be the} \\ \text{subhomomorphism afforded by } \rho: H \rightarrow \Sigma_2 \int S^1. \end{cases}$

The subhomomorphisms, ρ, fall into three classes in the following manner:

Case i: *im$(\rho) = \rho(H) \subset T^2 \subset \Sigma_2 \int S^1$ and $(\Sigma_2 \int S^1)^\rho = \Sigma_2 \int S^1$.*
This is the same as saying that for $k \geq k_0$,

(2.27a) $\mathrm{im}(\rho_k) \subset \mathrm{centre}(H_k)$ and $H_k^{\rho_k} = H_k$.

Case ii: $im(\rho) \subset T^2$ and $(\Sigma_2 \int S^1)^\rho = T^2$.
This is equivalent to $(k \geq k_0)$,

(2.27b) $$im(\rho_k) \subset B_k \quad \text{and} \quad H_k^{\rho_k} = B_k.$$

Case iii: $im(\rho) \not\subset T^2$
This means that there exists $g \in H$ such that

$$\rho_k(g) = \begin{bmatrix} 0 & a \\ b & 0 \end{bmatrix} \in H_k.$$

Therefore, if $\begin{bmatrix} z_1 & 0 \\ 0 & z_2 \end{bmatrix} = \sigma$ and $\sigma \rho_k(g) \sigma^{-1} = \rho_k(g)$, we must have $z_1 = z_2$.

Similarly, if $\sigma = \begin{bmatrix} 0 & z_1 \\ z_2 & 0 \end{bmatrix}$ and $\sigma \rho_k(g) \sigma^{-1} = \rho_k(g)$, then $z_1 b = z_2 a$. Therefore this
case is equivalent to $(k \geq k_0)$,

(2.27c) $$\begin{cases} im(\rho_k) \not\subset B_k \text{ and } H_k^{\rho_k} \text{ sits in a split extension} \\ \mathbb{Z}/2^k \cong \text{centre}(H_k) \rightarrowtail H_k^{\rho_k} \twoheadrightarrow \mathbb{Z}/2. \end{cases}$$

(2.28)   Now we examine, when G is a finite 2-group,

$$[\Sigma BG_+, \underset{k}{\text{holim}}\, Q((BH_k)_+)] \cong \underset{k}{\lim}\, \{\Sigma(BG_+),(BH_k)_+\}.$$

Arguing as in §1.47 (see also §§1.41–1.45), this group is isomorphic to

(2.29) $$\underset{k}{\lim} \bigoplus_{\rho_k} \pi_1^s((BW_{\rho_k})_+)\hat{}_2,$$

where (for $k \geq k_0$) ρ_k runs through subhomomorphisms associated, as in (2.26),
to $R_+(G, \Sigma_2 \int S^1)$-equivalence classes of $\{\rho: H \to \Sigma_2 \int S^1\}$. In fact the argument of
§1.47 shows that (2.29) equals

(2.30) $$\bigoplus_\rho \underset{k}{\lim}\, \pi_1^s((BW_{\rho_k})_+)\hat{}_2.$$

That is, for $k \geq k_0$, the inverse limit process of (2.29) respects the direct sum
decomposition.

In addition, by the argument of §1.47, each of the inverse limits (for a fixed
$\rho: H \to \Sigma_2 \int S^1$)

(2.31) $$\underset{k}{\lim}\, \pi_1^s((BW_{\rho_k})_+)\hat{}_2$$

is taken over transfer maps associated to the inclusions $W_{\rho_k} \subset W_{\rho_{k+1}}$ of finite groups. Since $\pi_1^S(S^0) \cong \mathbb{Z}/2$, (2.31) becomes

$$(2.32) \qquad \varprojlim_k ((W_{\rho_k})_{ab} \oplus \mathbb{Z}/2) \cong \varprojlim_k (W_{\rho_k})_{ab}.$$

The $\mathbb{Z}/2$-factor vanishes in $\varprojlim_k$ because the transfer induces multiplication by 2 on this factor (cf., §1.47).

From the extension of §1.44 (vii),

$$(2.33) \qquad H_k^{\rho_k} \rightarrowtail W_{\rho_k} \twoheadrightarrow V_\rho \qquad (k \geq k_0),$$

in which V_ρ is independent of k, we have an exact sequence [H-S, p. 202]

$$\to H_2(V_\rho) \to H_1(H_k^{\rho_k})_{V_\rho} \to (W_{\rho_k})_{ab} \twoheadrightarrow (V_\rho)_{ab} \to 0.$$

If we are in case (2.27a), then $\operatorname{im}(\rho)$ is central in $\Sigma_2 \int S^1$, V_ρ acts trivially on $H_1(H_k^{\rho_k}) = H_1(H_k)$. In fact, (2.33) is a product and hence split. Therefore we obtain an exact sequence

$$0 \to H_1(H_k) \rightarrowtail H_1(W_{\rho_k}) \twoheadrightarrow H_1(V_\rho) \to 0.$$

Taking $\varprojlim_k$ annihilates $H_1(V_\rho)$, by the argument of §1.47 (i.e., that the transfer is multiplication by an even integer on this factor). We therefore obtain an isomorphism

$$\varprojlim_k H_1(H_k) \cong \varprojlim_k H_1(W_{\rho_k}).$$

However, the inclusions

$$\mathbb{Z}/2^k \times \mathbb{Z}/2^k \subset H_k \quad \text{and} \quad \mathbb{Z}/2 = \left\langle \begin{bmatrix} 0 & 1 \\ 1 & 0 \end{bmatrix} \right\rangle \subset H_k$$

induce an epimorphism

$$H_1(\mathbb{Z}/2^k \times \mathbb{Z}/2^k) \oplus H_1(\mathbb{Z}/2) \longrightarrow H_1(H_k).$$

Since $\mathbb{Z}/2 \subset H_{k+1}$, the transfer multiplies by $|H_{k+1}:H_k| = 4$ on this factor. Similarly, the composition on the left in the diagram that follows.

$$\begin{array}{ccc} H_1(\mathbb{Z}/2^{k+1}) \otimes H_0(\mathbb{Z}/2^{k+1}) & \longrightarrow & H_1(H_{k+1}) \\ \Big\downarrow {\scriptstyle \iota'_* \otimes \iota''_*} & & \Big\downarrow {\scriptstyle t_*} \\ H_1(\mathbb{Z}/2^k) \otimes H_0(\mathbb{Z}/2^k) & \longrightarrow & H_1(H_k) \end{array}$$

has image which is 2-divisible since $t''_* = (2.\!—)$. Therefore

$$\operatorname{im}(H_1(H_{k+1}) \xrightarrow{t_*} H_1(H_k)) \quad \text{is 2-divisible in} \quad H_1(H_k).$$

This discussion implies that in case (2.27a),

(2.34) $$\varprojlim_k \pi_1^s((BW_{\rho_k})_+)_2^{\wedge} = 0.$$

In the case (2.27b), V_ρ again acts trivially on $H_1(H_k^{\rho_k}) = H_1(B_k) = B_k$ and (2.33) is split. A similar argument shows that in case (2.27b),

(2.35) $$\varprojlim_k \pi_1^s((BW_{\rho_k})_+)_2^{\wedge} = 0.$$

Finally, in case (2.27c),

$$H_k^{\rho_k} \cong \mathbb{Z}/2^k \times \mathbb{Z}/2 = (\text{centre } H_k) \times \left\langle \begin{bmatrix} 0 & a \\ b & 0 \end{bmatrix} \right\rangle.$$

Since this abelian V_ρ acts trivially on $H_1(H_k^{\rho_k})$, and in addition

$$H_1(W_{\rho_k}) \cong H_1(H_k^{\rho_k}) \oplus H_1(V_\rho) \cong \mathbb{Z}/2^k \oplus \mathbb{Z}/2 \oplus H_1(V_\rho).$$

The inverse limit of these groups kills the last two factors, whereas the first factor is easily seen to contribute a copy of $\hat{\mathbb{Z}}_2$.

Therefore we have, in case (2.27c),

(2.36) $$\varprojlim_k \pi_1^s((BW_{\rho_k})_+)_2^{\wedge} \cong \hat{\mathbb{Z}}_2.$$

(2.37) Theorem

Let G be a finite group. Then there is an isomorphism

$$\Phi_G : R_+\left(G, \Sigma_2 \int S^1\right)_{IA(G)}^{\wedge} \xrightarrow{\cong} \left\{ BG_+, B\Sigma_2 \int S_+^1 \right\}.$$

Proof. By §1.19, it suffices to assume G is a p-group.

When $p = 2$, Theorems 2.22 and 1.17 and the preceding discussion yield an exact sequence

(2.38) $$\cdots \to R_+(G, T^2)_2^{\wedge} \xrightarrow{\alpha'} \left\{ BG_+, B\Sigma_2 \int S_+^1 \right\} \to \bigoplus_\rho \hat{\mathbb{Z}}_2 \to 0.$$

Here ρ runs through the $R_+(G, \Sigma_2 \int S^1)$-equivalence classes of case (2.27c) and the right-hand zero is

$$\varprojlim_k \pi_1^s((BW_{\rho_k})_+)_2^{\wedge},$$

where $\{\rho_k\}$ is as in the proof in §1.47. However, that proof shows this to be zero.

From the commutative diagram,

$$R_+(G, T^2)_2^{\wedge} \xrightarrow{\alpha} R_+(G, \Sigma_2 \int S^1)_2^{\wedge}$$

$$\searrow_{\alpha'} \qquad \downarrow^{\Phi_G}$$

$$\{BG_+, B\Sigma_2 \int S^1_+\},$$

one sees that $\ker \alpha \subset \ker \alpha'$. However, from §2.24, $R_+(G, T^2)_2^{\wedge}/(\ker \alpha)$ is torsion free, and hence it is injected, via α', into $\{BG_+, B\Sigma_2 \int S^1_+\}$ as can be seen by composing with the transfer

$$t: \left\{BG_+, B\Sigma_2 \int S^1_+\right\} \to \{BG_+, BT^2_+\} \cong R_+(G, T^2)_2^{\wedge}.$$

Hence $\ker \alpha = \ker \alpha'$, and from the description in §2.24 combined with (2.38), we see that

$$\left\{BG_+, B\Sigma_2 \int S^1_+\right\} \cong R_+\left(G, \Sigma_2 \int S^1\right)_2^{\wedge}$$

abstractly. By arguments mentioned in [LMM], it is not hard to verify that this isomorphism is realized by Φ_G.

When $p \neq 2$ and G is a p-group, a transfer argument shows that

$$t: \left\{BG_+, B\Sigma_2 \int S^1_+\right\} \to \{BG_+, BT^2_+\} \cong R_+(G, T^2)_p^{\wedge}$$

embeds $\{BG_+, B\Sigma_2 \int S^1_+\}$ as the Σ_2-invariants of $R_+(G, T^2)_p^{\wedge}$. However, $R_+(G, \Sigma_2 \int S^1)_p^{\wedge} \cong (R_+(G, T^2)_p^{\wedge})^{\Sigma_2}$ in this case, from which it easily follows that Φ_G is an isomorphism. □

(2.39) We will now sketch a calculation, entirely analogous to that required to prove Theorem 3.37. We will need this later for an explicit example of the double-coset formula in action (see §3).

Let $U \subset \Sigma_2 \int S^1$ be the subgroup

(2.40)
$$U = \left\{ \begin{bmatrix} z & 0 \\ 0 & \pm z \end{bmatrix}, \begin{bmatrix} 0 & z \\ \pm z & 0 \end{bmatrix} \middle| z \in S^1 \right\}.$$

Inside U we have

(2.41)
$$V = \left\{ \begin{bmatrix} z & 0 \\ 0 & \pm z \end{bmatrix} \middle| z \in S^1 \right\} \cong \mathbb{Z}/2 \times S^1,$$

and $U = \mathbb{Z}/2 \ltimes (\mathbb{Z}/2 \times S^1) = \mathbb{Z}/2 \ltimes V$, a semidirect product in which $\mathbb{Z}/2$, generated by $\tau = \begin{bmatrix} 0 & 1 \\ 1 & 0 \end{bmatrix}$, acts on $\mathbb{Z}/2 \times S^1$ by

(2.42)
$$\begin{cases} \tau(a, z) = (a, -z) & (0 \neq a \in \mathbb{Z}/2, z \in S^1) \\ \tau(0, z) = (0, z). \end{cases}$$

As with the example of §3.37, we start by working at the prime 2. Accordingly, define subgroups

$$U_k = \left\{ \begin{bmatrix} z & 0 \\ 0 & \pm z \end{bmatrix}, \begin{bmatrix} 0 & z \\ \pm z & 0 \end{bmatrix} \middle| z \in S^1, z^{2^k} = 1 \right\}$$

and

$$V_k = \left\{ \begin{bmatrix} z & 0 \\ 0 & \pm z \end{bmatrix} \middle| z \in S^1, z^{2^k} = 1 \right\}.$$

Since the connected component of the identity in U and V consists of the centre, the representations, $\mathrm{Ad}(U)$ and $\mathrm{Ad}(V)$, are both one-dimensional trivial representations. Therefore we have transfer S-maps,

(2.43)
$$\Sigma((BU)_+) \xrightarrow{t} (BU_K)_+$$

$$\Sigma((BV)_+) \xrightarrow{t} (BV_k)_+.$$

A homology calculation, similar to that used to prove §§1.38–1.40 establishes the following theorem:

(2.44) Theorem

The S-maps of (2.43) induce homotopy equivalences

(i) $Q(\Sigma((BV)_+))^\wedge_2 \xrightarrow{\approx} \underleftarrow{\mathrm{holim}}_k Q((BV_k)_+),$

(ii) $Q(\Sigma((BU)_+)^\wedge_2 \xrightarrow{\approx} \underleftarrow{\mathrm{holim}}_k Q((BU_k)_+).$

(2.45)   Since V and V_k are abelian, if $\rho_k : H \to V_k \subset V$ is a subhomomorphism, then $N_{\rho_k} \cong M_{\rho_k} \times V_k$ in §1.44 (ii). By the arguments of §1.47, we find that, *when G is a finite 2-group,*

$$\{\Sigma^n(BG_+), BV_+\} \cong \varprojlim_k \bigoplus_\rho \pi^s_{n+1}((BW_{\rho_k})_+)\hat{}_2$$

$$\cong \bigoplus_\rho \varprojlim_k \pi^s_{n+1}((BV_k)_+)\hat{}_2,$$

where ρ runs through $R_+(G, V)$-equivalence classes of subhomomorphisms and, for each ρ with $\operatorname{im}(\rho) \subset V_k$, ρ_k is the resulting subhomomorphism, $\rho_k: H \to V_k$.

In particular, $\{BG_+, BV_+\} \cong \bigoplus_\rho \hat{\mathbb{Z}}_2 \cong R_+(G, V)\hat{}_2$, and we have the next proposition:

(2.46) Proposition

Let G be a finite group. There is an isomorphism

$$\Phi_G: R_+(G, V)\hat{}_{IA(G)} \xrightarrow{\cong} \{BG_+, BV_+\}.$$

Proof. The preceding discussion takes care of the case when G is a 2-group. When G is a p-group with p odd, $BV_+ \simeq BS^1$ (at the prime p) and the result follows from §1.17. $\qquad\square$

(2.47) Suppose now that we have a subhomomorphism, as in (2.26), $\rho: H \to U$ so that it gives, for $k \geq k_0$, $\rho_k: H \to U_k$. To evaluate $\varprojlim_k H_1(W_{\rho_k})\hat{}_2$, we divide the cases up according to the scheme of (2.27a–2.27c).

If $\operatorname{im}(\rho) \subset \operatorname{center}(U)$, then $U_k^{\rho_k} = U_k$ and, as in §2.28, the split extension $U_k \rightarrowtail W_{\rho_k} \twoheadrightarrow V_\rho$ yields

$$\varprojlim_k H_1(W_{\rho_k})\hat{}_2 \cong \varprojlim_k H_1(U_k)\hat{}_2$$

$$\cong \varprojlim_k (\mathbb{Z}/2 \oplus \mathbb{Z}/2^k)$$

$$\cong \hat{\mathbb{Z}}_2.$$

If $\operatorname{im}(\rho) \subset V$ and $U_k^{\rho_k} = V_k$, we obtain (as in (2.27b) and §2.28)

$$\varprojlim_k H_1(W_{\rho_k})\hat{}_2 \cong \varprojlim_k H_1(V_k)\hat{}_2 \cong \mathbb{Z}_2.$$

Finally, if $\operatorname{im}(\rho_k) \nsubseteq V_k$ (for $k \geq k_0$), then $U_k^{\rho_k}$ sits in a split extension

$$\mathbb{Z}/2^k \cong \operatorname{centre}(U_k) \rightarrowtail U_k^{\rho_k} \twoheadrightarrow \mathbb{Z}/2,$$

as in (2.27c). Arguing as in §2.28 (see (2.36)),

$$\varprojlim_{k} H_1(W_{\rho_k})^{\wedge}_2 \cong \hat{\mathbb{Z}}_2$$

in this case too.

Combining this discussion with the arguments (standard by now) of §2.37 (or §2.46), we obtain the following theorem:

(2.48) Theorem

Let G be a finite group and U as in (2.40). There is an isomorphism

$$\Phi_G: R_+(G, U)^{\wedge}_{IA(G)} \xrightarrow{\cong} \{BG_+, BU_+\}.$$

3.   THE DOUBLE-COSET FORMULA IN STABLE HOMOTOPY THEORY

(3.1)   In this section we will recall at some length the double-coset formula [Fe, §II.8], which evaluates a composition of S-maps

$$BK_+ \xrightarrow{Bj} BG_+ \xrightarrow{\tau} BH_+$$

in which τ is a transfer map, $j: K \rightarrow G$ is an inclusion and K, H are both *closed subgroups* of a *compact Lie group, G*.

(3.2)   Suppose that H, K, G are as in §3.1, and let G/H denote the space of left cosets on which K acts (smoothly) by left translation,

$$k(gH) = (kg)H \qquad (k \in K; g \in G).$$

This is a smooth action on a closed manifold. The orbit space (of *double cosets*), $K \backslash G/H$, is the disjoint union of differentiable manifolds (of possible differing dimensions). Each of these manifolds is a union of orbits of the same *orbit type*. An orbit is isomorphic to a homogeneous space K/L where L is the *stabilizer* of a point in the orbit, and the K-conjugacy class of L, (L), determines the orbit type. Denote by $M_{(L)}$ the set of cosets in G/H consisting of points of orbit-type (L). This manifold is locally closed, and its closure, $\bar{M}_{(L)}$, consists of orbit types, (L'), with L conjugate in K to a subgroup of L'. This gives a partial ordering on orbit types, namely, $(L) \geq (L')$ if and only if $kLk^{-1} \leq L'$ for some $k \in K$. Hence

$$\bar{M}_{(L)} = \bigcup_{(L') \leq (L)} M_{L'}.$$

Let $\{M_i; i \in I\}$ denote the set of K-orbit manifold components of $K \backslash G/H$, that is, connected components of the $K \backslash M_{(L)}$ described earlier.

Define the *internal Euler characteristic* of M_i by

$$(3.3) \qquad \aleph^{\#}(M_i) = \aleph(\bar{M}_i) - \aleph(\bar{M}_i - M_i).$$

In (3.3), $\aleph$ is the usual Euler characteristic [Spa],

$$\aleph(M) = \sum_n (-1)^n \dim_{\mathbb{Q}} (H_n(M; \mathbb{Q})).$$

The internal Euler characteristic is equal to the Euler characteristic defined by means of compactly supported, rational cohomology.

(3.4) Example

Let $K = \mathbb{Z}/2$ act on S^1 by $\tau(z) = \bar{z}$, complex conjugation. Thus

$$M_1 = (\mathbb{Z}/2) \backslash (S^1_{(1)}) \cong (-1, 1),$$

the open interval, whereas

$$S^1_{(K)} = \{\pm 1\}, \quad \text{two points, and}$$
$$M_2 = \{+1\}, M_3 = \{-1\}.$$

Hence $\aleph^{\#}(M_1) = \aleph([-1, 1]) - \aleph(S^0) = -1$, $\aleph^{\#}(M_2) = 1 = \aleph^{\#}(M_3)$.

(3.5) Choose $\{g_i \in G; i \in I\}$ such that $Kg_iH \in M_i$, one such for each $i \in I$ *with the convention that the double coset of the identity is represented by* $g_i = 1$.

Let $C_g : B(gHg^{-1}) \to BH$ denote the map induced by conjugation by g^{-1},

$$(g^{-1} \cdot \text{---} \cdot g) : gHg^{-1} \xrightarrow{\cong} H.$$

For each g_i, chosen as earlier, let ψ_i denote the S-map composite

$$(3.6) \qquad \begin{cases} \psi_i = Bs(C_{g_i})\tau, \\ BK_+ \xrightarrow{\tau} B(g_iHg_i^{-1} \cap K)_+ \xrightarrow{C_{g_i}} B(H \cap g_i^{-1}Kg_i)_+ \xrightarrow{Bs} BH_+ \end{cases}$$

In (3.6), τ is the Becker-Gottlieb transfer [B-G] for $g_iHg_i^{-1} \cap K \subset K$ and $s : H \cap g_iKg_i^{-1} \to H$ is the inclusion.

With the preceding notation, Feshbach's double-coset formula [Fe, §II.8] is as follows:

(3.7) Theorem

In $\{BK_+, BH_+\}$, the S-map $BK_+ \xrightarrow{Bj} BG_+ \xrightarrow{\tau} BH_+$ equals

$$\sum_{i \in I} \aleph^{\#}(M_i)\psi_i,$$

where $j: K \to G$ is the inclusion and τ is the transfer.

(3.8) Remark

In [Fe, §II.10] the important observation is made that $\tau: BG_+ \to BH_+$ is stably null-homotopic if the "Weyl group," $N_G(H)/H$, is not discrete.

(3.9)   TWO DIMENSIONAL EXAMPLES

Most of the remainder of this section will be devoted to consideration of the following example:

(3.10)
$$\begin{cases} K \subset \Sigma_2 \int S^1 \subset U(2) = G \supset \Sigma_2 \int S^1 = H \\ \text{with } K \text{ finite, } (U(2) \text{ is the unitary group of} \\ 2 \times 2 \text{ complex matrices}). \end{cases}$$

We must analyze the double-coset space $K \backslash U(2)/\Sigma_2 \int S^1$. Firstly, observe that the map

$$p: U(2) \to S^3, \quad p\begin{bmatrix} a & b \\ c & d \end{bmatrix} = \begin{bmatrix} a \\ c \end{bmatrix} = \begin{bmatrix} a & b \\ c & d \end{bmatrix}\begin{bmatrix} 1 \\ 0 \end{bmatrix}$$

induces a homeomorphism $U(2)/U(1) \cong S^3$, where

$$U(1) = \left\{ \begin{bmatrix} 1 & 0 \\ 0 & z \end{bmatrix} \middle| z \in S^1 \right\} = \operatorname{stab}\begin{bmatrix} 1 \\ 0 \end{bmatrix}.$$

Since $\begin{bmatrix} a & b \\ c & d \end{bmatrix}\begin{bmatrix} \alpha & 0 \\ 0 & \beta \end{bmatrix} = \begin{bmatrix} \alpha a & \beta b \\ \alpha c & \beta d \end{bmatrix}$, we find that

$$U(2)/T^2 \cong S^3 \middle/ \left(\begin{bmatrix} a \\ c \end{bmatrix} \sim \begin{bmatrix} \alpha a \\ \alpha c \end{bmatrix} \middle| \alpha \in S^1 \right)$$

$$\cong S^3/S^1$$

$$\cong \mathbb{C}P^1$$

$$\cong S^2.$$

Therefore $U(2)/\Sigma_2 \int S^1 \cong S^2/(\mathbb{Z}/2)$, where the $\mathbb{Z}/2$-action is described as follows: Let $\tau = \begin{bmatrix} 0 & 1 \\ 1 & 0 \end{bmatrix}$ so that $\begin{bmatrix} a & b \\ c & d \end{bmatrix} \tau = \begin{bmatrix} b & a \\ d & c \end{bmatrix}$. Consequently, in $S^2 = U(2)/T^2$, $\tau \begin{bmatrix} a \\ c \end{bmatrix} = \begin{bmatrix} b \\ d \end{bmatrix}$, where $|a|^2 + |c|^2 = 1 = |b|^2 + |d|^2$ and $\bar{a}b + \bar{c}d = 0$. In other words, $\begin{bmatrix} b \\ d \end{bmatrix}$ is the class in $\mathbb{C}P^1 = S^2$ of a unit vector perpendicular to $\begin{bmatrix} a \\ c \end{bmatrix}$. Since no vector is perpendicular to α times itself ($\alpha \in S^1$), we see that $\mathbb{Z}/2$ acts freely and consequently

$$(3.11) \qquad U(2)/\Sigma_2 \int S^1 \cong \mathbb{R}P^2.$$

We may see (3.11) in terms of the familiar models for S^2 and $\mathbb{R}P^2$ as follows: Write $S^2 = \mathbb{C} \cup \infty$, the one-point compactification of the complex numbers. Define

$$F: U(2)/T^2 \to \mathbb{C} \cup \infty \text{ by}$$

$$(3.12) \qquad F: \left[\begin{bmatrix} a & b \\ c & d \end{bmatrix} T^2 \right] = a/c \text{ if } c \neq 0$$

$$F: \left[\begin{bmatrix} 1 & 0 \\ 0 & 1 \end{bmatrix} T^2 \right] = \infty$$

The right action by τ therefore becomes

$$(3.13) \qquad \begin{cases} \tau: \mathbb{C} \cup \infty \to \mathbb{C} \cup \infty \\ \tau(\infty) = 0, \tau(0) = \infty, \tau(z) = -(z)^{-1} \text{ if } z \neq 0. \end{cases}$$

(3.13) transports to the standard antipodal action on S^2. Henceforth we will depict $\mathbb{R}P^2$ in the usual manner as

$$(3.14) \qquad \mathbb{R}P^2 = D^2/(z \sim -z \text{ if } z \in S^1),$$

where $D^2 = \{z \in \mathbb{C} \mid |z| \leq 1\}$.

Now consider the *left* action of $\Sigma_2 \int S^1$ on $\mathbb{R}P^2$ in (3.14). Firstly, since the central, diagonal, $\Delta(S^1)$, acts trivially on $U(2)/(\Sigma_2 \int S^1)$, we will study the action of $\Sigma_2 \int S^1/\Delta(S^1)$. There is an isomorphism

$$\Phi: \Sigma_2 \ltimes S^1 \xrightarrow{\cong} \Sigma_2 \int S^1/\Delta(S^1) \text{ given by}$$

$$(3.15) \qquad \Phi(\tau) = \begin{bmatrix} 0 & 1 \\ 1 & 0 \end{bmatrix}, \Phi(z) = \begin{bmatrix} z & 0 \\ 0 & 1 \end{bmatrix} \text{ for } z \in S^1,$$

where $\tau z \tau = \bar{z}$ in $\Sigma_2 \ltimes S^1$.

Since $\tau\begin{bmatrix} a & b \\ c & d \end{bmatrix} = \begin{bmatrix} c & d \\ a & b \end{bmatrix}$ for the left action of (3.12), this corresponds to
$\tau(z) = 1/z$, and in (3.14) to $\tau(z) = -\bar{z}$. Also, since $\begin{bmatrix} w & 0 \\ 0 & 1 \end{bmatrix}\begin{bmatrix} a & b \\ c & d \end{bmatrix} = \begin{bmatrix} wa & wb \\ c & d \end{bmatrix}$,
the left action of $w \in S^1$ on (3.14) is $w(z) = wz$. Therefore we have shown in (3.14)
that if $w \in S^1 \subset \Sigma_2 \ltimes S^1$ and $\Sigma_2 = \langle \tau \rangle$, then the left $\Sigma_2 \ltimes S^1$-action is given by

$$\begin{cases} w(z) = wz \\ w\tau(z) = -w\bar{z} \text{ for } z \in D^2. \end{cases}$$

Now we examine the double-coset space $K \backslash U(2)/\Sigma_2 \int S^1$, where K is the finite
group in (3.10). We will divide this examination into four cases. Let $\Delta(S^1)$ be
the multiples of the identity matrix in $\Sigma_2 \int S^1$, as in (3.15). Let W denote the
image of K in $(\Sigma_2 \int S^1)/\Delta(S^1) \cong \Sigma_2 \ltimes S^1$.

(3.16)

     Case i: $W \subset S^1 \subset \Sigma_2 \ltimes S^1$, $W \cong \mathbb{Z}/n$, with n odd.

     Case ii: $W \subset S^1$, $W \cong \mathbb{Z}/n$, with n even.

     Case iii: $W \not\subset S^1$, $W \cong D_{2n}$, with n odd.

     Case iv: $W \not\subset S^1$, $W \cong D_{2n}$, with n even.

As in Chapter 1, §(3.12), let x, y generate D_{2n}.

We will begin with Case iii, which is one of the more interesting cases. Let
$\xi_m = \exp(2\pi i/m)$. In this case a fundamental domain for the action on (3.14) by
x looks as shown in Figure 1.

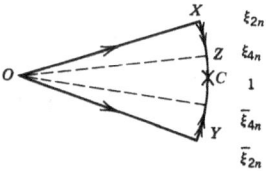

Figure 1

The W-stabilizer of a point on $(OZ]$ is $\{x^2y\}$ and of $(OY]$ is $\{xy\}$, which are
conjugate in W. Also, on ∂D^2, the W-stabilizer of X and of C is $\{y\}$, which is
conjugate to $\{xy\}$. Passing to orbit spaces Figure 1 becomes the picture shown in
Figure 2.

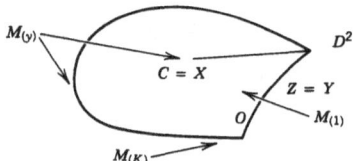

Figure 2. $K \backslash U(2)/\Sigma_2 \int S^1$ in Case iii

TABLE A

Case iii	$\aleph^{\#}(M_{(H)})$	$g \in U(2)$ such that $Kg\Sigma_2 \int S^1 \in M_{(H)}$
$O = M_{(K)}$	1	$\begin{bmatrix} 1 & 0 \\ 0 & 1 \end{bmatrix}$
$(OZYO) = M_{(y)}^1$	-1	$1/\sqrt{2}\begin{bmatrix} \xi_{4n} & 1 \\ 1 & -\bar{\xi}_{4n} \end{bmatrix}$
$X = M_{(y)}^2$	1	$1/\sqrt{2}\begin{bmatrix} 1 & -1 \\ 1 & 1 \end{bmatrix}$
$M_{(1)}$	0	$1/\sqrt{2}\begin{bmatrix} 1 & z \\ -\bar{z} & 1 \end{bmatrix}$

$$z = e^{i\theta}, \ \theta/\pi \text{ irrational}$$

Accordingly, we obtain the orbit-type components listed in Table A, and their internal Euler characteristics and double-coset representatives.

Now let us examine Case iv. A fundamental domain for the action of x on (3.14) looks as shown in Figure 3.

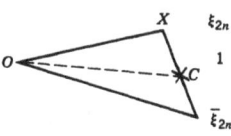

Figure 3

The W-stabilizer of an interior point of (OX) is $\{x^{m+1}y\}$, for (OC) it is $\{y\}$, for any interior point on XC the stabilizer is $\{x^m\}$. Passing to orbit spaces, we obtain the picture on D^2 shown in Figure 4.

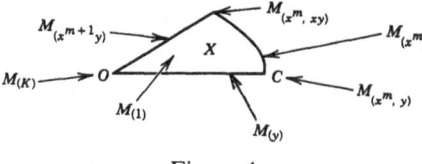

Figure 4

From Figure 4 we obtain Table B of orbit data.

Finally, we consider Cases i and ii. In (3.14), the action is free on the interior of D^2 except for the fixed point at the origin. Each point on the boundary, ∂D^2,

TABLE B

Case iv $(n = 2m)$	$\aleph^{\#}(M_{(H)})$	$g \in U(2)$ such that $Kg\Sigma_2 \int S^1 \in M_{(H)}$
$(O) = M_{(K)}$	1	$\begin{bmatrix} 1 & 0 \\ 0 & 1 \end{bmatrix}$
$(OX) = M_{(x^{m+1}y)}$	-1	$1/\sqrt{3}\begin{bmatrix} \xi_{2n} & -\sqrt{2} \\ \sqrt{2} & \bar{\xi}_{2n} \end{bmatrix}$
$(OC) = M_{(y)}$	-1	$1/\sqrt{3}\begin{bmatrix} 1 & -\sqrt{2} \\ \sqrt{2} & 1 \end{bmatrix}$
$(XC) = M_{(x^m)}$	-1	$1/\sqrt{2}\begin{bmatrix} 1 & e^{i\theta} \\ -e^{-i\theta} & 1 \end{bmatrix}$ $0 < \theta < \pi/m,$ θ/π irrational
$(X) = M^1_{(x^m, xy)}$	1	$1/\sqrt{2}\begin{bmatrix} \xi_{2n} & 1 \\ 1 & -\bar{\xi}_{2n} \end{bmatrix}$
$(C) = M^2_{(x^m, y)}$	1	$1/\sqrt{2}\begin{bmatrix} 1 & -1 \\ 1 & 1 \end{bmatrix}$
$M_{(1)}$	$\aleph(D^2) - \aleph(S^1) = 1$	$1/\sqrt{5}\begin{bmatrix} \xi_{4n} & 2 \\ 2 & -\bar{\xi}_{4n} \end{bmatrix}$

has stabilizer equal to $\{x^m\}$ if $n = 2m$, and trivial otherwise. The orbit space in Cases i and ii is given by one of the pictures, shown in Figures 5 and 6.

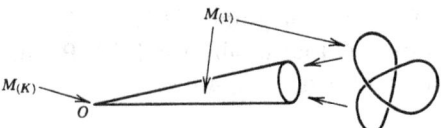

Figure 5. Case i

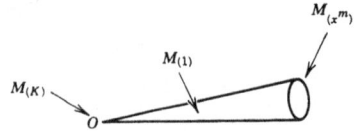

Figure 6. Case ii $(n = 2m)$

From Figures 5 and 6 we obtain Tables C and D of orbit data.

TABLE C

Case i	$\aleph^{\#}(M_{(H)})$	$g \in U(2)$ such that $Kg\Sigma_2 \int S^1 \in M_{(H)}$
$(0) = M_{(K)}$	1	$\begin{bmatrix} 1 & 0 \\ 0 & 1 \end{bmatrix}$
$M_{(1)}$	0	$1/2 \begin{bmatrix} \sqrt{3} & 1 \\ -1 & \sqrt{3} \end{bmatrix}$

TABLE D

Case ii	$\aleph^{\#}(M_{(H)})$	$g \in U(2)$ such that $Kg\Sigma_2 \int S^1 \in M_{(H)}$
$0 = M_{(K)}$	1	$\begin{bmatrix} 1 & 0 \\ 0 & 1 \end{bmatrix}$
$M_{(1)}$	0	$1/2 \begin{bmatrix} \sqrt{3} & 1 \\ -1 & \sqrt{3} \end{bmatrix}$
$M_{(x^m)}$	0	$1/\sqrt{2} \begin{bmatrix} 1 & -1 \\ 1 & 1 \end{bmatrix}$

(3.17)   We conclude this section by relating the Becker-Gottlieb transfer,

$$\tau: \left(B\Sigma_2 \int S^1 \right)_+ \to BU_+$$

(which occurred in the example summarized in §3.17) to the calculations of §2. To be precise, let

$$t: \Sigma T \left(\left(E\Sigma_2 \int S^1 \right) \times_{\Sigma_2 \int S^1} \mathbb{R}_\tau \right) \to \Sigma(BU_+)$$

denote the transfer, in the sense of §1.20, associated with $U \subset \Sigma_2 \int S^1$.

(3.18) Proposition

With the notation of §3.17 there is a homotopy commutative diagram of S-maps:

$$\Sigma(B\Sigma_2 \int S^1)_+ \xrightarrow{\ r\ } \Sigma T((E\Sigma_2 \int S^1) \times_{\Sigma_2 \int S^1} \mathbb{R}_\tau)$$

$$\tau \searrow \qquad \swarrow t$$

$$\Sigma(BU_+),$$

where r is induced by the zero-section map.

Proof. Suppose that M is a compact manifold on which G, a compact Lie group, acts freely. Let H be a subgroup of G. Let V be a finite-dimensional representation of G, and let $i: G/H \to V$ be a G-embedding with normal bundle, N_1 (cf., [Sn 3, Part I]). The Becker-Gottlieb transfer is formed from the composition of G-maps, of Thom spaces:

$$T(V) \xrightarrow{\gamma} T(N_1) \xrightarrow{i_1} T(N_1 \oplus \tau_{G/H}) = (G/H)_+ \wedge T(V),$$

in which i_1 is induced by the zero-section map of the tangent bundle $\tau_{G/H}$. To obtain τ, one crosses $i_1 \cdot \gamma$ with the identity on M, divides out by G, and stabilizes (in the sense of §1.20) by smashing with Thom spaces of a complementary bundle.

In our example $\tau_{G/H}$ is, as a $\Sigma_2 \int S^1$-space, $(\Sigma_2 \int S^1)/U \times \mathbb{R}_t = S^1 \times \mathbb{R}_t$. Furthermore, we can pre-compose with i_1 ($i_1 = r$ in our example) rather than post-compose, without changing τ.

However, if we use the embedding $i: G/H \to V$ to construct an embedding, $\hat{p}$,

$$M/H = M \times_G G/H \xrightarrow{1 \times i} M \times_G V \subset M/G \times \mathbb{R}^s$$

as in §1.20, we see at once that t is obtained by taking $\gamma: T(V) \to T(N_1)$, crossing with the identity on M, dividing out by G and stabilizing.

Therefore $t \cdot r = \tau$, as required, upon taking limits over compact submanifolds of $E\Sigma_2 \int S^1$. □

4. *AN ORTHOGONAL EXAMPLE/EXERCISE*

In this section we will prove the result of Theorem 4.1. The proof is very similar to that of Chapter 5, (2.37), and the diligent reader might like to treat this as an exercise.

(4.1) Theorem

Let G be a finite group. Then the homomorphism, which was introduced in Chapter 5, (1.15),

$$\varphi_G : R_+(G, O_2(\mathbb{R}))^{\wedge}_{IA(G)} \to \{BG_+, BO_2(\mathbb{R})_+\},$$

is an isomorphism.

(4.2)   To prove the theorem, we wish to show that Φ_G in (4.1) is an isomorphism. By Chapter 5, §1.19, it suffices to verify this when G is a finite p-group for each prime, p.

Suppose first that $p \neq 2$. Consider the extension

(4.3) $$S^1 \xrightarrow{i} O_2(\mathbb{R}) \xrightarrow{\det} \mathbb{Z}/2,$$

which is split by $s: \mathbb{Z}/2 \to O_2(\mathbb{R})$ given by $s(\tau) = \begin{bmatrix} 0 & 1 \\ 1 & 0 \end{bmatrix}$, $(1 \neq \tau \in \mathbb{Z}/2)$. Hence we may write $O_2(\mathbb{R}) \cong \mathbb{Z}/2 \ltimes S^1$, where $\tau z \tau = \bar{z}$ $(z \in S^1)$. Define homomorphisms

(4.4)
$$\begin{cases} O_2(\mathbb{R}) \cong \mathbb{Z}/2 \ltimes S^1 \underset{v}{\overset{u}{\rightleftarrows}} \Sigma_2 \int S^1 \text{ by} \\ u(\tau^\varepsilon z) = \tau^\varepsilon(z, \bar{z}) \text{ and} \\ v(\tau^\varepsilon(z_1, z_2)) = \tau^\varepsilon z_1 \bar{z}_2 \ (z, z_1, z_2 \in S^1). \end{cases}$$

Thus $vu(\tau^\varepsilon z) = \tau^\varepsilon z^2$, which implies that the induced map of suspension spectra

$$B(vu)_+ : \Sigma^\infty(BO_2(\mathbb{R})_+) \to \Sigma^\infty(BO_2(\mathbb{R})_+)$$

is a (stable) homotopy equivalence when p-localized, if $p \neq 2$. Hence, if G is a p-group and $p \neq 2$, we have the following commutative diagram in which u, u_* are split monic, v, v_* are split epic, and the right-hand isomorphism comes from Chapter 5, §2.37.

(4.5)
$$\begin{array}{ccc} R_+(G, O_2(\mathbb{R}))\hat{_p} & \underset{v}{\overset{u}{\rightleftarrows}} & R_+(G, \Sigma_2 \int S^1)\hat{_p} \\ \Phi_G \downarrow & & \downarrow \cong \\ \{BG_+, BO_2(\mathbb{R})_+\} & \underset{v_*}{\overset{u_*}{\rightleftarrows}} & \{BG_+, B\Sigma_2 \int S^1\}. \end{array}$$

To see that vu is an isomorphism on $R_+(G, O_2(\mathbb{R}))$, one merely observes that the image of any subhomomorphism $(G \supset H \xrightarrow{\rho} \mathbb{Z}/2 \ltimes S^1 \cong O_2(\mathbb{R}))$ must be in the p-primary torsion of $SO_2(\mathbb{R}) = S^1$ on which $vu(z) = z^2$ is an isomorphism.

Hence, from (4.5), we see that Φ_G *is an isomorphism in* (4.1) *when G is a finite p-group of odd order.*

(4.6) *For the remainder of this section, let G be a finite 2-group.* The adjoint representation, $\mathrm{Ad}(O_2(\mathbb{R}))$, is a copy of $\mathbb{R}$, the real numbers, acted upon antipodally via $\det: O_2(\mathbb{R}) \longrightarrow \mathbb{Z}/2$. Denote this representation by $\mathbb{R}_r$. Let $J_k = \mathbb{Z}/2 \ltimes (\mathbb{Z}/2^k)$, where $\tau g \tau = g^{-1}$ $(g \in \mathbb{Z}/2^k, 1 \neq \tau \in \mathbb{Z}/2)$, be the standard copy of $D_{2^{k+1}}$ within $O_2(\mathbb{R})$.

Applying to the extension (4.3) the arguments which in Chapter 5, §2 were applied to $T^2 \rightarrowtail \Sigma_2 \int S^1 \twoheadrightarrow \mathbb{Z}/2$, we obtain the following result (cf., Chapter 5, §2.22).

(4.7) Theorem

There is a fibration (up to homotopy)

$$Q(BO_2(\mathbb{R})_+)\hat{_2} \to \underset{k}{\mathrm{holim}} \, Q((BJ_k)_+) \xrightarrow{t} \underset{k}{\mathrm{holim}} \, Q((B\mathbb{Z}/2^k)_+),$$

in which t is induced by the transfers associated with $\mathbb{Z}/2^k \subset J_k$.

Mapping BG_+ into this fibration, we obtain the following commutative diagram from the analogue of Chapter 5, (2.21):

(4.8)

$$
\begin{array}{ccc}
R_+(G,S^1)^\wedge_2 & \xrightarrow{\ i\ } & R_+(G,O_2(\mathbb{R}))^\wedge_2 \\
\Phi\,\Big\downarrow\cong & & \Big\downarrow\Phi_G \\
\{BG_+, BS^1_+\} & \xrightarrow{\ i_*\ } & \{BG_+, BO_2(\mathbb{R})_+\} \to \varprojlim_k R_+(G,J_k)^\wedge_2 \xrightarrow{\ t\ } \varprojlim_k R_+(G,\mathbb{Z}/2^k)^\wedge_2.
\end{array}
$$

From Chapter 5, §1.42, we know that $t: R_+(G,J_k) \to R_+(G,\mathbb{Z}/2^k)$ is given by

(4.9) $$ t(G \supset H \xrightarrow{\ \rho\ } J_k) = (G \supset \rho^{-1}(\mathbb{Z}/2^k) \xrightarrow{\ \rho\ } \mathbb{Z}/2^k). $$

Also, by Chapter 5, §§1.42–1.43 (e.g., by the argument used in Chapter 5, §1.47),

(4.10)
$$
\begin{cases}
\varprojlim_k R_+(G,J_k)^\wedge_2 \cong R_+(G,O_2(\mathbb{R}))^\wedge_2 \ \text{and} \\[2ex]
\varprojlim_k R_+(G,\mathbb{Z}/2^k)^\wedge_2 \cong R_+(G,S^1)^\wedge_2.
\end{cases}
$$

Making these identifications,

(4.11) $$ t: R_+(G,O_2(\mathbb{R}))^\wedge_2 \to R_+(G,S^1)^\wedge_2 $$

is given by

(4.12) $$ t(G \supset H \xrightarrow{\ \rho\ } O_2(\mathbb{R})) = (G \supset \rho^{-1}(SO_2(\mathbb{R})) \xrightarrow{\ \rho\ } SO_2(\mathbb{R})). $$

However, by definition, there is an isomorphism

(4.13) $$ R_+(G,O_2(\mathbb{R}))^\wedge_2 \cong \bigoplus_{(\rho)} \hat{\mathbb{Z}}_2, $$

where (ρ) runs over the $R_+(G,O_2(\mathbb{R}))$ equivalence classes of subhomomorphisms. From this we see that t is injective on the subgroup corresponding to the (ρ), for which $\det \rho = 1$ and $\rho - i(t(\rho)) \in \ker(t)$. Hence the kernel of (4.11) is isomorphic to

$$ \bigoplus_{\substack{(\rho) \\ \det \rho \neq 1}} \hat{\mathbb{Z}}_2. $$

Accordingly, we obtain from (4.8) an exact sequence (since Φ_G is easily seen to be monic by a transfer argument):

$$(4.14) \qquad \operatorname{im}(i) \rightarrowtail \{BG_+, BO_2(\mathbb{R})_+\} \twoheadrightarrow \bigoplus_{\substack{(\rho) \\ \det \rho \neq 1}} \hat{\mathbb{Z}}_2.$$

From (4.14), we see that $R_+(G, O_2(\mathbb{R}))_2^{\hat{}}$ and $\{BG_+, BO_2(\mathbb{R})_+\}$ are abstractly isomorphic. By arguments mentioned in [LMM], it is not hard to show that Φ_G realizes this isomorphism.

We have now verified Theorem 4.1 for finite p-group when p is any prime, which suffices to complete the proof for all finite G, by Chapter 5, §1.19.

Chapter Six

Explicit Brauer Induction Theory

DANTON. *But we are the poor musicians, and our bodies the pitiful instruments.
Does the playing of our tuneless agony sound in the distance like a
sigh of delight to the almighty ear?*

—*GEORG BUECHNER,*
"Danton's Death" (1835)

*In this chapter we will derive a canonical form, which I have christened Explicit
Brauer Induction, in order to express a finite-dimensional representation of a finite
group, G, as a linear combination of monomial representations, after the manner
of the Brauer induction theorem. We do this for unitary, orthogonal (even
dimensional), and symplectic representations, and in these cases "monomial" means
of the type $Ind_H^G(\varphi:H \to X_1)$, where $X_1 = U(1)$, $O(2)$, or $Sp(1)$, depending on the
type of representation.*

*Once we have the canonical form, it is not hard to prove that it has the desired
properties. It is defined in terms of a canonical action of G on a symmetric space
associated to the representation. However, finding the formula is an important
step. We do this by using the stable homotopy results of Chapter 5 to establish
the formula in $R(G)\hat{\ }$, the completed representation ring of G. This is done in §1
and is stated in Theorem 1.16. In fact, the properties of the canonical form, as
refined in Theorem 3.13, may also be discovered and proved, after completion, by
means of the stable homotopy calculations of Chapter 5. In addition, for p-groups,
this suffices to establish all the results, since all the completions are just p-adic
completion in this case so that each of the torsion-free groups that occur maps
injectively into its completion. Historically, this is the route that I took to discover
the results of Chapter 6—namely, using stable homotopy results as a guide to
results in representation theory—and this is my justification for inflicting Chapter 5
upon the reader.*

Section 1 closes with a series of examples of the formula in the case of two-dimensional representations.

In §2, using the Lefschetz fixed-point theorem, we prove that the formula of Theorem 1.16 actually holds in $R(G)$, and not just in $R(G)^\wedge$. Once again the stable homotopy results lead the way, for Chapter 5 uses the stable homotopy transfer, which is essentially a fixed-point invariant for maps of fibre bundles. Naturally, therefore, the proofs of Chapter 5 suggest that one should use fixed-point theory to make the group character computations of §2. There is no more celebrated piece of fixed-point theory than the Lefschetz theorem, and it is clear from the context of §2 to which map the theorem is to be applied, namely, to translation by the group element at which we wish to calculate the character of our formula in Theorem 2.31.

Section 2 concludes with the examples of the two-dimensional octahedral and icosahedral representations.

Once one has a canonical form for Brauer induction, one may derive a presentation for $R(G)$ in which the generators are monomial representations. This is done in §3 (Theorems 3.13 and 3.22). Again we illustrate each of our formulae by an example. Theorem 3.13 refines the formula of §2 to give a map from representations to the Grothendieck group of monomial representations. This map is the basic ingredient in the presentation of $R(G)$, given in Theorem 3.22. Theorem 3.22 solves a problem of J-P. Serre [Ser 6, p. 71 (footnote)].

In Chapter 7 we use §§3.13 to 3.22 to give a new construction of the local root numbers of Galois representations in a manner which is fundamentally different from that of Deligne [De2] (see also [Ta]).

---

1. FINITE-DIMENSIONAL REPRESENTATION THEORY

(1.1)   In the theory of the finite-dimensional complex representations of a finite group, G, one of the most famous (and very useful) results is the Brauer induction theorem [Ser 6].

Let $R(G)$ denote the representation ring of G. It is defined as the quotient of the free abelian group on isomorphism classes, $[V]$, of complex representations, V, modulo the relation

$$[V \oplus W] = [V] + [W].$$

$R(G)$ inherits from direct sum and tensor product of representations a sum and product operation, making it into a ring. It is also the free abelian group on the isomorphism classes of irreducible representations.

Brauer's theorem states that, in $R(G)$, any element, x, may be written as a sum ($n_i \in \mathbb{Z}, H_i \subset G$),

(1.2) $$x = \sum_i n_i \operatorname{Ind}_{H_i}^G (L_i) \in R(G),$$

where $\dim L_i = 1$ for each i.

(1.3) Let us pause to illustrate the usefulness of (1.2). Suppose that A is an abelian group and that

$$\mu, \lambda: R(G) \to A$$

are two (additive) invariants of representations which are invariant under induction. An example of such a homomorphism is given by the Artin L-functions [Mar] (when G is the Galois group of a finite extension of number fields.) If μ, λ are equal on one-dimensional representations then (1.2) implies that $\mu = \lambda$ because

$$\mu(x) = \sum_i n_i \mu(\text{Ind}_{H_i}^G L_i)$$

$$= \sum_i n_i \mu(L_i)$$

$$= \sum_i n_i \lambda(L_i)$$

$$= \sum_i n_i \lambda(\text{Ind}_{H_i}^G L_i)$$

$$= \lambda(x).$$

This, for example, shows that there is only one way (see [Mar]) to extend the theory of abelian L-functions to a general theory of L-functions in the sense of Artin.

(1.4) The example in §1.3 illustrates the need for a canonical form for the equation (1.2). In fact, when G is soluble, Deligne and Langlands [De 2] obtained generators for the kernel of the natural map

$$b: R_+(G, S^1) \to R(G).$$

Here, as in Chapter 5, §1.9, $R_+(G, S^1)$ is the Grothendieck group of monomial homomorphisms (or representations), $\rho: G \to \Sigma_m \int S^1$. Since a monomial representation is evidently a sum of representations induced from one-dimensional representations (see Chapter 5, (1.1)–(1.6)), the Deligne-Langlands theorem gives the ambiguity in (1.2) when G is soluble. This result is used in the Deligne-Langlands construction of local constants for local Galois or Weil representations [De 2].

In this section we will work in the completed representation ring, $R(G)^{\hat{}}$. That is, if $\varepsilon: R(G) \to \mathbb{Z}$ is given by the dimension function and $I(G) = \ker \varepsilon$, then we set

(1.5) $$R(G)^{\hat{}} = \varprojlim_n R(G)/I(G)^n R(G).$$

When G is a p-group, $R(G)\hat{\ }\cong(I(G)\otimes\hat{\mathbb{Z}}_p)\oplus\mathbb{Z}$ (where $\hat{\mathbb{Z}}_p$ denotes the p-adic integers). In general, the kernel of

$$R(G)\to R(G)\hat{\ }$$

consists [At] *of characters which vanish on all elements of prime-power order.*

The advantage of working in $R(G)\hat{\ }$ is that we will be able, for all finite groups G, to find a canonical formula (Theorem 1.16) for Brauer induction by topological means. Once found, in the next section, we will develop its properties in $R(G)$ rather than in $R(G)\hat{\ }$.

(1.6)   Let $R_+(G,\Sigma_n\!\int S^1)$ denote the Grothendieck group of monomial homomorphisms, $\rho:G\to\Sigma_d\!\int\Sigma_n\!\int S^1$. By Chapter 5, (1.1)–(1.6), this is the Grothendieck group of subhomomorphisms

$$\left(G\supset H\xrightarrow{\ \rho'\ }\Sigma_n\!\int S^1\right),$$

and the homomorphism of Chapter 5, (1.15) and (1.16),

$$\Phi:R_+\left(G,\Sigma_n\!\int S^1\right)\hat{\ }_{IA(G)}\to\left\{BG_+,B\Sigma_n\!\int S^1_+\right\}$$

sends this subhomomorphism to the S-map

$$BG_+\xrightarrow{\ \tau\ }BH_+\xrightarrow{\ (B\rho')_+\ }B\Sigma_n\!\int S^1_+,$$

where τ is the transfer map associated with $H\subset G$. Incidentally, we have seen that the homomorphism Φ is an isomorphism when $n=1$ or 2. This is actually true for all n [Sn-Z].

Suppose now that $\chi:G\to U(n)$ is a homomorphism of the finite group, G, into the unitary group. The normalizer of the diagonal maximal torus in $U(n)$ is isomorphic to $\Sigma_n\!\int S^1$. Suppose that $g\in U(n)$ and that $H(g,\chi)$ is the subgroup of G given by

(1.7) $$H(g,\chi)=\chi^{-1}\left(g\left(\Sigma_n\!\int S^1\right)g^{-1}\right)\subset G.$$

(1.8) Lemma

With the notation of §1.6, the S-map given by

$$BG_+\xrightarrow{\ (B\chi)_+\ }B\chi(G)_+\xrightarrow{\ \tau\ }B\left(\left(g\left(\Sigma_n\!\int S^1\right)g^{-1}\right)\cap\chi(G)\right)_+$$

$$\xrightarrow{\ C_g\ }B\left(\left(\Sigma_n\!\int S^1\right)\cap(g^{-1}\chi(G)g)\right)_+\xrightarrow{\ (Bj)_+\ }B\Sigma_n\!\int S^1_+$$

is equal to the S-map

$$\Phi\left(G \supset H(g,\chi) \xrightarrow{g^{-1}\chi g} \Sigma_n \int S^1 \right) \in \left\{ BG_+, B\Sigma_n \int S^1_+ \right\}.$$

Here τ denotes a transfer map, and C_g, as in Chapter 5, §3.5, is induced by conjugation by g^{-1}.

Proof. We have a commutative diagram of group homomorphisms

$$\begin{array}{ccc}
H(g,\chi) & \xrightarrow{\chi} & (g(\Sigma_n \int S^1)g^{-1}) \cap \chi(G) \\
{\scriptstyle i_1}\downarrow & & \downarrow{\scriptstyle i_2} \\
G & \xrightarrow{\quad\chi\quad} & \chi(G),
\end{array}$$

in which the horizontal maps have the same kernel so that the spaces of cosets of the inclusions i_1 and i_2 are identified by χ. Hence, by naturality of the transfer, we obtain a homotopy commutative diagram of S-maps:

$$\begin{array}{ccc}
BG_+ & \xrightarrow{(B\chi)_+} & B\chi(G)_+ \\
{\scriptstyle \tau}\downarrow & & \downarrow{\scriptstyle \tau} \\
BH(g,\chi) & \xrightarrow{(B\chi)_+} & B((g(\Sigma_n \int S^1)g^{-1}) \cap \chi(G))_+.
\end{array}$$

Therefore, in $\{BG_+, B\Sigma_n \int S^1_+\}$,

$$(Bj)_+ (Cg)\tau(B\chi)_+ = (Bj)_+ (Cg)(B\chi_+)\tau$$

$$= (B(g^{-1}\chi g)_+)\tau$$

as required. □

(1.9) Suppose now that G is a finite group and that χ is a finite-dimensional representation of G. Since G is finite, we may assume that χ is unitary, and therefore corresponding to a homomorphism $\chi\colon G \to U(n)$.

This representation entitles us to an element, $[\chi]$, of $R(G)$. However, by virtue of the main theorem of [At] (see also [At-Se]), we may recover the image of $[\chi]$ in $R(G)\hat{\ }$ topologically by means of K-theory [At 2].

By [At; At-Se] the map that assigns to χ the vector bundle $\alpha(\chi)\colon EG \times_G \chi \to BG$ (see Chapter 4, (1.28)) induces a homomorphism $\alpha\colon R(G) \to K(BG)$, where $K(X)$ ($= K^0(X)$) is the (unitary) K-cohomology of X [At 2]. This homomorphism is continuous and extends to a homeomorphism

(1.10) $$\alpha\colon R(G)\hat{\ } \xrightarrow{\cong} K(BG).$$

If we set $\tilde{K}(X) = \ker(\dim: K(X) \to \mathbb{Z})$, then α induces an isomorphism

(1.11) $$\alpha: I(G)^{\wedge} \to \tilde{K}(BG).$$

An S-map, $f: X \to Y$, induces $f^*: \tilde{K}(Y) \to \tilde{K}(X)$ so that, in particular, we obtain a commutative diagram of homomorphisms:

(1.12)

$$
\begin{array}{ccc}
R(U(n)) & \xrightarrow{\;\chi^*\;} & R(G) \\
\downarrow{\scriptstyle\alpha} & & \downarrow{\scriptstyle\alpha} \\
K(BU(n)) & \xrightarrow{(B\chi)^*} & K(BG) \cong R(G)^{\wedge} \\
\downarrow{\scriptstyle\cong} & & \downarrow{\scriptstyle\cong} \\
\tilde{K}(BU(n)_+) & \xrightarrow{(B\chi_+)^*} & \tilde{K}(BG_+).
\end{array}
$$

Therefore, if we define

(1.13) $$\Gamma_n = [\mathbb{C}^n] \in R(U(n))$$

(where $[\mathbb{C}^n]$ is the natural action of $U(n)$ on complex n-space), then

(1.14) $$(B\chi_+)^*(\Gamma_n) = [\chi] \in R(G)^{\wedge} \cong \hat{K}(BG_+).$$

Suppose now that $H \subset G$ is an inclusion of finite groups and that

$$\tau: BG_+ \to BH_+$$

is the associated S-map transfer. In [K-P] a result of Boardman is recorded which implies that

(1.15) $$\tau^* = \operatorname{Ind}_H^G: K(BH) \cong R(H)^{\wedge} \to R(G)^{\wedge} \cong K(BG).$$

With these preliminaries out of the way, we are now ready to accomplish the objectives of §1.4.

(1.16) Theorem

Let $v: G \to U(n)$ be a representation of a finite group, G. Then, in $R(G)^{\wedge}$, we have

$$[v] = \sum_{g_i} \aleph^{\#}(M_i) \operatorname{Ind}_{H(g_i, v)}^G \left\{ \left[g_i^{-1} v g_i : H(g_i, v) \to \Sigma_n \int S^1 \to U(n) \right] \right\}.$$

(1.17) Remark

In Theorem 1.16 the sum is taken over orbit-type manifold representations, $g_i \in v(G) \backslash U(n)/(\Sigma_n \int S^1)$, as in the double-coset formula of Chapter 5, §3.7, with (K, G, H) replaced by $(v(G), U(n), \Sigma_n \int S^1)$. The $\aleph^{\#}(M_i)$ are the associated internal Euler characteristics of Chapter 5, §3.3.

Note that $g_i^{-1} v g_i$ is a sum of monomial representations, since its image lies in $\Sigma_n \int S^1$ and that the set of monomial representations is closed under induction.

Note also, that §1.16 is a formula for a representation, v, but not for a virtual representation. This is simply because the orbit structure for v and η would be different, so one must calculate the §1.16 separately for v and η to get the formulae for $v - \eta$.

(1.18) Proof of Theorem 1.16

We wish to calculate, by (1.13) and (1.14), $[v] = (Bv_+)^*(\Gamma_n) \in \tilde{K}(BG_+)$. However, the S-map

$$BU(n)_+ \xrightarrow{\tau} \left(B\Sigma_n \int S^1 \right)_+ \xrightarrow{(Bj_+)} BU(n)_+$$

is multiplication by the Euler characteristic, $\aleph(U(n)/\Sigma_n \int S^1) = 1$, on $\tilde{H}^*(-; \mathbb{Q})$ [B-G]. The Chern character gives an isomorphism $KU(BU(n)) \otimes \mathbb{Q} \cong H^{ev}(BU(n); \mathbb{Q})$ so that

$$\tau^*(Bj_+) = 1 : \tilde{K}U(BU(n)_+) \to \tilde{K}U(BU(n)_+),$$

since $\tilde{K}U(BU(n)_+)$ is torsion free.

Therefore we have, in $R(G)^{\hat{}}$,

$$[v] = (Bv_+)^*(\Gamma_n)$$

$$= (Bv_+)^* \tau^*(Bj_+)^*(\Gamma_n)$$

$$= \sum_{g_i} \aleph^{\#}(M_i) \psi_i^*(Bj_+)^*(\Gamma_n), \quad \text{by the double-coset formula of Chapter 5, §3.7,}$$

$$= \sum_{g_i} \aleph^{\#}(M_i) \tau_i^*(B(g_i^{-1} v g_i)_+)^*(Bj_+)^*(\Gamma_n),$$

by §1.18, where τ_i is the transfer map associated with $H(g_i, v) \subset G$, in the notation of §1.8. However, $j(g_i^{-1} v g_i)$ is the homomorphism $g_i^{-1} v g_i : H(g_i, v) \to \Sigma_n \int S^1 \to U(n)$ so that

$$(B(g_i^{-1} v g_i)_+)^*(Bj_+)^*(\Gamma_n) = [g_i^{-1} v g_i] \in R(H(g_i, v))^{\hat{}},$$

and the result follows from (1.15), which states that τ_i^* corresponds to $\text{Ind}_{H(g_i, v)}^G$. $\qquad\square$

(1.19)   In the next section we shall see that Theorem 1.16 is true in $R(G)$. However, we will pause to verify this fact on some two-dimensional examples, for which we calculated the internal Euler characteristic in Chapter 5.

(1.20) Examples

Let us consider the simplest nontrivial example of §1.16. Namely, let

$$v: D_{2n} \to \Sigma_2 \int S^1$$

be the inclusion homomorphism given by

$$v(y) = \begin{bmatrix} 0 & 1 \\ 1 & 0 \end{bmatrix}$$

$$v(x) = \begin{bmatrix} \xi_n & 0 \\ 0 & \bar{\xi}_n \end{bmatrix} \qquad (\xi_n = \exp(2\pi i/n)),$$

where D_{2n} is the dihedral group of order $2n$ generated by x and y as in Chapter 1, (3.12). From Chapter 5, §3.9 (in Cases iii and iv which are tabulated in Tables A and B), we see that the expression in §1.16 reduces to three terms in Case iii and to seven terms in Case iv.

Let us abbreviate $H(g_i, v)$ to $H(g)$. Firstly, I will give a résumé of the subgroups $\Sigma_2 \int S^1 \cap (g\Sigma_2 \int S^1 g^{-1})$ for each of the double-coset representatives listed in Chapter 5, §3 Tables A or B. We will need this information in order to examine $H(g) = D_{2n} \cap (g\Sigma_2 \int S^1 g^{-1})$. There are eight cases to list:

(1.21)

Case i: $g = \begin{bmatrix} 1 & 0 \\ 0 & 1 \end{bmatrix}$, $H(g) = \Sigma_2 \int S^1$.

Case ii: $g = 1/\sqrt{2} \begin{bmatrix} 1 & -1 \\ 1 & 1 \end{bmatrix}$, $H(g) = \left\{ \begin{bmatrix} a & 0 \\ 0 & \pm a \end{bmatrix}, \begin{bmatrix} 0 & a \\ \pm a & 0 \end{bmatrix}; a \in S^1 \right\}$.

Case iii: $g = 1/\sqrt{3} \begin{bmatrix} \xi_{2n} & -\sqrt{2} \\ \sqrt{2} & \bar{\xi}_{2n} \end{bmatrix}$, $H(g) = \left\{ \begin{bmatrix} b & 0 \\ 0 & b \end{bmatrix}, \begin{bmatrix} 0 & -b \\ \bar{\xi}_{2n}^2 b & 0 \end{bmatrix}; b \in S^1 \right\}$.

Case iv. $g = 1/\sqrt{3} \begin{bmatrix} 1 & -\sqrt{2} \\ \sqrt{2} & 1 \end{bmatrix}$, $H(g) = \left\{ \begin{bmatrix} a & 0 \\ 0 & a \end{bmatrix}, \begin{bmatrix} 0 & a \\ -a & 0 \end{bmatrix}; a \in S^1 \right\}$

Case v: $g = 1/\sqrt{2} \begin{bmatrix} 1 & z \\ -\bar{z} & 1 \end{bmatrix}$, $z = e^{i\theta}$, θ/π irrational;

$$H(g) = \left\{ \begin{bmatrix} b & 0 \\ 0 & \pm b \end{bmatrix}, \begin{bmatrix} 0 & zb \\ \pm \bar{z}b & 0 \end{bmatrix}; b \in S^1 \right\}.$$

Case vi: $g = 1/\sqrt{2}\begin{bmatrix} \xi_{4n} & 1 \\ 1 & -\bar{\xi}_{4n} \end{bmatrix}$ (n odd);

$$H(g) = \left\{ \begin{bmatrix} b & 0 \\ 0 & \pm b \end{bmatrix}, \begin{bmatrix} 0 & \xi_{4n}b \\ \pm \bar{\xi}_{4n}b & 0 \end{bmatrix}; b \in S^1 \right\}.$$

Case vii: $g = 1/\sqrt{2}\begin{bmatrix} \xi_{2n} & 1 \\ 1 & -\bar{\xi}_{2n} \end{bmatrix}$ (n even),

$$H(g) = \left\{ \begin{bmatrix} a & 0 \\ 0 & \pm a \end{bmatrix}, \begin{bmatrix} 0 & \pm \xi_{2n}a \\ \bar{\xi}_{2n}a & 0 \end{bmatrix}; a \in S^1 \right\}.$$

Case viii: $g = 1/\sqrt{5}\begin{bmatrix} \xi_{4n} & 2 \\ 2 & -\bar{\xi}_{4n} \end{bmatrix}$ (n even),

$$H(g) = \left\{ \begin{bmatrix} a & 0 \\ 0 & a \end{bmatrix}, \begin{bmatrix} 0 & a \\ -\bar{\xi}_{4n}^2 a & 0 \end{bmatrix}; a \in S^1 \right\}.$$

(1.22)   Now we will proceed to evaluate in $R(D_{2n})$, the expression in §1.16 when v is given as in §1.20. Our objective is to show that this expression is v.

We begin with the observation that conjugation by g_i is the identity on $R(D_{2n})$ so that we must show that

(1.23) $$\sum_{g_i \neq 1} \aleph^\#(M_i) \operatorname{Ind}_{H(g_i)}^{D_{2n}}(v) = 0 \qquad \text{in } R(D_{2n}).$$

We examine three cases separately

(1.24) $$\begin{cases} \text{Case i:} & n \text{ is odd, } n = 2s + 1. \\ \text{Case ii:} & n = 2(2s + 1). \\ \text{Case iii:} & n = 2^{\alpha-1}(2s + 1), \alpha \geq 3. \end{cases}$$

(1.25):   If $g(q)$ ($1 \leq q \leq 8$) is the qth $g \in U(2)$ in the cases of (1.21), then in (1.24) Case i we have, by Chapter 5, §3, Table A.

(1.26) $$\operatorname{Ind}_{H(g(2))}^{D_{2(2s+1)}}(v) - \operatorname{Ind}_{H(g(6))}^{D_{2(2s+1)}}(v) \in R(D_{2(2s+1)}).$$

However, from §1.21, Cases ii and vi, one easily verifies that $H(g(2))$ and $H(g(6))$ are each cyclic of order 2. Since they do not live in $\mathbb{Z}/(2s + 1)$, these two groups are conjugate in $D_{2(2s+1)}$. Furthermore in each case v restricts to the sum $1 + L$, where L is the nontrivial, one-dimensional representation. Hence (1.26) is zero, as required.

In (1.24), Case ii, $n = 2(2s + 1)$, and the image of $D_{4(2s+1)}$ in $(\Sigma_2 \int S^1)/\Delta(S^1) \cong \Sigma_2 \ltimes S^1$ is isomorphic to $D_{4(2s+1)}/(\pm I_2) \cong D_{2(2s+1)}$. Hence, by Chapter 5, §3, Table A, we must consider

(1.27) $$\operatorname{Ind}_{H(g(2))}^{D_{4(2s+1)}}(v) - \operatorname{Ind}_{H(g(6))}^{D_{4(2s+1)}}(v) \in R(D_{4(2s+1)}),$$

where $\tilde{H}(g(q))$ is the inverse image of $H(g(q))$ under the quotient map $D_{4(2s+1)} \to D_{2(2s+1)}$. Hence, by comparison with (1.24), Case i, $\tilde{H}(g(2))$ and $\tilde{H}(g(6))$ are conjugate copies of $\mathbb{Z}/2 \times \mathbb{Z}/2$ in $D_{4(2s+1)}$ and if L_1, L_2 are the two projections $\mathbb{Z}/2 \times \mathbb{Z}/2 \to \mathbb{Z}/2$ in each case v restricts to $L_1(1 + L_2)$ so that (1.27) vanishes as required.

Turning to (1.24), Case iii, the image of $D_{2n} = D_{2^\alpha(2s+1)} (\alpha \geq 3)$ in $\Sigma_2 \int S^1 / \Delta(S^1) \cong \Sigma_2 \ltimes S^1$ is D_{2N}, where $N = 2^{\alpha-2}(2s+1)$ is even. Hence, by Chapter 5, §3, Table B, we must consider

$$(1.28) \quad \left\{ \begin{array}{l} \mathrm{Ind}_{\tilde{H}(g(8))}^{D_{2n}}(v) + \mathrm{Ind}_{\tilde{H}(g(2))}^{D_{2n}}(v) + \mathrm{Ind}_{\tilde{H}(g(7))}^{D_{2n}}(v) \\[2mm] - \mathrm{Ind}_{\tilde{H}(g(3))}^{D_{2n}}(v) - \mathrm{Ind}_{\tilde{H}(g(4))}^{D_{2n}}(v) - \mathrm{Ind}_{\tilde{H}(g(5))}^{D_{2n}}(v). \end{array} \right.$$

In (1.28), $\tilde{H}(g(q))$ is the inverse image of $H(g(q))$ under the quotient map $D_{2n} \to D_{2N}$ (so that we replace n by N in (1.21), Cases i–viii). The details of this expression are as follows:

$$\tilde{H}(g(8)) = \{ \pm I_2 \} \cong \mathbb{Z}/2 \quad \text{and} \quad (v | \tilde{H}(g(8))) = 2L,$$

where L is the nontrivial, one-dimensional representation. However, $\tilde{H}(g(4))$ and $\tilde{H}(g(3))$ are both isomorphic to $\mathbb{Z}/2 \times \mathbb{Z}/2$, each one containing $\tilde{H}(g(8))$. In fact, in both cases, $H = \tilde{H}(g(3))$ and $\tilde{H}(g(4))$,

$$(v | H) \cong \mathrm{Ind}_{\tilde{H}(g(8))}^{H}(L),$$

with the result that the first, fourth, and fifth terms cancel in (1.28). Also each of $\tilde{H}(g(2))$ and $\tilde{H}(g(7))$ is isomorphic to a copy of D_8 containing the subgroup

$$\tilde{H}(g(5)) \cong \mathbb{Z}/4 \left\langle \begin{bmatrix} i & 0 \\ 0 & -i \end{bmatrix} \right\rangle.$$

In each case $(v | D_8)$ is the unique irreducible, two-dimensional representation of D_8. However, this irreducible as isomorphic to

$$\mathrm{Ind}_{\mathbb{Z}/4}^{D_8}(y) = \mathrm{Ind}_{\mathbb{Z}/4}^{D_8}(y^3),$$

where $y: \mathbb{Z}/4 \rightarrowtail S^1$ is an inclusion. Since $(v | \tilde{H}(g(5))) \cong y + y^3$ the second, third, and sixth terms in (1.28) cancel. Therefore (1.28) vanishes in $R(D_{2n})$, as required. $\qquad \square$

(1.29) Remark

From Tables C and D of Chapter 5, §3 one sees that if

$$v: K \to T^2 \subset U(2)$$

is a diagonal representation, then §1.16 merely states that $v = v$.

(1.30) A QUATERNIONIC EXAMPLE

If $\xi_m = \exp(\pi i/m)$, then setting

$$\eta(x) = \begin{bmatrix} \xi_{2n} & 0 \\ 0 & \bar{\xi}_{2n} \end{bmatrix}, \quad \eta(y) = \begin{bmatrix} 0 & -1 \\ 1 & 0 \end{bmatrix},$$

gives an embedding $\eta : Q_{4n} \to \Sigma_2 \int S^1$ of the quaternion group of Chapter 1, §(3.10)

$$Q_{4n} = \langle x, y \,|\, x^n = y^2, y^4 = 1, xyx = y \rangle.$$

Let us briefly examine §1.16 in this case. First, we observe that

$$\eta(x^n) = \eta(y^2) = \begin{bmatrix} -1 & 0 \\ 0 & -1 \end{bmatrix},$$

which is central so that the left action of Q_{4n} on $\mathbb{R}P^2 \cong U(2)/(\Sigma_2 \int S^1)$ factors through the quotient $Q_{4n} \to Q_{4n}/(y^2) \cong D_{2n}$. This means that the internal Euler characteristics of §1.16 may be read from Chapter 5, §3, Table A (if n is odd), and Table B (if n is even).

The discussion of (1.24), Cases i–iii, given in §1.25 lifts (by naturality of Ind_H^G under pullbacks) to verify that §1.16 holds when v is replaced by η; $Q_{4n} \to \Sigma_2 \int S^1$.

2. EXPLICIT BRAUER INDUCTION IN R(G)

2.1   In this section we will improve Theorem 1.16 so that it becomes an equation in $R(G)$. However, there is no additional difficulty if we formulate everything simultaneously for unitary, orthogonal, and symplectic representations. Therefore let us denote by X_n one of the following classical groups (unitary, orthogonal, or symplectic):

(2.2) $X_n = U(n), \quad O(2n), \quad \text{or} \quad Sp(n).$

We will also need the following subgroup, Y_n, of X_n. If $X_n = U(n)$, then Y_n will denote the normalizer of the standard, diagonal maximal torus, if $X_n = Sp(n)$, then Y_n will denote the normalizer of the diagonal subgroup, $X_1^n = Sp(1)^n$, and if $X_n = O(2n)$, then Y_n will denote the normalizer of the subgroup of diagonal 2×2 blocks (i.e., $O_2(\mathbb{R})^n = O(2)^n$ along the diagonal). In each case Y_n is given by a wreath product

(2.3) $Y_n = \Sigma_n \int X_1.$

These groups have been chosen for the following two basic properties:

(2.4) Lemma

If $\Lambda \subset \mathbb{C}$ is a subring in which $n!$ is invertible, then

$$H^*(X_n/Y_n; \Lambda) \cong H^*(\text{point}; \Lambda).$$

Proof. By calculations originally due to Borel [H], the inclusion of the diagonal torus, T^n, into $U(n)$ induces an isomorphism

$$(2.5) \qquad\qquad H^*(BU(n); \Lambda) \xrightarrow{\cong} H^*(BT^n; \Lambda)^{\Sigma_n},$$

where Σ_n permutes the factors in T^n. However, since $|\Sigma_n| = n!$, the Serre spectral sequence for the extension

$$T^n \rightarrowtail Y_n \twoheadrightarrow \Sigma_n$$

takes the form

$$E_2^{s,t} = \begin{bmatrix} H^t(BT^n; \Lambda)^{\Sigma_n} & \text{if } s = 0 \\ 0 & \text{if } s \neq 0 \end{bmatrix} \Rightarrow H^{s+t}(BY_n; \Lambda).$$

This means that restriction to T^n induces an isomorphism, compatible by naturality with (2.5)

$$H^*(BY_n; \Lambda) \xrightarrow{\cong} H^*(BT^n; \Lambda)^{\Sigma_n}.$$

Therefore in the unitary case the natural map

$$H^*(BX_n; \Lambda) \to H^*(BY_n; \Lambda)$$

is an isomorphism. However, the fibre of this natural map is X_n/Y_n, which therefore has trivial cohomology, by a simple argument using the Serre spectral sequence for the fibring

$$X_n/Y_n \to BY_n \to BX_n,$$
$$E_2^{s,t} = H^s(BX_n; H^t(X_n/Y_n; \Lambda)) \Rightarrow H^{s+t}(BY_n; \Lambda).$$

When X_n is $O(2n)$ or $Sp(n)$, the argument follows the same pattern, mutatis mutandis.   □

(2.6) Lemma

Let G be a finite group, and let $\rho: G \to Y_n$ be a homomorphism. Then, within Y_n,

ρ is conjugate to the matrix sum of monomial homomorphisms of the form

$$\lambda_i = \text{Ind}_{H_i}^G(\varphi_i): G \to Y_{n_i},$$

where $\varphi_i: H_i \to X_1$ and $\sum n_i = n$. Here λ_i is the "induced" monomial homomorphism constructed as in Chapter 5, §§(1.3)–(1.7). In addition, if ρ is varied by inner automorphisms of Y_n or of G, then, up to permutation, the $\{\varphi_i\}$ remain well-defined up to conjugation within G.

Proof. This is just a restatement, with the emphasis appropriate for this chapter, of the discussion of Chapter 5, §§(1.1)–(1.7). □

(2.7) Let G be a finite group, and let $R(G)$ denote its complex (unitary) representation ring. Let $RO(G)$ and $RH(G)$ denote, respectively, the Grothendieck groups of finite-dimensional orthogonal and symplectic representations of G.

We have maps, induced either by forgetting some structure or by tensoring up to a larger field:

(2.8) $$RO(G) \underset{c}{\overset{r}{\rightleftarrows}} R(G) \underset{q}{\overset{c'}{\rightleftarrows}} RH(G).$$

If V is a unitary G-representation, then $\bar{V}$ is the complex conjugate of V (i.e., same G-action but the complex conjugate vector space structure). We have the following relations between the homomorphisms of (2.8) [Ad, pp. 27–28]:

(2.9)
$$\begin{cases} rc(y) = 2y, \\ qc'(z) = 2z, \\ c'q(x) = x + \bar{x}, \\ \overline{c(y)} = c(y), \\ r(\bar{x}) = r(x), \\ \overline{c'(z)} = c'(z), \\ q(\bar{x}) = q(x), \\ (\bar{\bar{x}}) = x, \\ cr(x) = x + \bar{x}. \end{cases}$$

Suppose that V is a unitary representation of G, whose character

$$\aleph_V(g) = \text{Trace}\,((g\cdot-): V \to V)$$

is real valued. If V is *irreducible* among such real-valued characters, then V falls into one of the following (mutually exclusive) families [Mar p. 59; Ser 6]

(2.10)

  (i) $V = cr(W)$, where W irreducible and $\aleph_W$ is not real valued.
  (ii) $V = c(X)$, where X is an absolutely irreducible orthogonal representation.
  (iii) $V = c'(Z)$, where Z is an irreducible quaternionic representation.

A real representation is *absolutely irreducible* if its complexification is irreducible.

(2.11) Proposition [*Ad*, §3.50]

A unitary representation, V, of G is

  (i) real (i.e., $V = c(Y)$) if and only if there exists a symmetric, $\mathbb{C}$-linear, nondegenerate G-invariant, bilinear form on V.
  (ii) quaternionic (i.e., $V = c'(Z)$) if and only if there exists a skew-symmetric, $\mathbb{C}$-linear, nondegenerate, G-invariant bilinear from on V.

(2.12) Examples

(a) If $G \cong \mathbb{Z}/n = \langle g \rangle$ and n is even, then we have a homomorphism $\varepsilon \colon G \to \mathbb{Z}/2 \cong O_1(\mathbb{R})$. Also ε and 1 (i.e., $\mathbb{R}$ with trivial action) are the only one-dimensional real representations. The other irreducible representations are the realification of the action on $\mathbb{C}$, in which g acts by multiplication by $\exp((2\pi i)/n)^s$ for some $1 \le s \le n - 1$. When $s = 1$ call this representation y, and then $R(\mathbb{Z}/n) \cong \mathbb{Z}[y]/(y^n - 1)$. Note that $c \colon RO(\mathbb{Z}/2) \to R(\mathbb{Z}/2)$ is an isomorphism.

(b) Let $G = D_{2n} = \langle x, y; x^n = 1 = y^2, xyx = y \rangle$ be the dihedral group of Chapter 1, §3.12. We have one-dimensional real representations 1, x_1 and (if n is even) $x_2, x_1 x_2 \colon D_{2n} \to \{\pm 1\} = O_1(\mathbb{R})$ as in Chapter 1, (4.4). In addition, if y is the one-dimensional complex representation of $\mathbb{Z}/n = \langle x \rangle$, then

$$l_u = \mathrm{Ind}_{\mathbb{Z}/n}^{D_{2n}}(y^u) \qquad (1 \le u \le n - 1)$$

is a two-dimensional complex representation which is the complexification of the action on $\mathbb{R}^2$ given by

$$x = \begin{bmatrix} \cos(2\pi u/n) & \sin(2\pi u/n) \\ -\sin(2\pi u/n) & \cos(2\pi u/n) \end{bmatrix},$$

$$y = \begin{bmatrix} 0 & 1 \\ 1 & 0 \end{bmatrix}.$$

This is easily seen by observing that the character of $l_u = \mathbb{C}[D_{2n}] \otimes_{\mathbb{C}[\mathbb{Z}/n]} y^u$

is given by

(2.13)
$$\aleph_u(x^a y^\delta) = \begin{cases} 0 & \text{if } \delta = 1, \\ \zeta_n^{au} + \zeta_n^{-au} & \text{if } \delta = 0, \end{cases}$$

where $\zeta_n = \exp(2\pi i/n)$. Note that $l_u = l_{n-u}$.

To see that the set $\{1, x_1 \text{ (and } x_2, x_2x_1 \text{ if } n \text{ is even) } l_u \text{ for } 1 \le u < n/2\}$ are all the irreducible representations of D_{2n}, it suffices [Ser 6; C-R] to observe that the squares of their dimensions add up to $2n$.

We have shown that

(2.14)
$$c: RO(D_{2n}) \xrightarrow{\cong} R(D_{2n})$$

is an isomorphism, since it is surjective and $rc = 2$ by (2.9).

(2.15) Remark

Once we have shown that Theorem 1.16 is true in $R(G)$ (and its analogues in $RO(G)$ and $RH(G)$), we will have shown, as a corollary, that $R(G)$, $RO(G)$, and $RH(G)$ are generated by representations of the form

$$\text{Ind}_H^G(\varphi: H \to X_1)$$

(plus trivial representations in the orthogonal case). In the case of $R(G)$ this is Brauer's induction theorem. We are going to establish Explicit Brauer Induction formulae by studying the action of G on X_n/Y_n which is afforded by a representation, $v: G \to X_n$. However, I think the method may be introduced in a more satisfactory manner if we use it first to prove Brauer's result.

(2.16) Proposition (Brauer Induction)

Let G be a finite group. Then $R(G)$, $RH(G)$, and $RO(G)$ are generated by representations of the form

$$\text{Ind}_H^G(\varphi: H \to X_1),$$

together with the trivial representation in the orthogonal case. Here X_1 is as in (2.2).

Proof. Firstly, we remark that in the orthogonal case, $X_1 = O(2)$ so that we have to allow for adding a trivial representation in order to apply the following argument to even-dimensional representations in the orthogonal case.

Suppose that $v: G \to X_n$ is a representation of the appropriate type. If there exists $x \in X_n$ such that $xv(G)x^{-1}$ lies in Y_n, then we are done, by §2.6. Therefore we will suppose that this is not the case, and we will proceed by induction on the order of G to show that v is in the subgroup generated by induced representations as specified in the statement of the proposition.

The group G acts on X_n/Y_n by means of left translation, via v. Furthermore G has no fixed points because

(2.17)
$$\left[\begin{array}{l} v(g)yY_n = yY_n \text{ for all } g \in G \\ \text{if and only if } y^{-1}v(G)y \subset Y_n. \end{array}\right.$$

We may assume that the compact manifold, X_n/Y_n, has been triangulated so that G acts simplicially. This means that $g \in G$ either fixes an open simplex, σ, pointwise or it moves it to a disjoint simplex. Let $C_*(X_n/Y_n; \mathbb{C})$ denote the simplicial chain complex of X_n/Y_n. Each chain group, $C_i(X_n/Y_n; \mathbb{C})$, is a finite-dimensional G-representation so that the Euler characteristic

$$\sum_i (-1)^i C_i(X_n/Y_n; \mathbb{C}) \in R(G).$$

However, by a well-known and easy lemma, we can obtain the same Euler characteristic by passing to homology. By §2.4, X_n/Y_n has the complex homology of a point so that

$$1 = \sum_i (-1)^i C_i(X_n/Y_n; \mathbb{C}) \in R(G),$$

since the homology of a point can only have trivial G-action. On the other hand, because the action is simplicial, each $C_i(X_n/Y_n; \mathbb{C})$ is the sum of permutation representations $\mathrm{Ind}_H^G(1)$, where H is the stabilizer of some i-dimensional simplex. Moreover we have seen that $H \neq G$ so that

$$1 = \sum_{H \neq G} a_H \mathrm{Ind}_H^G(1) \in R(G).$$

Therefore, multiplying by v and applying Frobenius reciprocity,

$$v = \sum_{H \neq G} a_H \mathrm{Ind}_H^G(1) \otimes v$$

$$= \sum_{H \neq G} a_H \mathrm{Ind}_H^G(\mathrm{Res}_H^G(v)),$$

and the result follows by induction. $\square$

(2.18) *THE TOPOLOGY OF A REPRESENTATION*

Now we will examine the method of §2.17 more closely.

Let G denote a finite group, and let X denote a compact, connected Lie group with closed subgroup, Y. Suppose that

$$(2.19) \qquad\qquad v: G \to X$$

is a homomorphism. In subsequent applications, (2.19) will be a representation, with X being one of the classical groups $U(n)$, $SO(2n)$, or $Sp(n)$.

The homogeneous space, X/Y, is naturally a left G-space by means of the action

$$(2.20) \qquad\qquad g(xY) = v(g)xY \qquad (g\in G, x\in X).$$

The G-space, X/Y, admits a G-equivariant triangulation, making it into a finite G-simplicial complex in such a way that the orbit space, $M = G\backslash X/Y$, inherits the G-quotient simplicial structure. If H is a subgroup of G, then $M_{(H)}$ denotes the set of G-orbits in M which are G-isomorphic to G/H. Since M is a compact stratified space, it is possible to find a homotopy, H', from the identity map to a map

$$(2.21) \qquad\qquad \beta': M \to M = G\backslash X/Y,$$

having only a finite number of isolated fixed points. In fact, with care, one can ensure that H' and β' are well-behaved with respect to the orbit-type structure.

One merely moves M very slightly within each simplex. Thus one starts the inductive construction by leaving all the vertices fixed throughout the homotopy, H'. Now, on a one-simplex, σ, let H' be the homotopy which moves linearly within σ from the identity map to the map, β', with the graph shown in Figure 1.

Next, on each two-simplex (thought of as a disc), one superimposes this homotopy radially on top of the homotopy already given on the boundary—and so on, inductively.

For example, a suitable construction is made in [Fe, §§IX, X] whose properties I will summarize.

(2.22) Let $M = G\backslash X/Y$ have m-simplices $\{\Delta_i^m\}$, and let $b_{m,i}\in\Delta_i^m$ denote the barycentre, $\mathrm{Fix}(f)$ will denote the set of fixed points of a self-map, f.

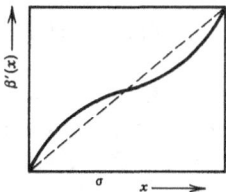

Figure 1

From [Fe, §§IX, X] there exists a homotopy

$$H': M \times I \to M,$$

such that $H'(x, 0) = x$ and $H'(x, 1) = \beta'(x)$, which satisfies the following properties:

(i) $\mathrm{Fix}(\beta') = \{b_{m,i}\}$, the set of barycentres of the triangulation.

(ii) H' preserves the triangulation of M, and therefore, inter alia, $H'(M_{(H)}, t) \subset M_{(H)}$ for each orbit type, (H) and for all $t \in I = [0, 1]$.

(iii) The fixed point index of $\beta' | \Delta_i^m$ is denoted by $I_{\beta' | \Delta_i^m}(b_{m,i})$ and equals $(-1)^m$.

(iv) Let M_α be a connected component (which is the same as a path component) of $M_{(H)}$. Denote by $\aleph^\#(M_\alpha)$ *the internal Euler characteristic* of M_α:

$$\aleph^\#(M_\alpha) = \aleph(\bar{M}_\alpha) - \aleph(\bar{M}_\alpha - M_\alpha),$$

where $\aleph$ denotes the usual Euler characteristic. With this notation (see [Fe, §XI.1])

$$\aleph^\#(M_\alpha) = \sum_{b_{m,i} \in M_\alpha} I_{\beta' | \Delta(m,i)}(b_{m,i}).$$

This formula follows by applying the Lefschetz fixed-point theorem to $\beta': M_\alpha \to M_\alpha$.

Let $\pi: X/Y \to M = G \backslash X/Y$ be the canonical projection.

(2.23) Proposition

Suppose that H', β', G, X, and Y satisfy the conditions of §2.22 (i)–(iv). Then there exists a G-equivariant homotopy

$$H: (X/Y) \times I \to X/Y$$

such that

(i) $H(xY, 0) = xY$ for all $x \in X$.

(ii) $\pi(H(xY, t)) = H'(\pi(xY), t)$ for all $x \in X$, $t \in I$.

(iii) If $\beta: X/Y \to X/Y$ is defined by $\beta(xY) = H(xY, 1)$, then $\mathrm{Fix}(\beta) = \{\pi^{-1}(b_{m,i})\}$.

Proof. From [Pa] one knows that $H': M \times I \to M$, satisfying §2.22 (ii) may be lifted to an equivariant homotopy starting at the identity. Clearly, for any such lifting, H, the map, $\beta = H(—, 1): X/Y \to X/Y$, will satisfy (i)–(ii), and in addition, $\mathrm{Fix}(\beta)$ will be a subset of $\{\pi^{-1}(\mathrm{Fix}(\beta'))\} = \{\pi^{-1}(b_{m,i})\}$. To ensure that these two sets are equal (to satisfy (iii)), we recall how H is constructed from H'. Firstly, by taking a barycentric subdivision, we may refine the triangulation to ensure that Fix (β') consists of zero cells of the triangulation. Next we may construct H by induction over the skeleton of this refined triangulation. Therefore it suffices to define the homotopy $H: \{\pi^{-1}(b_{m,i})\} \times I \to X/Y$ to satisfy (iii) and then extend it to H arbitrarily in such a way as to satisfy (i)–(ii). However, for each $b_{m,i}$ the path $H'(b_{m,i}, t)$ does not leave the interior, $\overset{\circ}{\Delta}{}^m_i$, of the simplex Δ^m_i of the original triangulation, by §2.22 (ii). Hence we may choose $x_{m,i} Y$ above $b_{m,i}$ and a slice, $S \subset X/Y$ through $x_{m,i} Y$ such that

(i) $\pi(S)$ is a neighbourhood of $b_{m,i}$ in Δ^m_i containing $H'(b_{m,i}, t)$ for all $t \in I$.

(ii) If $V = \operatorname{stab}(x_{m,i} Y)$ is the subgroup of G stabilizing $x_{m,i} Y$, then V fixes S and

$$\psi: G/V \times S \to X/Y$$

$$(gV, s) \mapsto gs$$

defines an equivariant homeomorphism onto a neighbourhood of $x_{m,i} Y$.

Given (i) and (ii), we choose a homeomorphism, $\hat{\pi}: S \to \pi(S)$, and define $(s \in S, g \in G, t \in I)$:

(2.24) $\qquad H(gs, t) = g(\hat{\pi}(H'(\pi(s), t))) = \psi(gV, \hat{\pi}(H'(\pi(s), t)))$.

This gives a well-defined, equivariant, lifted homotopy, H, on a neighbourhood of each $\{b_{m,i}\}$. Clearly, if $b_{m,i} = \pi(s)$, then $s = gx_{m,i} Y$ and

$$\begin{aligned}
\beta(s) &= H(s, 1) \\
&= H(\psi(gV, x_{m,i} Y), 1) \\
&= g(\hat{\pi}(\beta'(b_{m,i}))) \\
&= gx_{m,i} Y \\
&= s, \text{ as required.} \qquad \square
\end{aligned}$$

The following result is the main technical result of this section:

(2.25) Theorem

Let $v: G \to X$ be as in (2.19), and let Y be a closed subgroup of X. Then for any $x \in G$,

$$\aleph(X/Y) = \sum_{\alpha,(V)} \aleph^{\#}(M_\alpha)\#((G/V)^x) = \sum_{(V)} \aleph^{\#}(M_{(V)})\#((G/V)^x),$$

where $\#((G/V)^x)$ is the number of cosets gV such that $xgV = gV$. The first sum is taken over conjugacy classes (V) of subgroups, V, of G and over connected components M_α of orbit-type subsets, $M_{(V)} \subset (G\backslash X/Y) = M$, as in §2.22. The second sum is taken over orbit types, (V).

Proof. Consider the commutative diagram

$$
\begin{array}{ccc}
X/Y & \xrightarrow{x\beta} & X/Y \\
\pi \downarrow & & \downarrow \pi \\
G\backslash X/Y & \xrightarrow{\beta'} & G\backslash X/Y,
\end{array}
$$

where β', β are as in §§2.22 and 2.23. Since X is connected, $x\beta$ is homotopic to the identity map of X/Y. Consequently its Lefschetz number is equal to the Euler characteristic, $\aleph(X/Y)$. That is,

$$\aleph(X/Y) = L(x\beta, X/Y)$$

$$= \sum_n (-1)^n (\text{Trace}: (x\beta)_* : H_n(X/Y; \mathbb{Q}) \to H_n(X/Y; \mathbb{Q})).$$

However, by the Lefschetz fixed-point theorem, we have

(2.26)
$$L(x\beta, X/Y) = \sum_{s \in \text{Fix}\,(x\beta)} I_{x\beta}(s),$$

where $I_{x\beta}(s)$ is the local fixed-point index at a point $s \in X/Y$ such that $\beta(x) = s$. The only points in $\text{Fix}\,(x\beta)$ lie above points $b_{m,i} \in \text{Fix}\,(\beta')$, and if $V = \text{stab}\,(b_{m,i})$, there are precisely $\#((G/V)^x)$ points of $\text{Fix}\,(x\beta)$ above $b_{m,i}$. This is because β is the identity on $\pi^{-1}(b_{m,i})$ by §2.23 (iii). Furthermore the local fixed-point index $I_{x\beta}(s)$ at a point $s \in X/Y$ such that $\pi(s) = b_{m,i}$ is equal to $I_{\beta'|\Delta_i^m}(b_{m,i})$ because π is locally trivial at s. Therefore we may rewrite the right-hand side of (2.26) as

(2.27)
$$\sum_{\alpha,(V)} \sum_{b_{m,i} \in M_\alpha} (I_{\beta'|\Delta_i^m}(b_{m,i})) \#((G/V)^x),$$

where the sum is taken over conjugacy classes, (V), of subgroups, V, of G and over connected components, M_α, of $M_{(V)}$.

Substituting into (2.27) the expression for $\aleph^\#(M_\alpha)$ given in §2.22 (iv) completes the proof of the formula of §2.25.

The second formula follows from the first since the internal Euler characteristic is additive,

$$\sum \aleph^\#(M_\alpha) = \aleph^\#(M_{(V)}),$$

where M_α runs over the connected components of $M_{(V)}$. $\qquad\qquad\square$

(2.28) Remark

Ted Petrie showed me the following proof of Theorem 2.25, using compactly supported cohomology, H_c^*, with complex coefficients:

If $Z = (X/Y)$, we want to show that

$$(2.29) \qquad \sum_i (-1)^i H_c^i(Z) = \sum_{(V)} \aleph^\#(M_{(V)}) \, \mathrm{Ind}_V^G(1) \qquad \text{in } R(G).$$

However, Euler characteristics in H_c^* are additive on exact sequences. Hence the H_c^* exact sequences, resulting from the stratification by orbit type, decompose the left side of (2.29) into a sum of Euler characteristics, one for each orbit type. Hence we may suppose that Z has only one orbit type, (V). Further, since we are evaluating characters of representations, we may assume that G is cyclic. In this specialized case the cyclic group, G/V, acts freely on Z, and the result is obvious from the multiplicativity of the Euler characteristic in a covering space.

Independently, Jean-Pierre Serre suggested to me a proof using compactly supported cohomology along the lines of [Br, p. 235]. Serre also made the relevant observation that it is always H_c^*, rather than H^*, which occurs in the Deligne-Lusztig constructions.

In addition, there is a proof [Sn 4] which is valid for compact Lie groups, G, in which induction is defined analytically and in which the role of the Lefschetz fixed-point theorem is played by the Atiyah-Bott fixed-point theorem [A-B].

(2.30)   Now let X_n be as in (2.2), and let $v : G \to X_n$ be a representation. Let G act on X_n/Y_n, via v, by left translation, and set $M = {}_G \backslash (X_n/Y_n)$. For each conjugacy class of subgroups, (J), let $M_{(J)}$ denote the stratum of orbits of type (J) in M. Let $\aleph_{(J)}^\#$ denote the corresponding internal Euler characteristic,

$$\aleph_{(J)}^\# = \aleph(\bar{M}_{(J)}) - \aleph(\bar{M}_{(J)} - M_{(J)}).$$

Choose $g_J \in X_n$ lying above $M_{(J)}$ so that the stabilizer of $g_J Y_n$ will be

$$H(g_J, v) = v^{-1}(g_J Y_n g_J^{-1}),$$

as in Theorem 1.16.

(2.31) Theorem

With the notation of §2.30, let $v : G \to X_n$ be a representation. Then

   (i) $1 = \sum_{(J)} \aleph_{(J)}^\# \, \mathrm{Ind}_{H(g_J, v)}^G(1) \in R(G),$

   (ii) $v = \sum_{(J)} \aleph_{(J)}^\# \, \mathrm{Ind}_{H(g_J, v)}^G(\mathrm{Res}_{H(g_J, v)}^G(g_J^{-1} v g_J))$ *in* $R(G), RO(G),$ *or* $RH(G),$ *whichever is appropriate.*

Observe that, by §2.6, each representation

$$\text{Res}^G_{H(g_J,v)}(g_J^{-1}vg_J): H(g_J,v) \to Y_n \subset X_n$$

is (canonically) a sum of monomial representations of the form

$$\text{Ind}^G_H(\varphi: H \to X_1^?).$$

Proof. By §2.4, $\aleph(X_n/Y_n) = 1$ so that part (i) follows from the second form of Theorem 2.25. To obtain (ii), we multiply by v, apply Frobenius reciprocity, and then replace v by $g_J^{-1}vg_J$ in the term

$$\text{Ind}^G_{H(g_J,v)}(\text{Res}^G_{H(g_J,v)}(v)). \qquad \square$$

(2.32)   Let us conclude this section by examining two further examples which come from the icosahedral and octahedral groups.

(2.33)   *AN OCTAHEDRAL EXAMPLE*

Following [Sp 2, p. 92], we have the octahedral group, O, embedded as a subgroup of $SU(2)$ of order 48. It has the following set of generators:

$$a = \begin{bmatrix} \varepsilon & 0 \\ 0 & \varepsilon^7 \end{bmatrix}, \qquad (\varepsilon = \exp(\pi i/4) = (1+i)/(\sqrt{2})),$$

$$b = \begin{bmatrix} 0 & i \\ i & 0 \end{bmatrix}, \qquad (i^2 = -1),$$

$$c = (1/\sqrt{2}) \begin{bmatrix} \varepsilon^7 & \varepsilon^7 \\ \varepsilon^5 & \varepsilon \end{bmatrix},$$

$$O = \left\{ a, b, c \; \middle| \; \begin{matrix} bab^{-1} = a^{-1}, (ac)^2 = -a^2b = a^6b, cb = a^2c \\ b^2 = -1 = c^3 = a^4 \end{matrix} \right\}.$$

The elements of O are

$$a^h b^j c^m \qquad (0 \le h \le 7, 0 \le j \le 1, 0 \le m \le 2).$$

Associated to the representation

$$v: O \to SU(2),$$

we have the action of O on $U(2)/(\Sigma_2 \int S^1) = \mathbb{R}P^2$ through

$$O/(-I_2) = \left\{ a, b, c \; \middle| \; \begin{matrix} bab^{-1} = a^{-1}, (ac)^2 = a^2b, cb = a^2c \\ b^2 = a^4 = c^3 = 1 \end{matrix} \right\},$$

which is a group of order 24. This group is the symmetric group, Σ_4, which is seen by setting

$$a = (1234), \quad b = (14)(23), \quad \text{and} \quad c = (123).$$

The elements of Σ_4 are thus

$$a^h b^j c^m \qquad (0 \leq h \leq 3, 0 \leq j \leq 1, 0 \leq m \leq 2).$$

Plato and his colleagues have obligingly triangulated $\mathbb{R}P^2$ in such a way as to make the action of Σ_4 simplicial. Namely, we may represent $\mathbb{R}P^2$ as the surface of a cube with the antipodal points identified. The Platonic action is as follows: Σ_4 acts by permuting the diagonals of the cube. The full symmetry group of the cube is $\Sigma_4 \times \{\pm 1\}$ [R, p. 35] so that the full symmetry group of $\mathbb{R}P^2$ is Σ_4. $\mathbb{R}P^2$ is represented by the top half of the surface of the cube in Figure 2 with antipodal identifications along the equatorial boundary. Each face should, in addition, be triangulated in the following manner:

Let us number the diagonals

$$(AG) = D_1, \quad (BH) = D_2, \quad (CE) = D_3, \quad (DF) = D_4.$$

Clearly, $a = (1234)$ cyclically permutes the diagonals and rotates the top face through $\pi/4$. The full symmetry group of the top face (i.e., of a square) is the dihedral group of order 8, D_8. The action of b sends A to D or F, and as we are in projective space, we may suppose that it goes to D. Hence B must go to C and C to B. Therefore, in the symmetries of $\mathbb{R}P^2$, we have the subgroup

$$D_8 = \{a, b\} \qquad \text{whose action preserves } ABCD.$$

These observations quickly and easily yield the fact that a fundamental

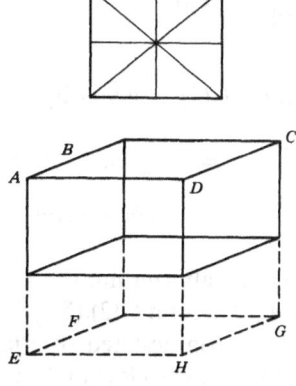

Figure 2. The cube

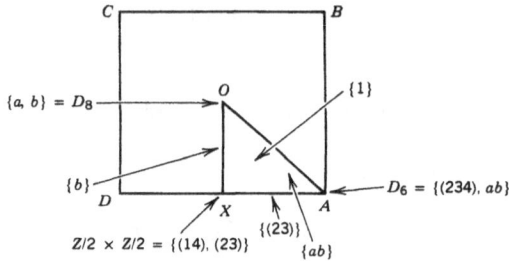

Figure 3. Fundamental domain and orbit-types for Σ_4

domain for the action of Σ_4 on $\mathbb{R}P^2$ is given by the triangle OXA on the top face. In other words, the orbit space of (2.30) is a triangle (Fig. 3).

Note that $ab = (24)$, which is of order 2, and conjugate to (23).

By Theorem 2.31 we obtain the following relation in $R(\Sigma_4)$:

(2.34)
$$1 = \text{Ind}_{D_8}^{\Sigma_4}(1) + \text{Ind}_{\{(14),(23)\}}^{\Sigma_4}(1) + \text{Ind}_{D_6}^{\Sigma_4}(1)$$

$$- \text{Ind}_{\{(13)(24)\}}^{\Sigma_4}(1) - 2\,\text{Ind}_{\{(24)\}}^{\Sigma_4}(1) + \text{Ind}_{\{1\}}^{\Sigma_4}(1).$$

(2.35)   AN ICOSAHEDRAL EXAMPLE

Let I denote the icosahedral group, and let

$$v: I \rightarrowtail SU(2)$$

denote the faithful representation given by the following formulae ($\eta = \exp(2\pi i/5)$):

$$v(a) = \begin{bmatrix} -\eta^3 & 0 \\ 0 & -\eta^2 \end{bmatrix}, \quad v(b) = \begin{bmatrix} 0 & 1 \\ -1 & 0 \end{bmatrix},$$

$$v(c) = (1/(\eta^2 - \eta^3))\begin{bmatrix} \eta + \eta^4 & 1 \\ 1 & -(\eta + \eta^4) \end{bmatrix}.$$

A presentation for I, within $SU(2)$, is given by

$$I = \{a, b, c \,|\, a^5 = b^2 = c^2 = -I_2, bab^{-1} = a^{-1}, bcb^{-1} = -c,$$

$$cac = acba, ca^2c = a^{-2}ca^{-2}\},$$

and $I/(\pm I_2)$ is isomorphic to the alternating group, A_5.

Thus, via its projection to A_5, I acts on $U(2)/(\Sigma_2 \int S^1) \cong \mathbb{R}P^2$, the real projective plane. This action has also been triangulated by Plato and his colleagues and is given by the classical action (e.g., see [R]) of I upon the dodecahedron with antipodal points identified.

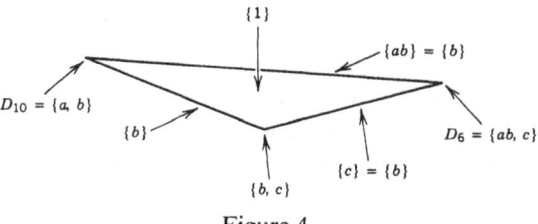

Figure 4

(2.36)   Consequently $M = I\backslash\mathbb{R}P^2$ is homeomorphic to a triangle on one of the pentagonal faces. Its A_5-orbit types are shown in Figure 4.

At once we obtain the equations

(2.37)
$$\begin{cases} 1 = \operatorname{Ind}_{D_{10}}^{A_5}(1) - 3\operatorname{Ind}_{\{b\}}^{A_5}(1) + \operatorname{Ind}_{D_6}^{A_5}(1) \\ \quad + \operatorname{Ind}_{\{b,c\}}^{A_5}(1) + \operatorname{Ind}_{\{1\}}^{A_5}(1) \end{cases}$$

in $R(A_5)$, and in $R(I)$,

(2.38)
$$\begin{cases} v = \operatorname{Ind}_{H_{20}}^{I}(v) - 3\operatorname{Ind}_{\{b\}}^{I}(v) + \operatorname{Ind}_{H_{12}}^{I}(v) \\ \quad + \operatorname{Ind}_{H_8}^{I}(v) + \operatorname{Ind}_{\{c^2\}}^{I}(v). \end{cases}$$

3.   A PRESENTATION FOR R(G)

(3.1)   The object of this section is to modify the Explicit Brauer Induction formula to obtain a presentation for the additive group, $R(G)$, of unitary representations of the finite group, G. This solves a problem which is posed in [Ser 6, p. 71 (footnote)]. As will become clear later in the book, a presentation for $R(G)$ by generators which are monomial homomorphisms is very useful when one wants to define refined invariants of Galois representations (local root numbers will be my main example), which one only has explicitly on one-dimensional representations (characters).

In this section we will only be concerned with unitary representations, which is the case $(X_n, Y_n) = (U(n), \Sigma_n \int S^1)$ in the notation of §2. We are going to refine Theorem 2.31 to give an element

(3.2)
$$\begin{cases} \tau_G(v) \in R_+(G, \Sigma_n \int S^1) \\ \text{for a representation, } v: G \to U(n). \end{cases}$$

We will establish some useful naturality properties for (3.2) and then use it to give the presentation for $R(G)$, which will be *natural* in G because $\tau_G(v)$ is natural for homomorphisms of G.

(3.3)　MORE CONCERNING $R_+(G, \pi)$

We will describe $R_+(G, \pi)$ of Chapter 5, §1.9 in terms of subhomomorphisms $(G \supset H \to \pi)$ as in Chapter 5, Theorem 1.10.

Suppose that $i: J \rightarrowtail G$ is an inclusion of a subgroup. Define the *restriction* homomorphism, between the groups introduced in Chapter 5, §1,

$$(3.4) \quad \begin{cases} \mathrm{Res}_J^G: R_+(G, \pi) \to R_+(J, \pi) \text{ by} \\[2mm] \mathrm{Res}_J^G(G \supset H \xrightarrow{\rho} \pi) \\[2mm] \quad = \displaystyle\sum_{x \in J \backslash G/H} (J \supset (J \cap (xHx^{-1})) \xrightarrow{\rho(x^{-1} - x)} \pi). \end{cases}$$

We may also define

$$(3.5) \quad \begin{aligned} &\mathrm{Ind}_J^G: R_+(J, \pi) \to R_+(G, \pi) \text{ by} \\[2mm] &\mathrm{Ind}_J^G(J \supset H \xrightarrow{\rho} \pi) = (G \supset H \xrightarrow{\rho} \pi). \end{aligned}$$

Specializing to the case $\pi = \Sigma_n \smallint S^1$, we obtain additional homomorphisms by considering a monomial homomorphism into $\Sigma_n \smallint S^1$ as an n-dimensional representation.

Define

$$(3.6) \quad \begin{aligned} &b_n: R_+(G, \Sigma_n \smallint S^1) \to R(G) \text{ by} \\[2mm] &b_n(G \supset H \xrightarrow{\rho} \Sigma_n \smallint S^1) = \mathrm{Ind}_H^G(\rho). \end{aligned}$$

The following result is clear from the definitions and the double coset formula for representations:

(3.7) Proposition

The following diagrams commute:

(i)

$$\begin{array}{ccc} R_+(G, \Sigma_n \smallint S^1) & \xrightarrow{\mathrm{Res}_J^G} & R_+(J, \Sigma_n \smallint S^1) \\ \downarrow{\scriptstyle b_n} & & \downarrow{\scriptstyle b_n} \\ R(G) & \xrightarrow{\mathrm{Res}_J^G} & R(J). \end{array}$$

(ii)

$$\begin{array}{ccc} R_+(J, \Sigma_n \smallint S^1) & \xrightarrow{\mathrm{Ind}_J^G} & R_+(G, \Sigma_n \smallint S^1) \\ \downarrow{\scriptstyle b_n} & & \downarrow{\scriptstyle b_n} \\ R(J) & \xrightarrow{\mathrm{Ind}_J^G} & R(G). \end{array}$$

(3.8)   The direct sum of matrices

$$\Sigma_n \int S^1 \times \Sigma_m \int S^1 \to \Sigma_{n+m} \int S^1,$$

which we met in Chapter 5, (1.8), induces an operation

(3.9) $\begin{cases} R_+(G_1, \Sigma_n \int S^1) \times R_+(G_2, \Sigma_m \int S^1) \to R_+(G_1 \times G_2, \Sigma_{n+m} \int S^1) \\ \text{denoted by } (x, y) \longmapsto (x * y). \end{cases}$

We may now state and prove the key result of this section. Let $v: G \to U(n)$ be a representation, and let G act, via v, by left translation on $U(n)/(\Sigma_n \int S^1)$ with orbit space, M, in the manner of §2.30. Let (J) denote a conjugacy class of subgroups of G, and let $\{M_{(J),\alpha}\}$ be the set of *connected components* of the orbit-type stratum, $M_{(J)}$ of §2.30. Set $\aleph_{(J),\alpha}^{\#}$ equal to the internal Euler characteristic

(3.10) $$\aleph_{(J),\alpha}^{\#} = \aleph(\bar{M}_{(J),\alpha}) - \aleph(\bar{M}_{(J),\alpha} - M_{(J),\alpha}).$$

In addition, for each pair $((J), \alpha) = \beta$, say, choose $g_\beta \in U(n)$ such that the orbit of g_β lies in $M_\beta = M_{(J),\alpha}$. The element, $\tau_G(v) \in R_+(G, \Sigma_n \int S^1)$ is defined by

(3.11) $$\tau_G(v) = \sum_\beta \aleph_\beta^{\#}(G \supset H(g_\beta, v) \xrightarrow{g_\beta^{-1} v g_\beta} \Sigma_n \int S^1),$$

where the sum extends over all pairs, $\beta = ((J), \alpha)$ and where, as in §2.30,

(3.12) $$H(g_\beta, v) = v^{-1}\left(g_\beta\left(\Sigma_n \int S^1\right)g_\beta^{-1}\right) \subset G.$$

(3.13) Theorem (*Explicit Brauer Induction—strong form*)

  (i) *Let $v: G \to U(n)$ be a representation as shown earlier,*

$$\tau_G(v) \in R_+\left(G, \Sigma_n \int S^1\right)$$

  is well-defined and depends only on the image of v in $R(G)$.
  (ii) *If b_n is the homomorphism of (3.6), then*

$$b_n(\tau_G(v)) = v \in R(G).$$

  (iii) *If $\mu_1: G_1 \to U(n)$ and $\mu_2: G_2 \to U(m)$ are unitary representations, then*

$$\tau_{G_1 \times G_2}(\mu_1 \oplus \mu_2) = \tau_{G_1}(\mu_1) * \tau_{G_2}(\mu_2).$$

(iv) *If* $i: J \to G$ *is an inclusion, then*

$$\operatorname{Res}_J^G(\tau_G(v)) = \tau_J(\operatorname{Res}_J^G(v)) \in R_+\left(J, \Sigma_n \int S^1\right).$$

Proof. To prove part (i), we observe firstly that the action on $U(n)/(\Sigma_n \int S^1)$, by means of which $\tau_G(v)$ is defined, sees only the image of v in $U(n)$. In addition

$$\operatorname{stab}\left(g_\beta\left(\Sigma_n \int S^1\right)\right) = \left\{x \in G \mid g_\beta^{-1} v(x) g_\beta \in \Sigma_n \int S^1\right\}$$

$$= H(g_\beta, v),$$

which does not change as we vary $g_\beta(\Sigma_n \int S^1)$ along a path in $U(n)/(\Sigma_n \int S^1)$ which covers a path in some component, $M_{(J),\alpha} = M_\beta$. On the other hand, if we change g_β to g'_β representing the same coset, the effect of the change is to conjugate

$$\left(H(g_\beta, v) \xrightarrow{\ g_\beta^{-1} v g_\beta\ } \Sigma_n \int S^1\right)$$

to

$$\left(H(g_\beta, v) \xrightarrow{\ (g'_\beta)^{-1} v(g'_\beta)\ } \Sigma_n \int S^1\right),$$

by the inner automorphism of $\Sigma_n \int S^1$ induced by $(g'_\beta)^{-1} g_\beta$. These remarks show that the equivalence class of $\tau_G(v) \in R_+(G, \Sigma_n \int S^1)$ depends only on v as a *homomorphism*. However, $\tau_G(v)$ depends only on v as a *representation* because, if we conjugate v to $\omega v \omega^{-1}$ for some $\omega \in U(n)$, there is an obvious commutative diagram relating the two actions of $g \in G$:

(3.14)

$$
\begin{array}{ccc}
U(n)/(\Sigma_n \int S^1) & \xrightarrow{\ v(g)\ } & U(n)/(\Sigma_n \int S^1) \\
\downarrow{\scriptstyle\omega} & & \downarrow{\scriptstyle\omega} \\
U(n)/(\Sigma_n \int S^1) & \xrightarrow{\ \omega v(g) \omega^{-1}\ } & U(n)/(\Sigma_n \int S^1).
\end{array}
$$

Consequently there is a bijection between the set of components of the orbit-type strata for the two actions. Under this bijection the stabilizers match up according to the rule

$$H(g, \omega v \omega^{-1}) = H(\omega^{-1} g, v).$$

Thus

$$\tau_G(\omega v \omega^{-1}) = \sum_\beta \aleph_\beta^\# \left(G \supset H(g_\beta, \omega v \omega^{-1}) \xrightarrow{g_\beta^{-1} \omega v \omega^{-1} g_\beta} \Sigma_n \int S^1 \right)$$

$$= \sum_\beta \aleph_\beta^\# \left(G \supset H(\omega^{-1} g_\beta, v) \xrightarrow{(g_\beta^{-1} \omega) v (g \omega^{-1} g_\beta)} \Sigma_n \int S^1 \right)$$

$$= \tau_G(v),$$

since the map induced by (3.14) on orbit spaces interchanges homeomorphically the connected components, $\{M_\beta\}$ of the two actions. This homeomorphism, of course, preserves internal Euler characteristics.

Part (ii) follows directly from Theorem 2.31, together with the equation (cf., §2.25 (proof))

$$\aleph_{(J)}^\# = \sum_{\beta = ((J), \alpha)} \aleph_\beta^\#,$$

where the sum runs over components of the stratum $M_{(J)}$.

Now we will prove part (iii), and we will do this by means of the relationship, explained in §2.18, between the internal Euler characteristics and local fixed-point indices. Consider the $(G_1 \times G_2)$-action, via $\mu_1 \oplus \mu_2$ on $U(n+m)/(\Sigma_{n+m} \int S^1) = X$, say. Within X the subspace $U(n)/(\Sigma_n \int S^1) \times U(m)/(\Sigma_m \int S^1) = W$, say, is stable under the action.

We wish to examine

$$\tau_{G_1 \times G_2}(\mu_1 \oplus \mu_2) = \sum_\gamma \aleph_\gamma^\# \left(G_1 \times G_2 \supset H(g_\gamma, \mu_1 \oplus \mu_2) \to \Sigma_{n+m} \int S^1 \right).$$

The formula we require is easily seen to be implied by the vanishing of the $\aleph_\gamma^\#$, which come from the orbit-type strata of $(X - W)$. However, I claim that there exists a $(G_1 \times G_2)$-equivariant map, f, of X, equivariantly homotopic to the identity map whose fixed points all lie within W. Therefore, on the orbit-space level, there exists a homotopy, preserving orbit types, from the identity to a map which only has fixed points within

$$(3.15) \qquad (G_1 \times G_2) \backslash W \cong G_1 \backslash \left(U(n) \Big/ \left(\Sigma_n \int S^1 \right) \right) \times G_2 \backslash \left(U(m) \Big/ \left(\Sigma_m \int S^1 \right) \right).$$

Applying the Lefschetz theorem to this map, restricted to a connected component of an orbit-type stratum, shows (by §2.18) that $\aleph_\gamma^\# = 0$ for strata outside (3.15).

It remains to construct the map, f. From [Sn 3, Chapter 1, §2.12], for example, there is a $(G_1 \times G_2)$-equivariant vector field, v, on X, obtained from the derivative of a translation map, whose zeros lie within W. Therefore we may construct f by the standard procedure of pushing a point, z, one (small) unit along $v(z)$ and projecting back to the nearest point in X (see Fig. 5).

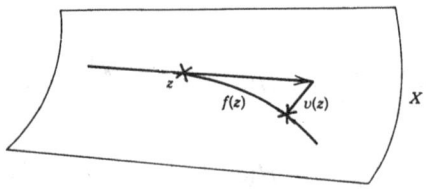

Figure 5

To prove part (iv), it is helpful to rewrite $\tau_G(v)$ in the following manner, which is justified by a further appeal to §2.18. Choose an isovariant map, $f: M \to M$, of the G-orbit space such that (a) the fixed points of f are isolated and (b) f is isovariantly homotopic to the identity map. Over each fixed point of f, choose $g \in U(n)$ so that $f(O(g\Sigma_n \int S^1)) = O(g\Sigma_n \int S^1)$, where $O(—)$ denotes this orbit. Let I_g denote the fixed-point contribution of f at $O(g(\Sigma_n \int S^1))$ (cf., §2.18) to $\aleph^{\#}_{(\beta)}$. That is, by §2.18, $\aleph^{\#}_{\beta} = \sum I_g$, where the sum runs over fixed points of f which lie in the connected component $M_\beta = M_{(V),\alpha}$ of the stratum $M_{(V)}$. Therefore we may write

$$\tau_G(v) = \sum_{\substack{g \\ O(g(\Sigma_n \int S^1)) \in \mathrm{Fix}(f)}} I_g\left(G \supset H(g,v) \xrightarrow{g^{-1}vg} \Sigma_n \int S^1\right).$$

By definition,

$$\mathrm{Res}^G_J(\tau_G(v)) = \sum_g I_g \sum_{\substack{x \in J \backslash G/V \\ (V)=(H(g,v))}} \left(J \supset J \cap xH(g,v)x^{-1} \xrightarrow{g^{-1}v(x^{-1} x)g} \Sigma_n \int S^1\right).$$

Next we observe that, for $x \in G$,

$$xH(g,v)x^{-1} = H(v(x)g,v).$$

Therefore we obtain

$$\mathrm{Res}^G_J \tau_G(v) = \sum_g \sum_x I_g\left(J \supset J \cap H(v(x)g,v) \xrightarrow{(v(x)g)^{-1}vv(x)g} \Sigma_n \int S^1\right).$$

On the other hand, by the same reasoning,

$$\tau_J(\mathrm{Res}^G_J(v)) = \sum_{\substack{h \\ O(h(\Sigma_n \int S^1)) \in \mathrm{Fix}(\tilde{f})}} I_h\left(J \supset H(h,v) \xrightarrow{h^{-1}vh} \Sigma_n \int S^1\right),$$

where this sum runs over $h \in U(n)$ chosen above a fixed point $O(h(\Sigma_n \int S^1))$ in the

J-orbit space, M', of a map $\tilde{f}: M' \to M'$ which is isovariantly homotopic to the identity and has isolated fixed points.

In §2.18 we saw how to manufacture f and $\tilde{f}$ so that $\tilde{f}$ induces f upon passage to G-orbits and such that above each fixed point of f on M representing an orbit, G/V, there lies in M' the set $J\backslash G/V$ which is pointwise fixed by $\tilde{f}$. At each point in this copy of $J\backslash G/V$ the fixed-point contribution, I_h, will be equal to that of the G-orbit, $O(g(\Sigma_n \int S^1))$, below it. That is, $I_h = I_g$ for all the $h \in J\backslash G/V$ lying over $O(g(\Sigma_n \int S^1)) \in \text{Fix}(f)$. Therefore, if we break up the sum for $\tau_J(\text{Res}_J^G(v))$ into a sum over G-orbit fixed points, followed by a sum over the double coset of $\tilde{f}$-fixed points, $J\backslash G/V$, above them, then it becomes apparent at once that

$$\tau_J(\text{Res}_J^G(v)) = \text{Res}_J^G(\tau_G(v)) \in R_+\left(J, \Sigma_n \int S^1\right),$$

as required.   □

(3.16) Remark

The connections with stable homotopy, which were described in §1, lead one to expect that Theorem 3.13 (iv) will be true. By §1, we have an isomorphism

$$\phi_G: R_+\left(G, \Sigma_n \int S^1\right)^{\wedge}_{IA(G)} \cong \left\{BG_+, B\left(\Sigma_n \int S^1\right)_+\right\}.$$

From Theorem 1.16 we know that the image in $\{BG_+, B(\Sigma_n \int S^1)_+\}$ of $\tau_G(v)$ is the S-map

$$BG_+ \xrightarrow{B(v_+)} BU(n)_+ \xrightarrow{\tau} B\left(\Sigma_n \int S^1\right)_+,$$

where τ is the Becker-Gottlieb transfer map. It is not difficult, by means of the double-coset formula, to verify that

$$\text{Res}_J^G: R_+\left(G, \Sigma_n \int S^1\right) \to R_+\left(J, \Sigma_n \int S^1\right)$$

corresponds to precomposition with

$$B(i_+): BJ_+ \to BG_+.$$

Hence, after completion, $\text{Res}_J^G \tau_G(v)$ corresponds to the S-map,

$$BJ_+ \xrightarrow{B(i_+)} BG_+ \xrightarrow{B(v_+)} BU(n)_+ \xrightarrow{\tau} B\left(\Sigma_n \int S^1\right)_+,$$

which also evidently corresponds to $\tau_J(\text{Res}_J^G(v))$.

This argument establishes Theorem 1.13 (iv) after completion, and if G is a p-group, in which case completion is just p-adic completion, the completed result implies that the formula holds in $R_+(G, \Sigma_n \int S^1)$ (i.e., without completion).

(3.17)   EXAMPLE OF THE NATURALITY FORMULA OF THEOREM 3.13 (iv)

As an illustration let us take the example

$$G = D_8 = \{x, y \mid x^4 = y^2 = 1, xyx = y\}$$

and $J = \{x^2, y\} \cong \mathbb{Z}/2 \times \mathbb{Z}/2$, with

$$v: D_8 \rightarrowtail U(2)$$

being given by the formulae (as in (1.20))

$$v(y) = \begin{bmatrix} 0 & 1 \\ 1 & 0 \end{bmatrix}$$

$$v(x) = \begin{bmatrix} i & 0 \\ 0 & -i \end{bmatrix} \qquad (i^2 = -1).$$

We depict $U(2)/(\Sigma_2 \int S^1) \cong \mathbb{R}P^2$, as in Chapter 5, as a disc with antipodal boundary points identified. With this convention the action is given by

(3.18) $$\begin{cases} x(z) = -z \\ y(z) = -\bar{z} \end{cases} \qquad (z \in D^2).$$

Let us establish some notation. Let U, V, W denote the following subgroups of $D_8 = G$:

$$U = \{x^2, xy\}, V = \{x\}, W = \{x^2\}.$$

In the following G-orbit and J-orbit diagrams the orbit-type components will

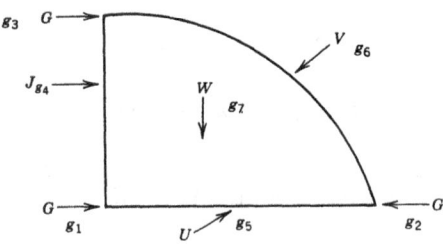

Figure 6. G-orbits

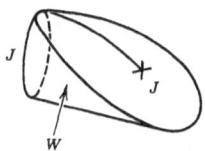 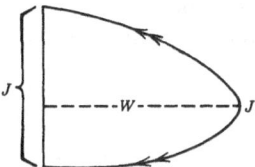

Figure 7. J-orbits

each be decorated with a group (e.g., V will denote a stratum in $M_{(V)}$) and a group element $g_i \in U(2)$ whose orbit represents a point of the stratum. Explicit g_i may be found from Table B, of Chapter 5, §3, but we will not need them.

Figure 6 is obtained from Chapter 5, §3, Figure 4 and Table B (otherwise known as Case iv when $n = 2$, $m = 1$, since D_8 acts on $\mathbb{R}P^2$ through its quotient $D_4 \cong \mathbb{Z}/2 \times \mathbb{Z}/2 \cong D_8/(x^2)$).

The J-orbit picture is a disc, as illustrated in the two diagrams of Figure 7. From Figure 6 we obtain

$$\tau_G(v) = (g_1^{-1}vg_1) + (g_2^{-1}vg_2) + (g_3^{-1}vg_3)$$

$$- (g_4^{-1}vg_4 | J) - (g_5^{-1}vg_5 | U)$$

$$- (g_6^{-1}vg_6 | V) + (g_7^{-1}vg_7 | W),$$

where $(\phi | H)$ denotes $(G \supset H \xrightarrow{\phi} \Sigma_2 \int S^1)$.

Above each G-orbit, G/H, are either one or two J-orbits corresponding to the cases $|J\backslash G/H| = 1$ or 2. Thus, from Figures 1 and 2, we find that $J\backslash D_8/J = \{1, x\} = J\backslash D_8/W$, whereas all other double cosets in the formula for $\operatorname{Res}^G_J \tau_G(v)$ have only one element. Since, for $x \in G$,

$$x^{-1}H(g, v)x = H(v(x)g, v),$$

we find that, in $R_+(J, \Sigma_2 \int S^1)$,

$$\operatorname{Res}^G_J \tau_G(v) = (g_1^{-1}vg_1) + (g_2^{-1}vg_2) + (g_3^{-1}vg_3)$$

$$- (g_4^{-1}vg_4) - ((v(x)g_4)^{-1}v(v(x)g_4))$$

$$- (g_5^{-1}vg_5 | W) - (g_6^{-1}vg_6 | W)$$

$$+ (g_7^{-1}vg_7 | W) + ((v(x)g_7)^{-1}v(x)g_7 | W).$$

However, if $g, h \in G$ have orbits in the same orbit-stratum component, then $(g^{-1}vg | H)$ equals $(h^{-1}vh | H)$. Also

$$v(x) \begin{bmatrix} a & b \\ c & d \end{bmatrix} = \begin{bmatrix} ia & ib \\ -ic & id \end{bmatrix}$$

so that, if g corresponds to $z \in D^2$, then $v(x)g$ corresponds to $-z$. From these remarks and the diagrams, one sees that the following cancellations occur:

first and fourth terms,
third and fifth terms,
seventh and eighth terms, and
the sixth and ninth terms.

Thus we have found that

$$\mathrm{Res}_J^G \tau_G(v) = \left(g_2^{-1} v g_2 : J \to \Sigma_2 \int S^1 \right).$$

This agrees with the formula for $\tau_J(\mathrm{Res}_J^G v)$, since the stratum, $M_{(J)}$, is the disjoint union of a circle ($\aleph_\beta^\# = 0$ for this) and the orbit of g_2 (for which $\aleph^\# = 1$), and the stratum, $M_{(W)}$, has

$$\aleph^\#(M_{(W)}) = \aleph^\#(\mathrm{disc}) - \aleph^\#(S^1 \cup (\text{point}))$$

$$= 1 - 1$$

$$= 0.$$

(3.19)   We are now ready to give a presentation for $R(G)$ for any finite group, G. The generators will be the elements of $R_+(G, \Sigma_n \int S^1)$ as n varies. Accordingly, we set

(3.20) $$\begin{cases} R_*(G) = \displaystyle\bigoplus_{n \geq 1} R_+(G, \Sigma_n \int S^1) \text{ and set} \\ B = \Sigma b_n : R_*(G) \longrightarrow R(G). \end{cases}$$

If $J \supset H$ is a subgroup of index d, and $\phi : J \to \Sigma_n \int S^1$ is a monomial homomorphism the construction which, in Chapter 5, §1.6, is denoted by λ gives an "induced" monomial homomorphism

$$\mathrm{Ind}_J^H(\phi) : H \to \Sigma_{dn} \int S^1.$$

This amounts to the process of considering ϕ as a representation with respect to the standard basis, $\{e_1, \ldots, e_n\}$, of $\mathbb{C}^n$, forming the representation $\mathrm{Ind}_J^H(\phi)$ and taking its matrix with respect to the basis $\{x_i \otimes e_j; 1 \leq i \leq d, 1 \leq j \leq n\}$ of $\mathbb{C}[H] \otimes_{\mathbb{C}[J]} \mathbb{C}^n$, where $\{x_i\}$ are chosen coset representatives in H/J.

(3.21)   Now define Λ to be the subgroup of $R_*(G)$ generated by the following elements:

(i) For $J \subset H \subset G$ and $d = [H:J]$, $((G \supset J \xrightarrow{\rho} \Sigma_n \int S^1) - (G \supset H \xrightarrow{\mathrm{Ind}_J^H(\rho)} \Sigma_{nd} \int S^1)) \in \Lambda$.

(ii) For $v : G \to \Sigma_n \int S^1$ and $\mu : G \to \Sigma_m \int S^1$, $((v \oplus \mu) - v - \mu) \in \Lambda$.

(iii) For $v: G \to \Sigma_n \int S^1$, $(\tau_G(v) - v) \in \Lambda$.

Notice that the effect of the relation (i) is to make all elements of $R_*(G)$ equivalent to a monomial homomorphism defined on the whole of G.

(3.22) Theorem

The homomorphism, B, of (3.20) induces an isomorphism

$$B: R_*(G)/\Lambda \xrightarrow{\cong} R(G).$$

Proof. We will define a homomorphism

$$t: R(G) \to R_*(G)/\Lambda,$$

and verify that $tB = 1 = Bt$. Our first inclination would be to define t by sending a representation, $v: G \to U(n)$, to $\tau_G(v)$. However, if v is not conjugate in $U(n)$ to a homomorphism into $\Sigma_n \int S^1$, we have not, at first sight, imposed sufficient relations to make this map into a homomorphism. However, once we have proved the result, it will follow that the above construction works, since $B\tau_G(v) = v$.

Instead, we take $x \in R(G)$ and write $x = \rho_1 - \rho_2$, where ρ_1, ρ_2 are *monomial homomorphisms*. Now define

$$t(x) = \tau_G(\rho_1) - \tau_G(\rho_2) \qquad (\mathrm{mod}\,\Lambda).$$

If $x = \lambda_1 - \lambda_2$, with λ_i given by monomial homomorphisms, then $\lambda_1 \oplus \rho_2 = \lambda_2 \oplus \rho_1$ in $R(G)$ so that

$$\tau_G(\lambda_1 \oplus \rho_2) = \tau_G(\lambda_2 \oplus \rho_1).$$

This ensures that $t(x)$ is well-defined, by relations (ii) and (iii). Therefore t is a homomorphism, and clearly,

$$Bt(x) = B\tau_G(\rho_1) - B\tau_G(\rho_2)$$

$$= \rho_1 - \rho_2$$

$$= x,$$

whereas

$$tB\left(G \supset H \xrightarrow{\rho} \Sigma_n \int S^1 \right) = t(\mathrm{Ind}_H^G \rho)$$

$$\equiv \tau_G(\mathrm{Ind}_H^G \rho) \qquad (\mathrm{mod}\,\Lambda)$$

$$\equiv \mathrm{Ind}_H^G(\rho), \qquad \text{by relation (iii),}$$

$$\equiv \left(G \supset H \xrightarrow{\rho} \Sigma_n \int S^1 \right), \qquad \text{by relation (i).} \qquad \square$$

(3.23)   We will close this chapter by recasting the presentation into two slightly different forms:

$$\text{Set } \mathscr{R}(G) = R_*(G)/(\Lambda_0),$$

where Λ_0 is the subgroup generated by relations of the form (3.21) (i) and (ii). Hence

(3.24) $$R(G) \cong \mathscr{R}(G) \Big/ \left\{ \tau_G(v) - v \,|\, v \colon G \to \Sigma_n \int S^1 \right\}.$$

$\mathscr{R}(G)$ is the Grothendieck group of monomial homomorphisms of the form $(n \geq 0)$,

$$v \colon G \to \Sigma_n \int S^1$$

under the equivalence relation induced by inner automorphisms of G and $\Sigma_n \int S^1$. Addition is induced by direct sum of matrices.

If $i \colon J \subset G$ is an inclusion, then

(3.25) $$\mathrm{Res}^G_J \colon \mathscr{R}(G) \to \mathscr{R}(J)$$

merely sends v to (vi), whereas

(3.26) $$\mathrm{Ind}^G_J \colon \mathscr{R}(J) \to \mathscr{R}(G)$$

is defined as in §3.19. In $\mathscr{R}(G)$, (3.25) and (3.26) satisfy the double-coset formula.

(3.27) $$\mathrm{Res}^G_J \, \mathrm{Ind}^G_H (v) = \sum_{x \in J \backslash G / H} \mathrm{Ind}^J_{J \cap xHx^{-1}} \, \mathrm{Res}^{xHx^{-1}}_{J \cap xHx^{-1}} (v(x^{-1} \_ x)).$$

(3.27) follows from the fact that the decomposition of $\mathbb{C}[G] \otimes_{\mathbb{C}[H]} \mathbb{C}^n$ as a J-space respects the (standard) bases used in §3.19 to define induction of monomial homomorphisms. That is, if $\{x_i\}$ is a set of coset representatives for G/H, then the basis $\{x_i \otimes e_j\}$ is acted upon by J through scalar multiplication and permutations. Therefore $\mathrm{Res}^G_J \, \mathrm{Ind}^G_H (v)$ decomposes into a sum of monomial homomorphisms, one for each double-coset element x, namely, that corresponding to

$$\{zx \otimes e_j \,|\, z \in J\} \subset \mathbb{C}[G] \underset{\mathbb{C}[H]}{\otimes} \mathbb{C}^n.$$

This decomposition yields (3.27).

Now for the second recasting of the presentation and the ρ-construction.

(3.28) THE ρ-CONSTRUCTION ON MONOMIAL HOMOMORPHISMS

Let $v: G \to \Sigma_n \int S^1$ be a monomial homomorphism, as previously, and let

$$\pi: \Sigma_n \int S^1 \to \left(\Sigma_n \int S^1 \right) \Big/ (S^1)^n \cong \Sigma_n$$

denote the canonical projection. From v we will define $\rho(v)$, another monomial homomorphism into $\Sigma_n \int S^1$, which is of the form

$$(3.29) \qquad \rho(v) = \sum_\alpha \operatorname{Ind}_{H_\alpha}^G(\rho_\alpha),$$

where $\{\rho_\alpha : H_\alpha \to S^1\}$ are homomorphisms.

Firstly, we will insist that

$$(3.30) \qquad \rho(v \oplus \mu) = \rho(v) \oplus \rho(\mu),$$

and therefore we may assume that G transitively permutes $\{1, 2, \ldots, n\}$ via the homomorphism $\pi v: G \to \Sigma_n$ (cf., Chapter 5, §1.7). In this case we define

$$\rho_1 : H \to S^1,$$

as in Chapter 5(1.3), by setting

$$(3.31) \qquad H = \{h \in G \,|\, \pi v(h)(1) = 1\}$$

$$(3.32) \qquad \begin{bmatrix} v(h) = \sigma(h)(\rho_1(h), \rho_2(h), \ldots) \in \Sigma_n \int S^1 \\ (h \in H, \sigma(h) \in \Sigma_n, \rho_i(H) \in S^1). \end{bmatrix}$$

With this notation we set

$$(3.33) \qquad \rho(v) = \operatorname{Ind}_H^G(\rho_1): G \to \Sigma_n \int S^1.$$

Notice that had we used the stabilizer of $j \in \{1, 2, \ldots, n\}$ in (3.31) and (3.32), the effect would have been to change $\{H, \rho_1\}$ by an inner automorphism of G (as was explained in Chapter 5, §§(1.7) and (1.8)). In other words, the set

$$(3.34) \qquad \rho(v): \{\rho_\alpha : H_\alpha \to S^1\}$$

of (3.29) is well-defined, depending only on the monomial homomorphism, up to permutation and inner automorphisms of G.

The following result summarizes the properties of $\rho(v)$:

(3.35) Lemma

With the notation just introduced, let $v: G \to \Sigma_n \int S^1$ be a monomial homomorphism.

(i) In (3.34) the set

$$\rho(v): \{\rho_\alpha: H_\alpha \to S^1\}$$

is well-defined up to the effect of inner automorphisms of G, depending only on the equivalence class of v.

(ii) If $v = \mathrm{Ind}_H^G(v')$, then $\rho(v) = \mathrm{Ind}_H^G(\rho(v'))$.

(iii) $\rho(\rho(v)) = \rho(v)$.

(iv) If $\rho(v): \{\rho_\alpha: H_\alpha \to S^1\}$, as in (3.34), and if J is a subgroup of G, then

$$\rho(\mathrm{Res}_J^G(v)): \{x_\alpha^*(\rho_\alpha): J \cap (x_\alpha H x_\alpha^{-1}) \to S^1\},$$

where x_α runs through $J \backslash G / H_\alpha$.

(v) As a representation $v = \rho(v) \in R(G)$.

Consequently, as a representation, v has a canonical form as a sum of monomial representations.

Proof. Part (i): The equivalence relation on monomial homomorphisms is given by inner automorphisms of G and of $\Sigma_n \int S^1$. This does not affect the G-orbit structure of $\{1, 2, \ldots, n\}$ acted upon via πv. Therefore we may assume that v is irreducible (i.e., G acts transitively on $\{1, 2, \ldots, n\}$ via πv. However, in this case we have verified, in Chapter 5, (1.7), et seq., that the construction of $\rho(v)$ is well-defined up to inner automorphisms of G.

Part (ii): Suppose that we have $v': H \to \Sigma_m \int S^1$ and that $y_1, y_2, \ldots, y_n$ are chosen coset representatives for G/H. Then

$$g y_i = y_{\tau(g)(i)} h(i, g) \qquad (g \in G, \tau(g) \in \Sigma_n, h(i, g) \in H)$$

and

$$\mathrm{Ind}_H^G(v')(g) = \tau(g)(v'(h(1, g)), v'(h(2, g)), \ldots, v'(h(n, g))).$$

Consider

$$\mu = \pi(\mathrm{Ind}_H^G(v')): G \to \Sigma_n \int \Sigma_m \to \Sigma_{nm},$$

and let us find the stabilizer of 1, $\mathrm{stab}_\mu(1)$, for this G-action on the set $\{1, 2, \ldots, nm\}$. Firstly we must have $\tau(g)(1) = 1$, for if we do not map the first $m \times m$ block to itself, we cannot fix the first entry via μ. That is, we must have

$$g y_1 H = y_1 H,$$

but, by convention, $y_1 = 1$ so that $g \in H$. In addition we must have that

$$\pi v'(g) \in \Sigma_m$$

fixes 1 so that, for example,

$$H \cap \text{stab}_{\pi v'}(1) = \text{stab}_\mu(1) = J.$$

For $g \in J$, $\text{Ind}_H^G(v')(g)$ looks like

$$\begin{bmatrix} v_1(x) \; 0 \; 0 \; 0 \ldots\ldots\ldots, \\ 0 \qquad \ldots\ldots\ldots\ldots \\ 0 \qquad \ldots\ldots\ldots\ldots \\ \vdots \qquad \end{bmatrix}$$

so that $\rho(\text{Ind}_H^G(v'))(g)$ will be a permutation multiplied, on the right, by a diagonal matrix consisting of values of $v_1(\text{—})$. If $x_1, \ldots, x_m$ are coset representatives for H/J, then $\{y_a x_b\}$ are coset representatives for G/J. Also

$$g y_a x_b = y_{\tau(g)(a)} h(a, g) x_b$$

$$= y_{\tau(g)(a)} x_{\sigma(h(a,g))(b)} j(b, h(a, g)).$$

On the other hand,

$$\text{Ind}_H^G(\rho(v'))(g) = \tau(g)(\rho(v'))(h(1, g)), (\rho(v'))(h(2, g)), \ldots)$$

$$= \tau(g)(\sigma(h(1, g)) v_1(j(1, h(1, g))), \ldots)$$

from which we see that

$$\rho(\text{Ind}_H^G(v')) = \text{Ind}_H^G(\rho(v')), \qquad \text{as required.}$$

Part (iii): By additivity, it suffices to consider the case when

$$\rho(v) = \text{Ind}_H^G(v_1) \quad \text{and} \quad H = \text{stab}_{\pi v}(1).$$

In this case we have

$$\rho(\rho(v)) = \rho(\text{Ind}_H^G(v_1))$$

$$= \text{Ind}_H^G(\rho(v_1)), \qquad \text{by (ii),}$$

$$= \text{Ind}_H^G(v_1), \qquad \text{since } \rho(v_1) = v_1 : H \to S^1,$$

$$= \rho(v), \text{ as required.}$$

Part (iv): If, as in (3.34),

$$\rho(v): \{\rho_\alpha: H_\alpha \to S^1\},$$

we wish to evaluate $\rho(\mathrm{Res}_J^G(v))$. We may suppose that G acts transitively on $\{1, 2, \ldots, n\}$ via $\pi v: G \to \Sigma_n$, which means that there is only one ρ_α in the expression for $\rho(v)$ with $H = \mathrm{stab}_{\pi v}(1)$. Next recall that there is a bijection of cosets

$$JxH/H \longleftrightarrow J/(xHx^{-1}) \qquad \text{given by}$$

$$jxH \longleftrightarrow jxHx^{-1}.$$

Let us look at the J-orbit of $x_a H$. This contributes a term to $\rho(\mathrm{Res}_J^G(v))$. Firstly, we must find the J-stabilizer of $x_a H$, which is the subgroup

$$J \cap (x_a H x_a^{-1}).$$

This means that the J-stabilizer, under $\pi v = \sigma$ say, of $q = \sigma(x_a)(1)$ is $J \cap (x_a H x_a^{-1})$, and if

$$v(x_a) = \sigma(x_a) \,\mathrm{diag}\,(\beta_1, \beta_2, \ldots, \beta_n),$$

then, if $h \in H$,

$$v(x_a h x_a^{-1})$$

$$= \sigma(x_a) \,\mathrm{diag}\,(\beta_1, \ldots) \begin{bmatrix} v_1(h)\ 0\ 0\ 0\ldots \\ 0 \qquad \ldots\ldots\ldots \\ 0 \qquad \ldots\ldots\ldots \\ \ldots\ldots\ldots\ldots \\ \ldots\ldots\ldots\ldots \\ \ldots\ldots\ldots\ldots \end{bmatrix} \mathrm{diag}\,(\beta_1^{-1}, \ldots)\sigma(x_a)^{-1},$$

which is a matrix lying within $\Sigma_n \int S^1$ and having (a, a)th entry equal to

$$\beta_1 v_1(h) \beta_1^{-1} \in S^1.$$

However, this is

$$\beta_1(x_a^*(v_1)(x_a h x_a^{-1}))\beta_1^{-1} = x_a^*(v_1)(x_a h x_a^{-1}) \in S^1.$$

This shows that the J-orbit of $x_a H$ contributes to $\rho(\mathrm{Res}_J^G(v))$ precisely the term

$$(x_a^*(v_1): J \cap (x_a H x_a^{-1}) \to S^1),$$

where $\rho(v) = \{v_1: H \rightarrow S^1\}$ in the notation of (3.34), which completes the proof.
Part (v):   Once again, we may suppose that G acts transitively upon $\{1, 2, \ldots, n\}$
via $\pi v: G \rightarrow \Sigma_n$. To show that $v = \rho(v)$ as representations, it suffices to evaluate
the characters of both sides on an element $g \in G$. For this purpose, let us write
$v_m(g)$ for the (m, m)th entry of $v(g)$, then we have

$$\text{Trace}(v(g)) = \sum_{\pi v(g)(m) = m} v_m(g)$$

$$= \sum_{gzH = zH} v_1(z^{-1}gz), \qquad \text{where } H = \text{stab}_{\pi v}(1),$$

$$= \{\text{the character of } \text{Ind}_H^G(v_1: H \rightarrow S^1) \text{ at } g\},$$

which completes the proof of Lemma 3.35.   □

Now we will assemble some useful commutative diagrams involving the
homomorphism

$$\rho: R_+\left(G, \Sigma_n \int S^1\right) \rightarrow R_+(G, S^1),$$

which sends $(G \supset H \xrightarrow{v} \Sigma_n \int S^1)$ to

$$\sum_\alpha (G \supset H_\alpha \xrightarrow{\rho_\alpha} S^1),$$

where, as in (3.34),

$$\rho(v): \{\rho_\alpha: H_\alpha \rightarrow S^1\}.$$

(3.36) Lemma

If $J \leq G$, then the following diagram commutes:

$$\begin{array}{ccc} R_+(J, \Sigma_n \int S^1) & \xrightarrow{\rho} & R_+(J, S^1) \\ {\scriptstyle\text{Ind}_J^G}\downarrow & & \downarrow{\scriptstyle\text{Ind}_J^G} \\ R_+(G, \Sigma_n \int S^1) & \xrightarrow{\rho} & R_+(G, S^1). \end{array}$$

Proof. If $v: H \rightarrow \Sigma_n \int S^1$ is a monomial homomorphism such that

$$\rho(v): \{\rho_\alpha: H_\alpha \rightarrow S^1\}$$

in (3.34), then

$$\rho\left(\operatorname{Ind}_J^G\left(J \supset H \xrightarrow{v} \Sigma_n \int S^1 \right) \right) = \rho\left(G \supset H \xrightarrow{v} \Sigma_n \int S^1 \right)$$

$$= \sum_\alpha (G \supset H_\alpha \xrightarrow{\rho_\alpha} S^1)$$

$$= \sum_\alpha \operatorname{Ind}_J^G (J \supset H_\alpha \xrightarrow{\rho_\alpha} S^1)$$

$$= \operatorname{Ind}_J^G (\rho(v)), \qquad \text{as required.} \qquad \square$$

(3.37) Lemma

If $J \leq G$, then the following diagram commutes:

$$
\begin{array}{ccc}
R_+(G, \Sigma_n \int S^1) & \xrightarrow{\rho} & R_+(G, S^1) \\
\downarrow {\scriptstyle \operatorname{Res}_J^G} & & \downarrow {\scriptstyle \operatorname{Res}_J^G} \\
R_+(J, \Sigma_n \int S^1) & \xrightarrow{\rho} & R_+(J, S^1).
\end{array}
$$

Proof. Suppose that $\rho(v)$: $\{\rho_\alpha : H_\alpha \to S^1\}$ is as in (3.34), where $v : H \to \Sigma_n \int S^1$ is a monomial homomorphism. By definition,

$$\operatorname{Res}_J^G (\rho(v)) = \operatorname{Res}_J^G \left(\sum_\alpha (G \supset H_\alpha \xrightarrow{\rho_\alpha} S^1) \right)$$

$$= \sum_\alpha \sum_{x \in J \backslash G / H_\alpha} (J \supset J \cap (x H_\alpha x^{-1}) \xrightarrow{x^*(\operatorname{Res}_{(x^{-1}Jx)\cap H_\alpha}^{H}(\rho_\alpha))} S^1).$$

However, for the other route, we have

$$\rho\left(\operatorname{Res}_J^G\left(G \supset H \xrightarrow{v} \Sigma_n \int S^1 \right) \right)$$

$$= \sum_{y \in J \backslash G / H} \rho(J \supset J \cap (y H y^{-1}) \xrightarrow{y^*(\operatorname{Res}_{(y^{-1}Jy)\cap H}^{H}(v))} S^1).$$

By §3.35 (iv),

$$\rho(y^*(\operatorname{Res}_{(y^{-1}Jy)\cap H}^{H}(v)))$$

$$= \sum_\alpha y^* \left(\sum_{z \in (y^{-1}Jy) \backslash H / H_\alpha} ((y^{-1}Jy) \cap H \cap (z H_\alpha z^{-1}) \xrightarrow{z^*(\rho_\alpha)} S^1 \right)$$

so that

$$\rho\left(\operatorname{Res}^G_J\left(G \supset H \xrightarrow{\nu} \Sigma_n \int S^1 \right) \right)$$

$$= \sum_{\alpha,\, y \in J \backslash G/H} \left(\sum_{z \in (y^{-1}Jy) \backslash H/H_\alpha} (J \supset J \cap (yzH_\alpha(yz)^{-1}) \xrightarrow{(yz)^*(\rho_\alpha)} S^1) \right).$$

However, for fixed α, with y, z, y', z' being double-coset representatives as previously,

$$yz = y'z' \qquad \text{in } J \backslash G/H_\alpha$$

implies that

$$y = y' \qquad \text{in } J \backslash G/H, \text{ and then}$$
$$yz = yz' \qquad \text{in } J \backslash G/H_\alpha \text{ implies that}$$
$$z = z' \qquad \text{in } (y^{-1}Jy) \backslash H/H_\alpha.$$

The converse is also true so that the triple sum reduces to the required formula for $\operatorname{Res}^G_J(\rho(v))$. $\qquad\square$

Immediately from §3.35 (v) we obtain the following result:

(3.38) Lemma

The diagram

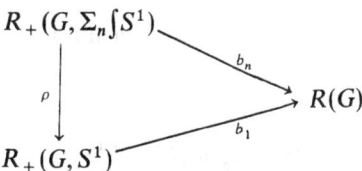

commutes.

(3.39) By means of the ρ-construction of (3.28), we may easily recast the presentation of Theorem 3.22 entirely in terms of $R_+(G, S^1)$. Write b for any of the homomorphisms

$$b_n : R_+\left(G, \Sigma_n \int S^1 \right) \to R(G).$$

Let $\mathcal{T} \le R_+(G, S^1)$ be the subgroup generated by the following two types of elements:

(i) For $(v: G \supset H \to S^1)$, $\rho(\tau_G(b(v))) - v \in \mathcal{T}$.
(ii) If $(\mu: G \supset J \to S^1)$ is a second element of $R_+(G, S^1)$, then $\rho(\tau_G(b(v) \oplus b(\mu))) - \rho(\tau_G(b(v))) - \rho(\tau_G(b(\mu))) \in \mathcal{T}$.

(3.40) Theorem

The homomorphism, b, induces an isomorphism

$$b: R_+(G, S^1)/\mathcal{T} \xrightarrow{\cong} R(G).$$

(3.41) Remark

If $v: G \to \Sigma_n \int S^1$ is a monomial homomorphism such that

$$\rho(v): \{\rho_\alpha: H_\alpha \to S^1\}$$

in (3.34), then

$$\rho(v) = \sum_\alpha \mathrm{Ind}_{H_\alpha}^G(\rho_\alpha): G \to \Sigma_n \int S^1$$

is also a well-defined monomial homomorphism. However, as I warned in Chapter 5, §1.8, v and $\rho(v)$ are presumably not generally equivalent as monomial homomorphisms.
  For example, if $G = \mathbb{Z}/3 = \langle x \rangle$, we may define

$$v: \mathbb{Z}/3 \to \Sigma_3 \int S^1 \qquad \text{by}$$

$$v(x) = [\mathrm{diag}\,(1, \xi, \xi^2)] \qquad \{\text{the permutation } (1, 2, 3)\},$$

where $\xi = \exp(2\pi i/3)$. Hence

$$\rho(v): \{H_\alpha = \{1\} \to S^1\}$$

and

$$\rho(v)(x) = \begin{bmatrix} 0 & 0 & 1 \\ 1 & 0 & 0 \\ 0 & 1 & 0 \end{bmatrix},$$

whereas

$$v(x) = \begin{bmatrix} 0 & 0 & 1 \\ \xi & 0 & 0 \\ 0 & \xi^2 & 0 \end{bmatrix}.$$

Actually, in this case

$$XvX^{-1} = \rho(v),$$

where

$$X = \begin{bmatrix} 0 & 0 & 1 \\ 1 & 0 & 0 \\ 0 & \xi^2 & 0 \end{bmatrix},$$

but this only works because G is cyclic. I expect that some fairly simple nonabelian groups will yield examples for which v and $\rho(v)$ are inequivalent.

Chapter Seven

Applications of Explicit Brauer Induction to Artin Root Numbers and Local Root Numbers

Grässträcts barn, sä lika sin fader,
Vära och dö; oàndliga rader;
Morgon blir afton; evig ring.
Tiden är ingenting.

—*F. M. FRANZÉN* (19th century)*

In this chapter we will begin with a brief review of the L-functions of Artin. For a more extensive review, we refer the reader to an excellent survey article [Mar]. *However, the review in this chapter will provide the reader with an idea of the type of mathematical behaviour that one can expect from an L-function, as well as acquaint the reader with the Artin root number and its properties on one-dimensional Galois representations.*

We will then proceed to construct the local root numbers (also called local constants) of [De 2; Ta]. *The construction could fairly justly be described as fundamentally different from the previous construction* [De 2]. *The parts of the picture and the various properties of root numbers emerge in a completely different order in the two approaches. Our approach uses the Explicit Brauer Induction method developed in Chapter 6.*

The local root number is an invariant of a local Galois representation which is to be taken very seriously. For example, in the theory of the structure of a ring

---

*The children of the blade of grass; so like their father;
Grow and die in endless ranks together;
Dawn turns to dusk in an eternal ring.
Time is an inconsequential thing.

of algebraic integers as a Galois module, the local root numbers determine whether or not this projective module is free, in the case of a tame extension [T]. In the Langlands programme, which speculates about bijections between Galois representations and other categories of representations [Car; Gel], the local root number plays an important role as part of the detection machinery in the local conjecture.

Our construction will be first to establish, by direct construction, the local root numbers of orthogonal representations. This construction is entirely local and gives at once a formula, due to Deligne [De], which relates orthogonal local root numbers to the second Stiefel-Whitney class. In view of the results of Chapter 3, namely the connection between Hasse-Witt classes and the second Stiefel-Whitney class, it is not surprising that Witt groups of bilinear forms enter into the construction. Actually in my original construction, I was unaware that the group, Y_K, which I introduced to make the construction, was equal to the augmentation ideal in the Witt ring of the local field, K. This observation, which is due to Pierre Conner, is a very useful insight. In particular, from the results on local root numbers, we obtain some information on the Weil character of the Witt ring. All this is to be found in §§1 and 2, together with sample local root number calculations. In §3 we use the results of Chapter 6, together with the existence of orthogonal local root numbers, to deduce the existence of a local root number function which is related to the Artin root number by a local/global factorization. All constructions up to this point use only local methods (i.e., representation theory methods). However, from our presentation for R(G), in §3 we show finally that, at least in some cases, the local/global factorization implies that our local root number function is actually a homomorphism with the required inductive properties. Previously, the local/global factorization would be derived as a consequence of the other properties. To complete the last step, I have used, for simplicity, the existence of tame local root numbers, which is established in [F-T] by local techniques.

---

1   THE ARTIN L-FUNCTIONS

Let L/K be a finite normal extension of number fields with Galois group, G. Let

$$\rho: G \to GL(V)$$

be a finite-dimensional, complex representation of G. If we wish we may put a G-invariant inner product on V (by averaging an inner product over G) so that ρ may, without loss of generality, be considered as a unitary representation, $\rho: G \to U(n)$, where dim $V = n$, of the type treated in Chapter 6.

We are now going to recall some details concerning the Artin L-function of ρ and the "extended" Artin L-function of ρ. Artin introduced these functions in [A2], extending the notion to its present-day form in [A3]. One cannot hope to improve upon the survey article [Mar]. On the other hand, a brief review of L-functions seems necessary here, for the reader's convenience, just as a text on Leonardo da Vinci cannot hope to get far on the topic of the *Mona Lisa* without assuming firsthand familiarity with the work but can get nowhere before revealing that it is a portrait of a lady.

As is customary a *finite prime* of K will mean a prime ideal, $P \lhd O_K$, of the (algebraic) integers, O_K, of K or equivalently the *place* given by the inclusion of K into the complete local field, K_p, obtained by P-adically completing K. An *infinite* or *Archimedean prime* means an embedding of K into a complete local field which is isomorphic to $\mathbb{R}$ or $\mathbb{C}$, the real or complex numbers. Let Q be a prime of L over P (written $Q|P$; Q *divides* P). In terms of places this means that the following diagram commutes:

(1.1)
$$\begin{array}{ccc} K & \longrightarrow & K_P \\ \downarrow & & \downarrow \\ L & \longrightarrow & L_Q \end{array}.$$

In terms of ideals it means that $P \cdot O_L = Q \lhd O_L$. G acts on L, preserving the integers of L and permuting the places. A *decomposition group* for P is defined (up to conjugation in G) by

(1.2)
$$D_P = \{g \in G | gQ = Q\}.$$

For finite primes, the *residue fields* of K and L are, respectively,

(1.3)
$$k = (O_K)/P \quad \text{and} \quad l = (O_L)/Q.$$

We have an evident map, which is surjective,

(1.4)
$$D_P \twoheadrightarrow G(l/k)$$

whose kernel is the *inertia group*, I_P. The Galois group, $G(l/k)$, is cyclic since k and l are finite. If $N(P)$ is the absolute norm of P (i.e., the positive integer which generates the ideal in the rational integers generated by the norms from P), then the generator, $\tilde{\sigma}_P$, satisfies

(1.5)
$$\tilde{\sigma}_P(x) = x^{N(P)} \qquad \text{in } l.$$

This is the Frobenius element. It may be lifted to

(1.6)
$$\begin{cases} \sigma_P \in D_P \subset G \text{ characterized (mod } I_P) \text{ by} \\ \sigma_P(z) \equiv z^{N(P)} \pmod{Q} \text{ for } z \in O_L. \end{cases}$$

Returning to the representation, ρ, define

(1.7)
$$V_P = \{v \in V | \rho(g)v = v \text{ for all } g \in I_P\},$$

the subspace of I_P-invariants of V.

The *Artin L-function* of ρ is defined as

(1.8)
$$\begin{cases} L_K(s,\rho) = \prod_{\substack{P \lhd O_K \\ \text{finite}}} (\det_{V_P}(1 - N(P)^{-s}\sigma_P))^{-1} \\ \text{for } s \text{ complex and } Re(s) > 1. \end{cases}$$

The L-function satisfies the following properties [Mar, p. 9]:

(1.9) Proposition

(i) $L_K(s, \rho_1 \oplus \rho_2) = L_K(s, \rho_1)L_K(s, \rho_2)$.

(ii) If $K \subset L \subset N$ is a chain of finite normal extensions and $\pi: G(N/K) \to G(L/K)$ is the natural epimorphism, then

$$L_K(s, \rho) = L_K(s, \rho\pi).$$

In other words, $L_K(s, \text{—})$ is defined on continuous, finite-dimensional representations of the absolute Galois group, Ω_K, of Chapter 1, §(1.24).

(iii) If F is an intermediate field of L/K and $\psi: G(L/F) \to GL((W))$ is a representation, then

$$L_F(s, \psi) = L_K(s, \text{Ind}_{F/K}(\psi)).$$

(1.10)   When $\dim(\rho) = 1$ the L-function specializes to give the L-series of Dirichlet, and when $\rho = 1$, we obtain the Dedekind zeta function,

(1.11)
$$L_K(s, 1) = \zeta_K(s).$$

According to [Car, §1.2], Artin introduced the L-function, inter alia, to analyze multiplicative relations between Dedekind zeta functions of number fields. The following theorem of Artin reduces the problem to representation theory. It is proved by means of the Cebotarev density theorem.

(1.12) Theorem (E. Artin)

Let $L/\mathbb{Q}$ be a finite normal extension, then

$$L_\mathbb{Q}(s, \text{—}): R(G(L/\mathbb{Q})) \to \begin{bmatrix} \text{multiplicative group of} \\ \text{nonzero meromorphic} \\ \text{functions} \end{bmatrix}$$

is an injective homomorphism.

(1.13) Corollary

Let $L/\mathbb{Q}$ be a finite normal extension. All relations between zeta functions of intermediate fields, $\mathbb{Q} \subset F \subset L$, are obtained by applying $L_\mathbb{Q}(s, \text{—})$ to the kernel of *the natural map*

$$b: A(G(L/\mathbb{Q})) \to R(G(L/\mathbb{Q})),$$

where $A(G)$ is the Burnside ring, $A(G) = R_+(G, \{1\})$.

Proof. Set $G = G(L/\mathbb{Q})$, if

$$0 = b(\sum a_H(G \supset H \to \{1\})$$
$$= \sum a_H \operatorname{Ind}_H^G(1),$$

then

$$1 = \prod L_{\mathbb{Q}}(s, \operatorname{Ind}_H^G(1))^{a_H}$$
$$= \prod_H \zeta_{L(H)}(s)^{a_H}$$

where $L(H)$ is the fixed field of H. By (1.12), all such relations are generated in this manner. □

(1.14) Example

Let $G = G(L/\mathbb{Q})$ be as in §1.13. Suppose that $v: G \to U(n)$ is a representation and that

$$1 = \sum_{(J)} \aleph_{(J)}^{\#} \operatorname{Ind}_J^G(1) \in R(G)$$

is the equation of Chapter 6, (2.31), then

$$\zeta_{\mathbb{Q}}(s) = \prod_{(J)} \zeta_{L(J)}(s)^{\aleph^{\#}(J)}.$$

(1.15) Remark

Our presentation for $R(G)$, derived in Chapter 6, §3, codifies in a geometrical manner the way in which all the relations in (1.13) arise. Consider the presentation in the form of Chapter 6, (3.23),

$$R(G) = \mathscr{R}(G)/\{\tau_G v - v \mid v: G \to \Sigma_n \int S^1 \text{ some } n\}.$$

The map from $A(G)$ to $\mathscr{R}(G)$, induced by the inclusion of Σ_n into $\Sigma_n \int S^1$, is clearly split injective. Hence, in $\mathscr{R}(G)$, we have

$$(1.16) \qquad \operatorname{Ker}(b: A(G) \to R(G)) = A(G) \cap \left\{ \tau_G(v) - v \mid v: G \to \Sigma_n \int S^1 \right\}.$$

Alternatively, from Theorem 3.40, we may obtain all the relations in (1.13) *by applying* $L_{\mathbb{Q}}(s, —)$ *to*

$$\mathrm{Ker}\,(b: A(G) \to R(G)) = A(G) \cap \mathcal{T} \subset R_{+}(G, S^1)$$

as G varies through finite Galois groups of the form $G = G(L/\mathbb{Q})$.

(1.17)   THE ANALYTIC CLASS NUMBER FORMULA

The Artin L-function owes its importance in number theory to the remarkable manner in which it encodes vital number-theoretic information. An excellent example of this is the analytic class number formula

(1.18) $$\lim_{s \to 1+} (s - 1)\zeta_K(s) = 2^{r+t}\pi^t Rh/m(\sqrt{D}).$$

In (1.18), r is the number of real places of K, $2t$ is the number of complex places, R is the regulator, D the discriminant, m is the order of $\mu(K)$, the roots of unity in K^*, and h is the order of the class group of K.

(1.19)   THE CLASS GROUP

I am not going to go further into the details of regulators and discriminants. For these and further information, I refer the reader to [B-S, p. 313]. The discriminant is computed by means of an integer basis for $\mathcal{O}_K$. The regulator is more complicated and involves a *fundamental system of units*, that is, a basis for the (multiplicative) free abelian group of units in $\mathcal{O}_K$, $\mathcal{O}_K^*$, modulo the subgroup of roots of unity in K. I am going to concentrate on the class number, h, although similar discussions apply for each invariant appearing in the right side of (1.18).

The *ideal class group* of K is defined [Co, p. 49] to be the group of all *fractional ideals* of $\mathcal{O}_K$, under multiplication, modulo the principal fractional ideals. Its role in life is to measure how far $\mathcal{O}_K$ is from being a unique factorization domain. It is also known as the *Picard group* of $\mathcal{O}_K$. We will denote it by

$$Cl(K).$$

It is a finite group.

If we let $K_0(\mathcal{O}_K)$ denote the Grothendieck group of isomorphism classes of finitely generated projective $\mathcal{O}_K$-modules (i.e., summands of modules of the form $\mathcal{O}_K^N$), then there is an exact sequence (see [M2])

$$0 \to Cl(K) \rightarrowtail K_0(\mathcal{O}_K) \xrightarrow{rk} \mathbb{Z} \to 0,$$

where rk sends a module to its rank. This is because every finitely generated projective module has the form

$$\mathcal{O}_K^w \oplus P$$

with $P \lhd \mathcal{O}_K$ a prime ideal, and there are isomorphisms of $\mathcal{O}_K$-modules

$$P_1 \oplus P_2 \equiv \mathcal{O}_K \oplus P_1 P_2.$$

Suppose now that L/K is a finite Galois extension of number fields. Tensoring a projective $\mathcal{O}_K$-module, M, with $\mathcal{O}_L$ yields a projective $\mathcal{O}_L$-module, $M \otimes_{\mathcal{O}_K} \mathcal{O}_L$, and there results a natural homomorphism,

$$(1.20) \qquad\qquad i_* : Cl(K) \to Cl(L).$$

In addition $\mathcal{O}_L$ is a finitely generated $\mathcal{O}_K$-module so that considering an $\mathcal{O}_L$-module as an $\mathcal{O}_K$-module yields a *transfer map*, or *norm homomorphism*,

$$(1.21) \qquad\qquad i^* : Cl(L) \to Cl(K).$$

The homomorphisms i^* and i_* enjoy properties analogous to the cohomology transfer and restriction map of Chapter 1, (2.45). In particular,

$$(1.22) \qquad \begin{cases} i_* i^*(x) = \sum\limits_{g \in G(L/K)} g_*(x) \in Cl(L) \quad \text{and} \\ i^* i_*(y) = [L:K] y \in Cl(K). \end{cases}$$

Consequently i_* induces an isomorphism:

$$(1.23) \qquad \begin{cases} i_* : Cl(K)[1/d] \xrightarrow{\;\cong\;} (Cl(L)[1/d])^{G(L/K)}, \quad \text{where} \\ d = [L:K] = |G(L/K)|. \end{cases}$$

(1.24) We will use (1.23) later to derive class number relations from relations in $R(G(L/K))$ between permutation representations. These relations were discovered by Brauer [Bra], who used analytic methods together with §1.13 and the analytic class number formula. Brauer's results are recovered purely algebraically in [Wa]. Before giving the latter proof, I would like to give a topological example/exercise—an exercise because it involves a little more topology than we have been assuming—which gives further insight into how the geometrical principles of Chapter 6 fit into the picture.

(1.25) Exercise/Example

Let G be a finite group, and let $v : G \to X_n$ ($X_n = U(n), Sp(n),$ or $O(2n)$) be a representation. Let

$$1 = \sum_{(J)} \aleph_{(J)}^{\#}, \operatorname{Ind}_J^G(1) \in R(G)$$

be the relation of Chapter 6, (2.31) (where I have replaced $H(g_J, v)$ by its conjugate, J, for notational convenience). Let M be any $\mathbb{Z}[1/d][G]$-module where $d = |G|(n!)$. Then, in the sense of Euler characteristics,

$$(1.26) \qquad\qquad M^G = \sum_J \aleph_{(J)}^{\#} M^J.$$

This means that there is an exact sequence whose terms are sums of the M^J's and whose Euler characteristic gives (1.26). As we shall see, one can remove the $n!$ from d, but for the moment let it remain.

To see (1.26), set $A = X_n/Y_n$, as in Chapter 6, (2.31). We may assume that A is triangulated as a finite simplicial complex on which G acts simplicially. Now consider the Bredon cohomology groups [Bre]:

$$(1.27) \qquad\qquad H_G^*(A, \underline{M}).$$

I claim that the right side of (1.26) is equal to the Euler characteristic

$$(1.28) \qquad\qquad \sum (-1)^i H_G^i(A; \underline{M})$$

for any G-module, M. To see this, observe that the cochain complex from which (1.27) is computed,

$$(1.29) \qquad\qquad 0 \to C_G^0(A, \underline{M}) \to C_G^1(A, \underline{M}) \to \cdots,$$

has

$$C_G^n(A, \underline{M}) = \bigoplus_{\sigma^n} M^{\mathrm{stab}(\sigma^n)},$$

where the direct sum is over orbit representatives, σ^n, of the n-simplices of A.

However, if $|G|$ is invertible in M, then we may replace $M^{\mathrm{stab}(\sigma)}$ by $H^*(\mathrm{stab}(\sigma); M)$. This shows that the double complex [H-S; C-E; Mac],

$$\bigoplus_{m} \quad \bigoplus_{\substack{\sigma^n \\ \text{orbit} \\ \text{representative}}} \quad C^m(\mathrm{stab}(\sigma^n); M),$$

has total cohomology equal to $H_G^*(X; \underline{M})$. This follows from the spectral sequence of a double complex.

On the other hand, it is not hard to see that, in dimension $m + n$ the preceding double complex is just

$$C^{m+n}(EG \times_G A; M),$$

the simplicial cochain group for the G-orbit space

$$EG \underset{G}{\times} A = (EG \times A)/G,$$

where G acts diagonally.

Since $|G|$ is invertible in M the spectral sequence for

$$A \to EG \underset{G}{\times} A \to BG$$

shows that

$$H^*_G(EG \underset{G}{\times} A; M) \cong H^*(A; M)^G,$$

the G-invariants of the cohomology of A with (simple) coefficients in M. If $n!$ is invertible also, then, by Chapter 6, (2.4),

$$H^*(A; M) = \begin{cases} M & \text{if } * = 0, \\ 0 & \text{if } * \neq 0. \end{cases}$$

This means that we can make (1.29) exact by replacing the left-hand trivial map by the evident diagonal map,

$$\mathscr{E}: M^G \to \bigoplus_{\sigma^0} M^{\text{stab}(\sigma^0)}.$$

(1.30) Example

The following example is due to Dirichlet (1842) and may also be found in [Co, p. 253].

Let $m > 0$ be a square-free integer, and set $F_1 = \mathbb{Q}(\sqrt{m})$, $F_2 = \mathbb{Q}(\sqrt{-m})$, $N = \mathbb{Q}(\sqrt{m}, \sqrt{-m})$. Let $L = \mathbb{Q}(a, i)$, where $i^2 = -1$ and $a^2 = \sqrt{m}$. Hence $G(L/\mathbb{Q}) \cong D_8$, by Chapter 3, (2.2) and (2.3), where $D_8 = \{x, y \mid x^4 = y^2 = 1, xyx = y\}$ and

(1.31)
$$\begin{cases} x(i) = i, & x(a) = ia, \\ y(i) = -i, & y(a) = a. \end{cases}$$

From Case iv in Chapter 5 (i.e., Table B and Figure 4 of Chapter 5, § (3.16)), the orbit-space stratification of $D_8 \backslash U(2)/(\Sigma_2 \int S^1) = D_8 \backslash \mathbb{R}P^2$ looks as shown in Figure 1.

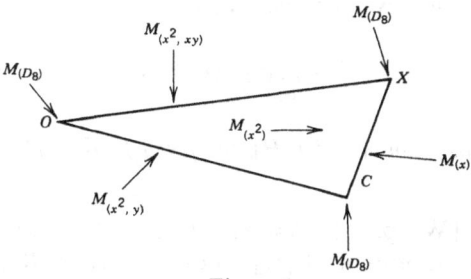

Figure 1

Since x^2 acts trivially on $\mathbb{R}P^2$, one finds that the resulting relation originates from the following relation in $R(D_8/\{x^2\}) = R(G(N/\mathbb{Q}))$:

$$2 + \operatorname{Ind}_{\{1\}}^{G(N/\mathbb{Q})}(1) = \operatorname{Ind}_{\{y\}}^{G(N/\mathbb{Q})}(1) + \operatorname{Ind}_{\{x\}}^{G(N/\mathbb{Q})}(1) + \operatorname{Ind}_{\{xy\}}^{G(N/\mathbb{Q})}(1).$$

The fixed fields in $N/\mathbb{Q}$ are

$$N^{G(N/\mathbb{Q})} = \mathbb{Q}, \, N^{\{x\}} = \mathbb{Q}(i), \, N^{\{xy\}} = F_2$$

and

$$N^{\{y\}} = F_1.$$

Applying §1.25 to the example $G = G(L/\mathbb{Q})$ and $M = Cl(L)[\frac{1}{2}]$, by (1.23) we obtain a relation between class numbers (h_K denotes $|Cl(K)|$):

(1.32) $\{h_N = h_{F_1} \cdot h_{F_2}$ up to powers of 2$\}$,

since $\mathbb{Z}$ and $\mathbb{Z}[i]$ are unique factorization domains.

Notice that the power of 2 is an essential ambiguity. When m is an odd prime, the exact result is [Co, p. 253, §19.8 (c)]

$$h_N = \begin{cases} (h_{F_1} \cdot h_{F_2})/2 & \text{if } m \equiv 1 \ (\mathrm{mod} \ 4), \\ h_{F_1} \cdot h_{F_2} & \text{if } m \equiv 3 \ (\mathrm{mod} \ 4). \end{cases}$$

I have included §1.25 and §1.30 because of their evident connection with the Explicit Brauer Induction of Chapter 6. However, in this instance, following [Wa], one can do better by a simple algebraic argument.

(1.33) Proposition

Let L/K be a finite, normal extension of number fields with group G. Suppose that

$$0 = \sum_{(H)} a_H \operatorname{Ind}_H^G(1) \in R(G),$$

then, up to powers of primes appearing in $|G|$,

$$1 = \prod_{(H)} (h_{L(H)})^{a_H} \in \mathbb{Q},$$

where $h_{L(H)}$ is the class number of $L(H)$, the fixed field of H.

Proof. According to [Wa, p. 35] if $\hat{\Lambda} = \mathbb{Z}[1/|G|]$, then any isomorphism between permutation representations as $\mathbb{C}[G]$-modules (i.e., in $R(G)$) is also true as $\Lambda[G]$-modules for Λ equal to a *finite* quotient of $\hat{\Lambda}$. Now set $C(L) = Cl(L)[1/|G|]$, which is a $\Lambda[G]$-module, for an appropriate Λ. By the preceding remark,

$$0 = \sum_{(H)} a_H \operatorname{Hom}_{\Lambda[G]}\left(\Lambda[G] \underset{\Lambda[H]}{\otimes} \Lambda, C(L) \right),$$

in the sense of Euler characteristics. The result now follows from the iso-morphisms,

$$\operatorname{Hom}_{\Lambda[G]}\left(\Lambda[G] \underset{\Lambda[H]}{\otimes} \Lambda, Cl(L) \right) \cong \operatorname{Hom}_{\Lambda[H]}(\Lambda, Cl(L))$$

$$\cong Cl(L)^H$$

$$\cong Cl(L(H)), \qquad \text{by (1.23).}$$

(1.34) THE EXTENDED L-FUNCTIONS AND ARTIN ROOT NUMBERS

Let L/K be a finite, normal extension of number fields with Galois group G. Let

$$\rho: G \to GL(V)$$

be a finite-dimensional representation, as in (1.8).

The *extended Artin L-function* has the form

(1.35) $$\Lambda_K(s, \rho) = A(\rho)^{s/2} \gamma_\rho(s) L_K(s, \rho), \qquad \operatorname{Re}(s) > 1.$$

The extra terms, which are described in detail in [Mar, pp. 11–17] are innocuous functions given in terms of Γ-functions, the Artin conductor, the absolute discriminant of K, and the Archimedean places of K. With these extra factors in place, Hecke had shown that $\Lambda_K(s, \rho)$ had an analytic continuation to the whole plane when $\dim \rho = 1$, $\rho \neq 1$. When $\rho = 1$, $\Lambda_K(s, 1)$ has a mero-morphic continuation.

In addition $\Lambda_K(s, \rho)$ enjoys the same properties as are listed in §1.9 for $L_K(s, \rho)$. Hence, by Brauer induction, from the one-dimensional case we find that $\Lambda_K(s, \rho)$ has a meromorphic continuation to the whole plane. If $\rho \neq 1$ and ρ is irreducible, a famous *conjecture of Artin* states that $\Lambda_K(s, \rho)$ (and $L_K(s, \rho)$, equivalently) will be *holomorphic*.

The meromorphic continuation satisfies the following functional equation, which again follows from Hecke's work when $\dim(\rho) = 1$:

(1.36) $$\Lambda_K(1 - s, \rho) = W_K(\rho) \Lambda_K(s, \bar\rho),$$

where $\bar\rho$ is given by the complex conjugate of ρ.

$W_K(\rho)$ is a complex number lying on the unit circle; it is called the *Artin root number*. From §1.9 the following properties for $W_K(\rho)$ are evident:

(1.37) Proposition

With the preceding notation, $W_K(\rho) \in S^1$ and

(i) $W_K(\rho_1 \oplus \rho_2) = W_K(\rho_1)W_K(\rho_2)$.

(ii) If $K < L < N$ is a chain of finite, normal extensions and $\pi: G(N/K) \to G(L/K)$ is the natural epimorphism, then

$$W_K(\rho) = W_K(\rho \cdot \pi).$$

(iii) If F is an intermediate field of L/K and if $\psi: G(L/F) \to GL:(W)$ is a representation, then

$$W_F(\psi) = W_K(\text{Ind}_{F/K}(\psi)).$$

(1.38)   Let L/K again be a finite, normal extension of number fields with group, $G(L/K)$. Let P, Q be primes of K and L, respectively, with Q dividing P. When P is finite, we have the decomposition group, D_P, of (1.2). When P is Archimedean, then D_P will mean the cyclic group of order 2 (generated by complex conjugation) if the pair of completions $K_P \subset L_Q$ is isomorphic to $\mathbb{R} \subset \mathbb{C}$ and D_P will be trivial in the other Archimedean cases. With this definition the completions, L_Q/K_P, form a Galois extension of local fields, with

(1.39) $G(L_Q/K_P) \cong D_P$.

The group, D_P, is defined up to conjugation as a subgroup of $G(L/K)$.

Every local Galois group, G, has a finite, descending chain of normal subgroups $\{1\} = G_m \subset \cdots \subset G_2 \subset G_1 \subset G_0 \subset G$. These are the *ramification groups* of G (see [Ser 2]). If $G = D_P$, where P is a finite prime as in (1.2), then

(1.40)
$$
\begin{aligned}
&G_0 = I_p, \text{ the inertia group,}\\
&G_1 = \{\text{the first wild ramification group}\} \text{ is a}\\
&p\text{-group } (p = \text{residue characteristic})\\
&G_0/G_1 \text{ is cyclic of order prime to } p \text{ and}\\
&G(D_P/I_P) \cong G(l/k) \text{ in (1.3) is cyclic.}
\end{aligned}
$$

When P is Archimedean and if $D_P \neq \{1\}$, then, we set $I_P = \{1\}$ so that (1.40) is still true.

Now suppose that

$$\rho: G(L/K) \to GL(V)$$

is a global Galois representation, as in (1.34); then, for each place, v, of K we may choose a place, w, of L over v and, by restricting ρ to a choice of decomposition group, obtain a local Galois representation

(1.41) $$\rho_v \colon G(L_w/K_v) = D_v \to GL(V).$$

The *local root number* will be an invariant of Galois representations of local fields, satisfying properties similar to §1.37 (but not identical) which will be made precise in the next section. The local root number will take values in the complex numbers of unit norm, and its most important property will be to give a local/global factorization of the Artin root number of the form

(1.42) $$W_K(\rho) = \prod_{\substack{v \text{ place} \\ \text{of } K}} W_{K_v}(\rho_v).$$

The product in (1.42), taken over all places of K, will actually be a finite product because $W_{K_v}(\rho_v)$ will equal one for all but finitely many primes, namely, those at which ρ ramifies.

In the abelian case (i.e., $\dim(\rho) = 1$) the factorization of (1.42) is due to Hecke, and the rest of this chapter will be concerned with promoting the one-dimensional invariants to an invariant of all local Galois representations. In preparation for that we will have to review the abelian theory. But first, we will deal with the local root numbers in the Archimedean case. This is not strictly necessary because in the Archimedean case $\Omega_K = \{1\}$ or $\mathbb{Z}/2$ and all representations are orthogonal, so that we could incorporate this into our later treatment of orthogonal root numbers. The latter approach is the shorter but the more contrived, so I have chosen the former.

(1.43)   ARCHIMEDEAN ROOT NUMBERS

Let K be an Archimedean local field, and let

$$\rho \colon \Omega_K = G(\bar{K}/K) \to GL(V)$$

be a representation (automatically continuous). If $K \cong \mathbb{C}$, this representation is trivial, and so we set

(1.44) $$W_K(\rho) = 1 \qquad \text{if } K \cong \mathbb{C}.$$

If $K \cong \mathbb{R}$, then Ω_K is cyclic of order 2, and if y is the nontrivial, one-dimensional representation, we have

$$\rho = a + by \qquad (a + b = \dim(V)),$$

and [Mar, p. 32] we set

(1.45) $$W_K(\rho) = i^{-b} \qquad \text{if } K \cong \mathbb{R}, \text{ where } i^2 = -1.$$

(1.46) SYMPLECTIC ROOT NUMBERS AT INFINITY AND THE FIRST PONTRJAGIN CLASS

Suppose now that L/K again be a finite, normal extension of number fields, and let

$$(1.47) \qquad \rho: G(L/K) \to Sp(n) \subset U(2n)$$

be a symplectic representation.

The cohomology of $BSp(n)$ is given by a polynomial ring on the Pontrjagin classes (cf., Chapter 1, (3.16) and (4.22), vis à vis Stiefel-Whitney and Chern classes):

$$(1.48) \qquad \begin{cases} H^*(BSp(n); \mathbb{Z}) \cong \mathbb{Z}[p_1, p_2, \ldots, p_n], \deg(p_i) = 4i, \\ H^*(BSp(n); \mathbb{Z}/2) \cong \mathbb{Z}/2[p_1, p_2, \ldots, p_n]. \end{cases}$$

Therefore we have the first Pontrjagin class of ρ

$$(1.49) \qquad p_1(\rho) = \rho^*(p_1) \in H^4(K; \mathbb{Z}/2).$$

However, if v is a real place of K, then $H^4(K_v; \mathbb{Z}/2) \cong H^4(\Omega_{K_v}; \mathbb{Z}/2) \cong H^4(\mathbb{Z}/2; \mathbb{Z}/2) \cong \{\pm 1\}$, and the real places induce an isomorphism

$$(1.50) \qquad H^4(K; \mathbb{Z}/2) \xrightarrow{\cong} \prod_{\substack{v \text{ real} \\ \text{place of } K}} H^4(K_v; \mathbb{Z}/2) \cong \prod_{\substack{v \\ \text{real}}} \{\pm 1\}.$$

Let $p_1(\rho)_v \in \{\pm 1\}$ denote the vth coordinate of $p_1(\rho)$ in terms of (1.50). Since symplectic representations, ρ, restrict to $\Omega_{K_v} \subseteq \mathbb{Z}/2$ to give representations of the form $2\alpha + 2\beta y$, we see that, by (1.45),

$$W_{K_v}(\rho_v) \in \{\pm 1\}$$

also. Thus the following result is hardly surprising:

(1.51) Theorem

With the notation of §1.46,

$$p_1(\rho) = \{W_{K_v}(\rho_v)\} \in \prod_{\substack{v \text{ real} \\ \text{place of } K}} \{\pm 1\} \cong H^4(K; \mathbb{Z}/2).$$

We will prove this theorem, using the symplectic induction theorem of Chapter 6, § (2.31). We will need the following result on one-dimensional symplectic Galois representations:

(1.52) Lemma

Let L/K be a finite, normal extension of number fields whose group, $G(L/K)$, is a subgroup of $Sp(1)$ $(= S^3$, the quaternions of unit length).
Let

$$\rho: G(L/K) \rightarrowtail Sp(1) \subset U(2)$$

be the resulting symplectic representation. Then

$$p_1(\rho) = \{W_{K_v}(\rho_v)\} \in \prod_{\substack{v \text{ real} \\ \text{place of} \\ K}} \{\pm 1\},$$

if $G(L/K)$ is cyclic or generalized quaternion.

Proof. Let us assume that $G(L/K) = Q_{4n}$, and let $K \rightarrowtail K_v \cong \mathbb{R}$ be a real place of K. Then we have a commutative diagram of Galois groups:

(1.53)

$$\mathbb{Z}/2 \cong \Omega_{K_v} \rightarrowtail \Omega_K$$
$$\searrow_{\varphi} \qquad \downarrow$$
$$Q_{4n}$$

In (1.53), if φ is trivial, then complex conjugation acts trivially on L, and L lies in $K_v \cong \mathbb{R}$ so that $p_1(\rho)$ vanishes in $H^4(K_v; \mathbb{Z}/2)$, as does $W_{K_v}(\rho_v)$.
Otherwise, there is a commutative diagram

$$\begin{array}{ccc} K & \rightarrowtail & L \\ {\scriptstyle v}\downarrow & & \downarrow{\scriptstyle w} \\ \mathbb{R} & \rightarrowtail & \mathbb{C} \end{array}$$

In this case φ maps Ω_{K_v} isomorphically to the central subgroup of order 2 in Q_{4n}. But, by Chapter 1, (4.40), this means that φ^* detects nontrivially the Pontrjagin class, $p_1(\rho) \in H^4(Q_{4n}; \mathbb{Z}/2)$ so that $p_1(\rho)_v = -1 \in \{\pm 1\}$. However, this equals $W_{K_v}(\rho_v)$ in this case, since ρ_v restricts to 2_y on $\Omega_{K_v} \cong \mathbb{Z}/2$.
When $G(L/K)$ is cyclic, then, as a complex representation, $\rho = \psi + \bar{\psi}$, where ψ is one dimensional. In this case, restricted to $\Omega_{K_v} \cong \mathbb{Z}/2$, $\rho_v = 2y$ and $W_{K_v}(\rho_v) = -1$, but also $p_1(2y) = c_1(y)^2 \neq 0 \in H^4(\mathbb{Z}/2; \mathbb{Z}/2)$, where c_1 is the first Chern class. This completes the proof. $\square$

(1.54) Proof of Theorem 1.51
Suppose that $v: G(L/F) \rightarrowtail Sp(m)$ is a symplectic representation, where $K \subset F \subset L$ and that

$$\rho = \text{Ind}_{F/K}(V).$$

Since p_1, the first Pontrjagin class, is the first k-invariant of BSp, it commutes with induction. This comes from a general result about transfers in K-cohomology theories. We will not go into details. However, we have met the analogous fact, that SW_1 commutes with transfer and, in §2.21, will meet the fact that SW_2 commutes with transfer for special orthogonal representations. These facts correspond to SW_1 and SW_2 being the first k-invariants of BO and BSO, respectively.

Granting this, we have

$$p_1(\rho) = p_1\left(\mathrm{Ind}_{F/K}(v)\right) = i_*(p_1(v)) \in H^4(K; \mathbb{Z}/2).$$

Therefore, if we can show that the right side of (1.51) commutes with transfer, then the symplectic case of Chapter 6, (2.31) will reduce the proof to the cases which were dealt with in §1.52 (cf., [Mar, p. 76]) since these are the basic $Sp(1)$-cases.

Accordingly, let us evaluate the cohomology transfer

$$i_* : H^4(F; \mathbb{Z}/2) \to H^4(K; \mathbb{Z}/2)$$

considered as

$$i_* : \bigoplus_{\substack{u \text{ real} \\ \text{place of} \\ F}} \{\pm 1\} \to \bigoplus_{\substack{v \text{ real} \\ \text{place of} \\ K}} \{\pm 1\}.$$

If we have a diagram of places

$$
\begin{array}{ccccc}
K & \rightarrowtail & F & \rightarrowtail & L \\
\downarrow & & \downarrow & & \downarrow \\
\mathbb{R} \cong K_v & \overset{=}{\rightarrowtail} & F_u & \rightarrowtail & L_w
\end{array},
$$

then i_* is the identity from $H^4(F_u; \mathbb{Z}/2)$ to $H^4(K_v; \mathbb{Z}/2)$. From this we see that

$$p_1(\rho)_v = p_1(\mathrm{Ind}_{F/K}(v))_v = \prod_{\substack{u | v \\ \text{real}}} p_1(v)_u \in \{\pm 1\}.$$

On the other hand, let us consider how $W_{K_v}(\rho_v)$ is related to the $\{W_{F_u}(v_u)\}$. Let $\mathbb{Z}/2 = \Omega_{K_v}$; then, by the double-coset formula

$$\mathrm{Res}_{\mathbb{Z}/2}^{G(L/K)}(\mathrm{Ind}_{F/K}(v)) = \sum_{x \in (\mathbb{Z}/2) \backslash G(L/K)/G(L/F)} \mathrm{Ind}_{H_x}^{\mathbb{Z}/2}(\mathrm{Res}_{H_x}^{G(L/F)}(v)),$$

where $H_x = (xG(L/F)x^{-1}) \cap (\mathbb{Z}/2)$.

Observe that $G(L/K)/G(L/F)$ corresponds to the set of conjugates of F in L. If $H_x = \mathbb{Z}/2$, this corresponds to a place, $F_u \cong \mathbb{R}$, containing K_v. This term contributes $W_{F_u}(v_u)$ to $W_{K_v}(\rho_v)$. If, on the other hand, $H_x = \{1\}$, then

$\text{Ind}_{H_x}^{Z/2}(\text{Res}_{H_x}^{G(L/F)}(v)) = (\dim_{\mathbb{C}} v)(1 + y)$ which contributes $(-1)^{(\dim_{\mathbb{C}}(v))/2}$. But these terms correspond to complex places, F_u, which occur in pairs. Each such pair contributes $(-1)^{\dim_{\mathbb{C}}(v)} = 1$, as v is even dimensional. Thus, as required, we find that

$$W_{K_v}(\rho_v) = \prod_{\substack{u|v \\ \text{real}}} W_{F_u}(v_u) \in \{\pm 1\}. \qquad \square$$

(1.55)   ABELIAN LOCAL ROOT NUMBERS

Let us consider the abelian case of (1.42)—that is, when $\dim(\rho) = 1$. In this case a factorization of the (Artin) root number into local terms was given by Hecke. From that factorization we know what the abelian local root numbers should be, and we will now proceed to review their definition.

Let K be a local field over $\mathbb{Q}_p$. Let $P \lhd O_K$ denote the maximal ideal of the integers, O_K, of K. Set $U_K^0 = O_K^*$, the units in O_K and, for $i \geq 1$,

$$U_K^i = 1 + P^i \cdot O_K.$$

Hence $U_K^0 \supset U_K^1 \supset U_K^2 \supset \cdots$ is a descending series of groups, whose quotients $(U_K^i)/(U_K^{i+1})$ are given by the theory of ramification groups [Ser 2].

Let π be a generator of P (called a *uniformizing parameter* of O_K), and let $\mathscr{D}$ denote the *different* of $K/\mathbb{Q}_p$. Thus $\mathscr{D}^{-1}$ (the *codifferent*) is the greatest fractional ideal of K, on which

$$\psi: K \to S^1 \subset \mathbb{C}^*$$

is trivial, where ψ is equal to the composition

$$(1.56) \qquad K \xrightarrow{\text{Trace}} \mathbb{Q}_p \to \mathbb{Q}_p/\mathbb{Z}_p \xrightarrow{\exp(2\pi i\,\_)} S^1.$$

The *conductor* of any representation

$$\theta: G(L/K) \to GL(V)$$

is defined to be $\pi^{n(\theta)}$, where

$$(1.57) \qquad n(\theta) = \sum_{0 \leq i} |G_i|/|G_0| \operatorname{codim}_{\mathbb{C}}(V^{G_i}),$$

where $\{G_i\}$ are the ramification groups of §1.38 and $V^{G_i} = \{v \in V | \theta(g)v = v$ for all $g \in G_i\}$. The conductor is well-defined as an ideal of O_K, $f(\theta)$.

Observe that $\psi = \exp(2\pi i(\text{Trace}(\text{---})))$ is an exponential homomorphism and for this reason is called an *additive character* of K.

If $\dim \rho = 1$ and ρ is a representation of $G(L/K)$ for some L/K, then, by local class field theory, this corresponds precisely to a (multiplicative) character

(1.58) $\rho: K^* \to \mathbb{C}^*$

such that $\mathrm{im}(\rho)$ is finite. This is because local class field theory gives us a dense injection of the form

$$K^* \to (\Omega_K)_{ab}.$$

Let c be a generator of $\mathcal{D}_\rho = f(\rho)\mathcal{D}$, and let $n = n(\rho)$, then the following Gauss sum is well-defined and nonzero:

(1.59) $\tau(\rho) = \sum_{x \in (U_K)/U_K^n} \rho(x/c)\psi(x/c).$

Set

(1.60) $W_K(\rho) = \tau(\bar{\rho})/\sqrt{N(f(\rho))},$

where the denominator is the positive square root of the absolute norm (i.e., down to $\mathbb{Q}_p$) of the conductor of ρ.

$W_K(\rho)$ lies in the unit circle, and with this definition, (1.42) is true for one-dimensional representations. That is, the abelian local root numbers, in the non-Archimedean case, are to be given by (1.59) and (1.60). The remainder of this chapter will be concerned with promoting the one-dimensional local root number to a function on all local Galois representations in such a manner that (1.42) is true in general.

(1.61)   Examples

(a) If $K/\mathbb{Q}_p$ is unramified, then $\tau(\rho) = W_K(\rho) = 1$. More generally, if ρ is unramified, then $n(\rho) = 0$, and

$$\tau(\rho) = \rho(\mathcal{D}^{-1}),$$

where this means the value of ρ on a generator of the fractional ideal $\mathcal{D}^{-1}$.

(b) In Chapter 3, § (3.32) we examined a local field with Galois group isomorphic to Q_{16} which was constructed in [Ser 5, §4, p. 413]. We will conclude this section by evaluating an abelian local root number associated to this extension. The extension has the form

$$\mathbb{Q}_2 \rightarrowtail \mathbb{Q}_2(i) = K \rightarrowtail L,$$

where $i^2 = -1$ and

$$G(L/K) \cong \mathbb{Z}/8\langle x \rangle, \quad G(K/\mathbb{Q}_2) \cong \mathbb{Z}/2\langle y \rangle,$$

and $G(L/\mathbb{Q}_2) \cong Q_{16} = \{x, y \mid x^4 = y^2, y^4 = 1, xyx = y\}.$

Let $\theta: G(L/K) \to S^1$ be the representation which sends x to $\xi_8 = \exp(\pi i/4)$. We will calculate

$$W_K(\theta) \in S^1.$$

As a multiplicative character, θ is given by the composition

$$K^* \longrightarrow K^*/A \cong \mathbb{Z}/8 \overset{\theta}{\longrightarrow} S^1,$$

where, if $\pi = i - 1$ is taken as the uniformizing parameter, then $\theta(\pi) = \xi_8$ and

$$A = \{2^n u \mid n \in \mathbb{Z} \text{ and } v_\pi(1 - u) \geq 3\}.$$

The chain of ramification groups is given in [Ser 5, p. 414] as

$$G_4 = \{1\}, G_3 = \{x^4\} = G_2, \, G_1 = \{x^2\} = G_0.$$

Hence $n(\theta)$, from (1.57), is given by

$$n(\theta) = 1 + 1 + (1/2) + (1/2) = 3.$$

Also $\psi(1/\pi^2) = \psi(-1/2i) = 1$ so that

$$\mathscr{D}^{-1} = \{1/\pi^2\}, \quad \mathscr{D} = \{\pi^2\},$$

and $f(\theta)\mathscr{D} = \{\pi^5\} = \{8/(i+1)\}$. Take $c = 8/(i+1)$, then $\theta(1/c) = -\bar{\xi}_8$. Hence

$$\tau(\theta) = \sum_{z \in U_{K/U_K^3}} \theta(z/c)\psi(z/c)$$

$$= (-\bar{\xi}_8)\{\theta(i)\psi(i/c) + \theta(-1)\psi(-1/c) + \theta(-i)\psi(-i/c) + \theta(1)\psi(1/c)\},$$

and since $\theta(-i) = i$, $\theta(-1) = -1$, $\theta(i) = -i$ and $\psi(i/c) = -i$, $\psi(1/c) = i$, we find that

$$\tau(\theta) = (-\bar{\xi}_8)\{(-i)(-i) + (-1)(-i) + i(i) + 1(i)\}$$

$$= -2\sqrt{2}i.$$

However, the norm of π is 2 so that

$$W_K(\theta) = -2\sqrt{2}i/2\sqrt{2} = -i.$$

2. ORTHOGONAL ROOT NUMBERS

Throughout this section K will denote a local field over $\mathbb{Q}_p$. Let L/K be a finite Galois extension, and let ρ be an orthogonal representation of $G(L/K)$

(2.1) $$\rho: G(L/K) \to O_n(\mathbb{R}) \subset U(n).$$

As in Chapter 6, (2.7) we will sometimes consider ρ as a representation given by the homomorphism into the orthogonal group, $O_n(\mathbb{R})$, and sometimes we will follow the practice of [Mar; Ta, etc] and consider ρ in terms of its complexification (i.e., the composite of (2.1)).

We recall that $c: RO(G) \to R(G)$ is injective, which justifies identifying ρ with $c(\rho)$. From this point of view what we are going to do in this section is to construct the *restriction* of the local root number

(2.2) $$W_K: R(G(L/K)) \to S^1$$

on $RO(G(L/K))$. Let me pause to explain why this is necessary as a first step toward the construction of (2.2). To see this, we will need to see the axioms for W_K.

(2.3)   LOCAL ROOT NUMBER AXIOMS (cf., Chapter 7, (1.37))

The local root number of (2.2) will be required to satisfy the following properties:

(i) $W_K(\rho_1 \oplus \rho_2) = W_K(\rho_1) W_K(\rho_2)$.

(ii) If $K \subset L \subset N$ is a chain of finite, normal extensions and $\pi: G(N/K) \twoheadrightarrow G(L/K)$ is the natural epimorphism, then $W_K(\rho) = W_K(\rho\pi)$.

(iii) If F is an intermediate field of L/K and if $\psi: G(L/F) \to GL(W)$ is a (complex) representation, then $W_F(\psi) = W_K(\mathrm{Ind}_{F/K}(\psi - \dim \psi))$.

(iv) If ρ is a one-dimensional representation of $G(L/K)$ and $\rho: K^* \to \mathbb{C}^*$ is the associated character of finite order, then $W_K(\rho)$ is given by the formula of §§(1.55)–(1.60). In particular, $W_K(1) = 1$.

(2.4) Remarks

The axioms of (2.3) are suggested by the properties of the Artin root number of Chapter 6, (1.27). By Chapter 6, (2.16) there can only be one local root number, W_K, satisfying the axioms of (2.3). This is because, as we shall see in this section, the axioms determine W_K on $RO(G(L/K))$, which means that W_K is determined on permutation representations and thence on all monomial representations, by (2.3) (iii).

The inductivity axiom (iii) and Hecke's proof of (1.42) in the one-dimensional case suffice to yield (1.42) in general. (cf., Chapter 7, §3). This is the motivation for the axioms. In fact, we will prove (1.42) before establishing the axioms of §2.3, and will then derive them from (1.42).

In [Dw] W_K is constructed up to sign, and it is shown that W_K cannot be fully inductive—in other words, that the fourth roots of unity, $W_K(\mathrm{Ind}_{F/K}(1))$, must sometimes be nontrivial. The sign is, however, very important, just as it is in the case of the quadratic Gauss sums which we are about to calculate. Gauss spent four years attempting to determine their sign [I-R, pp. 71–75]. Once one has *control of the $W_K(\text{Ind}_{F/K}(1))$'s, one can derive the existence of W_K by building upon* [Dw]. *This was done in a very long, unpublished essay by R. P. Langlands. Our method of obtaining the $W_K(\text{Ind}_{F/K}(1))$'s is much more succinct—we will simply construct W_K for all orthogonal representations—but, since we have the Explicit Induction results of Chapter 6, which were not available previously, we will not need to appeal to* [Dw] *because it will follow very simply that W_K exists, as a function, and satisfies (1.42), from which all of (2.3) also follows, at least in some cases (see §3.13).*

We will begin by computing W_{Q_p} on one-dimensional orthogonal representations. These results are originally due to C. F. Gauss.

(2.5) QUADRATIC CHARACTERS

A continuous, one-dimensional orthogonal Galois representation is just a homomorphism:

$$\psi:\Omega_K \longrightarrow \{\pm 1\}.$$
$$\searrow \qquad \nearrow$$
$$G(L/K)$$

Hence ψ represents a class

$$l(a)\in H^1(K;\mathbb{Z}/2) \cong K^*/(K^*)^2$$

in the notation of Chapter 2, (3.3). Alternatively, $l(a)$ may be described as a character

(2.6) $$l(a): K^* \to \{\pm 1\} \subset \mathbb{C}^*$$

by class field theory. If we observe that

$$H^2(K;\mathbb{Z}/2) \cong \{\pm 1\},$$

then (2.6) is simply given by the formula

(2.7) $$l(a)(b) = l(a)\cup l(b) = (a,b),$$

the cohomology cup-product, or the Hilbert symbol, according to taste.

Accordingly, we may evaluate

$$W_K(l(a))\in\{\pm 1, \pm \sqrt{(-1)}\},$$

where K is a non-Archimedean local field and $a\in K^*/(K^*)^2$, by the formula of Chapter 7, (1.60).

$$(2.8) \quad W_{\mathbb{Q}_p}(l(a))$$

First, from [Ser 8, pp. 15f], $\mathbb{Q}_p = \mathbb{Z}_p[1/p]$, where $\mathbb{Z}_p$ denotes the p-adic integers, so that

$$\mathbb{Q}_p^* = \mathbb{Z}\langle p\rangle \times U^0, \qquad U^0 \cong \mathbb{Z}_p^*$$

and

$$U^n = 1 + p^n \mathbb{Z}_p = \ker(U \xrightarrow{\varepsilon_n} (\mathbb{Z}/p^n)^*).$$

Note that [Ser 8] uses U_i for $U^i_{\mathbb{Q}_p}$ of Chapter 7, (1.55). Also $\varepsilon_1 : U^1 \to \mathbb{F}_p^*$ is split (via the Teichmüller map) so that

$$U \cong \mathbb{F}_p^* \times U^1.$$

When $p \neq 2$, $U^1 \cong \mathbb{Z}_p$ (via the logarithm), whereas, if $p = 2$, $U \cong \{\pm 1\} \times U^2$ and $U^2 \cong \mathbb{Z}_2$.

This discussion quickly yields the following result:

(2.9) Proposition [Ser 8, pp. 17–18]

(i) If $p \neq 2$,

$$Q_p^*/(\mathbb{Q}_p^*)^2 \cong \mathbb{Z}/2 \times \mathbb{Z}/2\langle u, p\rangle,$$

where $u \in U$ and the Legendre symbol $\left[\dfrac{u}{p}\right] = -1$ (i.e., u is a nonsquare mod p).

(ii) $\mathbb{Q}_2^*/(\mathbb{Q}_2^*)^2 \cong \mathbb{Z}/2 \times \mathbb{Z}/2 \times \mathbb{Z}/2\langle -1, 5, 2\rangle.$

When $p = 2$, define [Ser 8, p. 20]

$$\omega, \varepsilon : U/U^3 \to \mathbb{Z}/2 \text{ by}$$

$$(2.10) \qquad \varepsilon(z) \equiv ((z-1)/2)\,(\mathrm{mod}\,2) = \begin{cases} 0 & \text{if } z \equiv 1 \,(\mathrm{mod}\,4) \\ 1 & \text{if } z \equiv 3 \,(\mathrm{mod}\,4) \end{cases}$$

and

$$\omega(z) = ((z^2 - 1)/8)\,(\mathrm{mod}\,2) = \begin{cases} 0 & \text{if } z \equiv \pm 1 \,(\mathrm{mod}\,8) \\ 1 & \text{if } z \equiv \pm 5 \,(\mathrm{mod}\,8). \end{cases}$$

(2.11) Proposition [Ser 8, p. 20]

If $a = p^\alpha u$, $b = p^\beta v$ with $u, v \in U$, the Hilbert symbol is given, in terms of Legendre symbols, by

$$(i) \quad (a, b) = (-1)^{\alpha\beta\varepsilon(\rho)} \left[\dfrac{u}{p}\right]^\beta \left[\dfrac{v}{p}\right]^\alpha \qquad \text{if } p \neq 2.$$

(ii) $(a,b)=(-1)^{\varepsilon(u)\varepsilon(v)+\alpha\omega(v)+\beta\omega(u)}$ *if* $p=2$.

From (2.10) and (2.11) we obtain Tables A and B of cup-products, as in (2.7). In Table A, $u\in U$ is as in §2.9.

TABLE A. Cup-products on $\mathbb{Q}_p^*/(\mathbb{Q}_p^*)^2$ ***if*** $p\neq 2$

	1	u	p	up
1	1	1	1	1
u	1	1	-1	-1
p	1	-1	$(-1)^{\varepsilon(p)}=\left[\frac{-1}{p}\right]$	$(-1)^{\varepsilon(p)+1}$
up	1	-1	$(-1)^{\varepsilon(p)+1}$	$(-1)^{\varepsilon(p)}$

TABLE B. Cup-products on $\mathbb{Q}_2^*/(\mathbb{Q}_2^*)^2$

	1	−1	5	−5	2	−2	10	−10
1	1	1	1	1	1	1	1	1
−1	1	−1	1	−1	1	−1	1	−1
5	1	1	1	1	−1	−1	−1	−1
−5	1	−1	1	−1	−1	1	−1	1
2	1	1	−1	−1	1	1	−1	−1
−2	1	−1	−1	1	1	−1	−1	1
10	1	1	−1	−1	−1	−1	1	1
−10	1	−1	−1	1	−1	1	1	−1

From Chapter 3, (2.5) we know that, if $a\in\mathbb{Q}_p$, $l(a)\cup l(x)=(a,x)$ is trivial if and only if $x\in N_{L/\mathbb{Q}_p}(L^*)$, where $L=\mathbb{Q}_p(\sqrt{a})$. Therefore the conductor of

$$l(a)\colon \mathbb{Q}_p^*/(\mathbb{Q}_p^*)^2\to\{\pm 1\}$$

equals p^f where f is the least integer such that $U^f=1+p^f\mathbb{Z}_p$ lies in $N_{L/\mathbb{Q}_p}(L^*)$.

Let us consider the case $p=2$. If $a=-1$, then, from Table B, 2 and 5 are norms from L^* but not -1 so that $l(-1)$ has conductor 2^2, since 5 generates U^2. Therefore $c=4$ in Chapter 7, (1.60) and, by definition,

$$W_{\mathbb{Q}_2}(l(-1))=1/\sqrt{(2^2)}\sum_{x=\pm 1}(-1,x/4)i^x=2i/2=i,$$

where $i^2=-1$.

If $a=2$, then Table B implies that -1 is a norm but not 5, so that $f>2$, but 5^2 is a norm and generates U^3 so that $l(2)$ has conductor 2^3 and $c=8$. Thus, if $\xi_8 = (1 + i)/\sqrt{2}$,

$$W_{Q_2}(l(2)) = 1/2\sqrt{(2)} \sum_{x=\pm 1, \pm 3} (2, x/8)\xi_8^x$$

$$= (\xi_8 + \bar{\xi}_8 - (\xi_8^3 + \bar{\xi}_8^3))/(2\sqrt{2})$$

$$= 8/(2\sqrt{2})^2$$

$$= 1.$$

Finally, if $a = 5$, Table B shows that $l(5)$ has trivial conductor so that

$$W_{Q_2}(l(5)) = 1.$$

One may continue in this manner to complete Table C. Alternatively, one may use the relation

(2.12) $$W_K(l(ab)) = W_K(l(a))W_K(l(b))(a, b).$$

We will prove (2.12) (see §2.45) as an easy corollary (following [Ta, p. 126, Cor. 2]) of the calculation of two-dimensional dihedral local root numbers.

TABLE C. $W_{Q_2}(l(a)) \in \{\pm 1, \pm i\}, \ a \in Q_2^*/(Q_2^*)^2$

a	1	-1	2	-2	5	-5	10	-10
$W_{Q_2}(l(a))$	1	i	1	i	1	i	-1	$-i$

Turning to the case when $p \neq 2$ and $a = p$, we see from Table A that u is not a norm, but $U^1 \cong \mathbb{Z}_p$ is 2-divisible so that $l(p)$ has conductor equal to p. Hence $c = p$, and if $\xi_p = \exp(2\pi i/p)$,

$$W_{Q_p}(l(p)) = (1/\sqrt{p})\left(\sum_{x \in \mathbb{F}_p^*} (p, x/p)\xi_p^x \right)$$

$$= (1/\sqrt{p})(-1)^{\varepsilon(p)} \sum_{x \in \mathbb{F}_p^*} \left[\frac{x}{p} \right] \xi_p^x$$

$$= \begin{cases} -i & \text{if } p \equiv 3 \ (\mathrm{mod}\ 4) \\ 1 & \text{if } p \equiv 1 \ (\mathrm{mod}\ 4) \end{cases}$$

by calculations of Gauss [I-R, pp. 71–75].

Hence we obtain Table D, since, from Table A, $(u, u) = 1$ and $l(u)$ is unramified.

TABLE D. $W_{\mathbb{Q}_p}(l(a)) \in \{\pm 1, \pm i\},\, a \in \mathbb{Q}_p^* / (\mathbb{Q}_p^*)^2$

a	1	u	p	up
$W_{\mathbb{Q}_p}(l(a))$ $p \equiv 3 \pmod 4$	1	1	$-i$	i
$W_{\mathbb{Q}_p}(l(a))$ $p \equiv 1 \pmod 4$	1	1	1	-1

(2.13) A CONSTRUCTION WITH ORTHOGONAL REPRESENTATIONS

Let G be a finite group, and consider the augmentation ideal $IO(G) \lhd RO(G)$. Set

$$(2.14) \qquad J_G = \{x \in IO(G) \mid SW_1(x) = 0 = SW_2(x)\}.$$

(2.15) Lemma

J_G is an ideal of $RO(G)$.

Proof. We will prove the stronger statement that if X is any connected space,

$$J_X = \{x \in KO(X) \mid \dim(x) = 0 = SW_1(x) = SW_2(x)\},$$

then J_X is an ideal in $KO(X)$. The result will follow by taking $X = BG$ and recalling that there is an isomorphism $KO(BG) \cong RO(G)\widehat{}$, induced by Chapter 4, (1.28).

By the splitting principle of Chapter 4, (1.32), we see that $SW_1(x \otimes y)$ and $SW_2(x \otimes y)$ are homogeneous polynomials in $SW_1(x), SW_1(y), SW_2(x)$, and $SW_2(y)$ if $\dim(x) = 0$. The splitting principle implies that it is sufficient to show this when x and y are formal sums of line bundles. Therefore, if $\dim x = 0 = SW_1(x) = SW_2(x)$, the same is true for $x \otimes y$. $\qquad \square$

Hence $IO(G)/J_G$ is a ring, and so is

$$(2.16) \qquad Y_K = \varinjlim_{L/K} \{IO(G(L/K))/(J_{G(L/K)})\},$$

where the limit is take over finite Galois extensions, which lie within a fixed choice of separable closure, of the local field K.

If $a \in K^*/(K^*)^2$, then we define

$$(2.17) \qquad P_K(a) = l(a) - 1 \in Y_K.$$

(2.18) Theorem

Let K be a non-Archimedean local field.

(i) There is an exact sequence

$$H^2(K/\mathbb{Z}/2) \cong \{\pm 1\} \overset{i}{\rightarrowtail} Y_K \overset{\pi}{\longrightarrow} K^*/(K^*)^2 \cong H^1(K; \mathbb{Z}/2).$$

(ii) $\pi(p(a)) = l(a)$ for $a \in K^*/(K^*)^2$.
(iii) If $a, b \in K^*/(K^*)^2$, then $p(ab) = p(a) + p(b) + i\{(a,b)\}$.

Proof. By definition, any $x \in IO(G)/J_G$ is detected by $SW_1(x)$ and $SW_2(x)$. Therefore any $x \in Y_K$ is detected by $SW_1(x)$ or $SW_2(x)$ in $H^*(K; \mathbb{Z}/2)$. This yields the exact sequence if we define $\pi(x) = SW_1(x)$, and it also implies that (ii) is true. Part (iii) follows from the fact that

$$SW_1(l(ab)) = SW_1(l(a)) + SW_1(l(b)) \in K^*/(K^*)^2$$

so that

$$p(ab) - p(a) - p(b) = l(ab) - l(a) - l(b) + 1$$

lies in image (i). Hence it is equal to

$$SW_2(l(ab) - l(a) - l(b)) = l(a)l(b) = (a, b),$$

by an easy exercise with the Cartan formula.   □

(2.19) Corollary

Let $\mu_4 = \{\pm 1, \pm i\}$, where $i^2 = -1$; then there is a homomorphism

$$\varphi_p : Y_{\mathbb{Q}_p} \to \mu_4 \qquad \text{given by}$$

$$\varphi_p(p(a)) = W_{\mathbb{Q}_p}(l(a)).$$

Proof. This is easily verified from Tables C and D.   □

The following result was pointed out to me by P. E. Conner:

(2.20) Corollary

Let $GW(K)$ denote the Grothendieck-Witt ring of symmetric, nondegenerate bilinear forms over the local field K. If $IGW(K)$ denotes the augmentation ideal, there is an isomorphism

$$\psi_K : IGW(K) \overset{\cong}{\longrightarrow} Y_K$$

given by

$$\psi_K(\langle a \rangle - \langle 1 \rangle) = p(a),$$

where $a \in K^*/(K^*)^2$ and $\langle a \rangle : K \times K \to K$ sends (x, y) to axy.

Proof. The Hasse-Witt classes, HW_1 and HW_2 are known to detect $IGW(K)$, from which one easily deduces the analogue of Theorem 2.18 for $IGW(K)$. From this it is clear that ψ is well-defined and an isomorphism.

We will return to $GW(K)$ later in this section. $\qquad\square$

(2.21) Lemma

Let G be a finite group, and let $J_G \lhd RO(G)$ be the ideal defined in (2.14). If $H < G$, then

$$\mathrm{Ind}_H^G(J_H) \subset J_G.$$

Proof. As in the proof of Lemma 2.15 we consider, more generally, $J_X \lhd KO(X)$, where X is a connected space. The homotopy classes of maps from X to BSO satisfy an isomorphism

$$[X, BSO] \cong \{x \in KO(X) | \dim(x) = 0 = SW_1(x)\}.$$

Thus $J_X = \ker\{[X, BSO] \to H^2(X; \mathbb{Z}/2)\}$, where the homomorphism is induced by SW_2, the second Stiefel-Whitney class—considered as a map of the form $SW_2 : BSO \to K(\mathbb{Z}/2, 2)$.

The preceding map is the first nontrivial k-invariant of BSO (i.e., it induces an isomorphism on the second homotopy groups). These spaces are infinite loopspaces and, from general theory, SW_2 is infinitely deloopable. Also from the general theory of infinite loopspaces, this means that SW_2 commutes with transfer for finite coverings, because of the homotopy theoretic description of the transfer in [K-P]. However, as mentioned in [K-P], the transfer on $[X, BSO]$ defined by the infinite loopspace construction coincides, by a result due to J. M. Boardman, with the K-theory transfer given by the induced bundle construction, which completes the proof. $\qquad\square$

Put succinctly, "first k-invariants commute with transfer." In fact, we used this fact for BSp in §1.54.

(2.22) Now let K be a local field over $\mathbb{Q}_p$. Let $\rho: \Omega_K \to O_n(\mathbb{R})$ be a continuous, orthogonal Galois representation. The induction homomorphism, by §(2.21), induces a homomorphism

(2.23) $$\mathrm{Ind}_{K/\mathbb{Q}_p}: Y_K \to Y_{\mathbb{Q}_p}$$

of the groups of (2.16), and consequently we may define

$$(2.24) \qquad \Gamma_K(\rho) = \varphi_p(\mathrm{Ind}_{K/\mathbb{Q}_p}[\rho - n]) \in \mu_4 \subset S^1,$$

where φ_p is as in §2.19. It will be shown later in this section that $\Gamma_K(\rho)$ is equal to the inevitable value on orthogonal representations of a local root number, satisfying the conditions of (2.3). To see that, we must first deduce some properties of $\Gamma_K(\rho)$. We will need the following commutative diagram, in which $IO(K) = \varinjlim_{L/K} IO(G(L/K))$ and λ_K is the canonical quotient map:

$$(2.25)$$

$$
\begin{array}{ccccccc}
& & IO(K) \xrightarrow{\lambda_K} & & & & \\
H^2(K;\mathbb{Z}/2) & \overset{i}{\rightarrowtail} & & Y_K & \xrightarrow{\pi} & K^*/(K^*)^2 & \\
& & \downarrow{\scriptstyle \mathrm{ind}_{K/\mathbb{Q}_p}} & & & & \\
\cong \Big\downarrow {\scriptstyle \mathrm{Tr}_{K/\mathbb{Q}_p}} & & IO(\mathbb{Q}_p) \xrightarrow{\lambda_{\mathbb{Q}_p}} & \Big\downarrow {\scriptstyle \mathrm{Ind}_{K/\mathbb{Q}_p}} & & \Big\downarrow {\scriptstyle N_{K/\mathbb{Q}_p}} & \\
\{\pm 1\} \cong H^2(\mathbb{Q}_p;\mathbb{Z}/2) & \underset{i}{\rightarrowtail} & & Y_{\mathbb{Q}_p} & \underset{\pi}{\longrightarrow} & \mathbb{Q}_p^*/(\mathbb{Q}_p^*)^2. &
\end{array}
$$

(2.26) Theorem

Let $\Gamma_K : RO(K) = \varinjlim_{L/K} R(G(L/K)) \to \mu_4$ be the homomorphism of (2.24).

(i) Let F/K be a finite extension, and let $\rho : \Omega_F \to O_n(\mathbb{R})$ be a continuous, orthogonal representation, then

$$\Gamma_K(\mathrm{Ind}_{F/K}(\rho)) = \Gamma_F(\rho) \cdot \Gamma_K(\mathrm{Ind}_{F/K}(1))^n \in \mu_4.$$

(ii) If $a, b \in K^*$, then

$$\Gamma_K(l(ab)) = \Gamma_K(l(a))\Gamma_K(l(b))(a, b) \in \mu_4.$$

(iii) For all continuous, orthogonal representations, $\rho : \Omega_K \to O_n(\mathbb{R})$,

$$\Gamma_K(\rho) = \Gamma_K(\det \rho) SW_2(\rho) \in \mu_4,$$

where $SW_2(\rho) \in H^2(K;\mathbb{Z}/2) \cong \{\pm 1\}$.

Proof. Part (i) is a formality, by construction. For we have

$$\Gamma_K(\mathrm{Ind}_{F/K}(\rho)) = \varphi_p(\mathrm{Ind}_{K/\mathbb{Q}_p}(\mathrm{Ind}_{F/K}(\rho) - nd)), \qquad \text{where } d = [F:K],$$

$$= \varphi_p(\mathrm{Ind}_{F/\mathbb{Q}_p}(\rho - n))\varphi_p(\mathrm{Ind}_{F/\mathbb{Q}_p}(n) - \mathrm{Ind}_{K/\mathbb{Q}_p}(nd)), \qquad \text{by §2.19,}$$

$$= \Gamma_F(\rho)\varphi_p(\mathrm{Ind}_{K/\mathbb{Q}_p}(\mathrm{Ind}_{F/K}(n) - nd))$$

$$= \Gamma_F(\rho)\Gamma_K(\mathrm{Ind}_{F/K}(1))^n, \qquad \text{by definition and §2.19.}$$

Part (ii) is the special case of part (iii) in which one sets $\rho = l(a) \oplus l(b)$, for then $\det \rho = l(ab)$ and, by the Cartan formula,

$$SW_2(\rho) = SW_1(l(a))SW_1(l(b)) = (a, b) \in \{\pm 1\}.$$

Finally, part (iii) is clear because if $x = \rho - \det \rho - n + 1 \in IO(K)$, then $\pi(x) = \det \rho - \det(\rho)$ is trivial in (2.25) so that the class of x in Y_K is detected by $SW_2(x)$ and is equal to the image of $SW_2(x)$ under i. However, the Cartan formula (written multiplicatively) yields

$$SW_2(x) = SW_2(\rho - \det \rho)$$

$$= SW_2(\rho)\{SW_1(\rho) \cup SW_1(\rho)\}SW_2(-\rho)$$

$$= SW_2(\rho)\{SW_1(\rho) \cup SW_1(\rho)\}^2, \quad \text{since } SW_1(\rho) = SW_1(\det \rho),$$

$$= SW_2(\rho), \quad \text{as required.} \qquad \square$$

We will also have need of the following "uniqueness" result, which is a companion to Theorem 2.26:

(2.27) Proposition

Let $\Gamma_K: RO(K) \to \mu_4$ be any homomorphism which satisfies §2.26 (i) and (ii). If §2.26 (iii) holds for $\rho: \Omega_K \to O_n(\mathbb{R})$ with $n \leq 2$, then §2.26 (iii) holds for all ρ.

Proof. Throughout this proof we will abbreviate $SW_2(\rho)$ to $s(\rho) \in \{\pm 1\}$.

By §2.26(i), $\Gamma_K(1) = 1$, so we may study $\Gamma_K(\rho - n)$, where $n = \dim \rho$. By §2.26 (ii) we may replace ρ by $\rho \oplus \det \rho$ if necessary, since if $\Gamma_K(\rho \oplus (\det \rho)) = SW_2(\rho \oplus (\det \rho))$, then

$$SW_2(\rho)\Gamma_K(\det \rho) = \Gamma_K(\rho)\Gamma_K(\det \rho)^2(SW_1(\rho) \cup SW_1(\rho))$$

$$= \Gamma_K(\rho)\Gamma_K((\det \rho)^2)$$

$$= \Gamma_K(\rho)\Gamma_K(1)$$

$$= \Gamma_K(\rho).$$

Hence we may consider only $\rho: \Omega_K \to SO_{2m}(\mathbb{R})$, $m \geq 2$.

Assume now that we have proved §2.26 (iii) for such ρ of dimension at most $2m - 2$.

By the orthogonal case of Chapter 6, (2.31),

$$\rho - 2m = \sum_\alpha n_\alpha(\text{Ind}_{F_\alpha/K} \text{Res}_{\Omega_{F_\alpha}}^{\Omega_K}(\rho) - 2m)),$$

where F_α/K is such that the restriction of ρ to Ω_{F_α} may be assumed to map into the normalizer, NT^m, of $T^m = SO_2(\mathbb{R})^m$ in $SO_{2m}(\mathbb{R})$.

On special orthogonal virtual representations of dimension zero SW_2 is a homomorphism which commutes with induction (see §2.21 (proof)). Therefore, if 2.26 (iii) holds for $\mathrm{Res}^{\Omega_K}_{\Omega_{F_\alpha}}(\rho)$, then

$$s(\rho - 2m) = \prod_\alpha s(\mathrm{Ind}_{F_{\alpha/K}}(\mathrm{Res}^{\Omega_K}_{\Omega_{F_\alpha}}(\rho) - 2m))^{n_\alpha}$$

$$= \prod_\alpha s(\mathrm{Res}^{\Omega_K}_{\Omega_{F_\alpha}}(\rho) - 2m)^{n_\alpha}$$

$$= \prod_\alpha \Gamma_{F_\alpha}(\mathrm{Res}^{\Omega_K}_{\Omega_{F_\alpha}}(\rho) - 2m)^{n_\alpha}$$

$$= \prod_\alpha \Gamma_K(\mathrm{Ind}_{F_{\alpha/K}}(\mathrm{Res}^{\Omega_K}_{\Omega_{F_\alpha}}(\rho) - 2m))^{n_\alpha}$$

$$= \Gamma_K(\rho - 2m).$$

Therefore we are reduced to studying ρ of the form $\rho: \Omega_K \to NT^m \subset SO_{2m}(\mathbb{R})$.

Let $\Sigma_m \int O_2(\mathbb{R})$ denote the wreath product generated by the diagonal $O_2(\mathbb{R})^m$ in $O_{2m}(\mathbb{R})$ and the permutation matrices which permute the diagonal 2×2 blocks, $O_2(\mathbb{R})^m$. Suppose that

$$\rho: \Omega_K \to NT^m \subset \Sigma_m \int O_2(\mathbb{R}) \twoheadrightarrow \Sigma_m$$

acts on $\{1, 2, \ldots, m\}$ with orbits of size $m_1, m_2, \ldots, m_r$, then ρ can be written, by Chapter 6, §(2.6), as

$$\rho = \sum_{j=1}^r \mathrm{Ind}_{N_j/K}(\Omega_{N_j} \xrightarrow{\rho_j} O_2(\mathbb{R})),$$

where $[N_j:K] = m_j$. Write ε_j for the homomorphism

$$\begin{cases} \varepsilon_j: \Omega_{N_j} \longrightarrow O_2(\mathbb{R}) \\ g \longmapsto \begin{bmatrix} \det(\rho_j(g)) & 0 \\ 0 & 1 \end{bmatrix} \end{cases}.$$

Hence

$$\Gamma_K(\rho) = \Gamma_K\left(\rho - \sum_j \mathrm{Ind}_{N_j/K}(\varepsilon_j) + \sum_j \mathrm{Ind}_{N_j/K}(\varepsilon_j)\right)$$

$$= (\Pi_j \Gamma_{N_j}(\rho_j - \varepsilon_j))(\Pi_j \Gamma_K(\mathrm{Ind}_{N_j/K}(\varepsilon_j)))$$

$$= (\Pi_j(s(\rho_j - \varepsilon_j)))(\Pi_j \Gamma_K(\mathrm{Ind}_{N_j/K}(\varepsilon_j)))$$

$$= (\Pi_j(s(\mathrm{Ind}_{N_j/K}(\rho_j - \varepsilon_j))))(\Pi_j \Gamma_K(\mathrm{Ind}_{N_j/K}(\varepsilon_j)))$$

$$= \left(s\left(\rho - \sum_j \mathrm{Ind}_{N_j/K}(\varepsilon_j)\right)\right)(\Pi_j \Gamma_K(\mathrm{Ind}_{N_j/K}(\varepsilon_j)))$$

since $\det(\rho_j) = \det(\varepsilon_j)$.

Now, as a representation, $\varepsilon_j = 1 + d_j$, where d_j is given by the determinant of ε_j so that, for example,

$$\mathrm{Ind}_{N_j/K}(\varepsilon_j) = \mathrm{Ind}_{N_j/K}(1) \oplus \mathrm{Ind}_{N_j/K}(d_j) = A_j \oplus B_j.$$

Notice that each of A_j and B_j has dimension less than or equal to $2m - 2$. Hence, by induction,

$$\begin{aligned}
\Gamma_K(\mathrm{Ind}_{N_j/K}(\varepsilon_j)) &= s(A_j)s(B_j)\Gamma_K(\det A_j)\Gamma_K(\det B_j) \\
&= s(A_j)s(B_j)\Gamma_K(A_j \oplus B_j)[(\det A_j \cup \det B_j)], \qquad \text{by §2.26 (ii),} \\
&= s(A_j)s(B_j)\Gamma_K(A_j \oplus B_j)[(SW_1(A_j) \cup (SW_1(B_j))] \\
&= s(A_j \oplus B_j)\Gamma_K(A_j \oplus B_j),
\end{aligned}$$

by the Cartan formula.

Similarly, by the iterated application of §2.26 (ii) and the Cartan formula,

$$\begin{aligned}
\Pi\Gamma_K(\mathrm{Ind}(\varepsilon_j)) &= s\left(\sum_j \mathrm{Ind}(\varepsilon_j)\right)\Gamma_K\left(\det\left(\sum_j \mathrm{Ind}(\varepsilon_j)\right)\right) \\
&= s\left(\sum_j \mathrm{Ind}(\varepsilon_j)\right),
\end{aligned}$$

since the preceding determinant is equal to that of ρ, which is trivial, since ρ is special orthogonal. This completes the proof. $\qquad\square$

(2.28) Remark

Next we are going to show that if the local root numbers exist and satisfy the conditions of (2.3), then, on orthogonal representations, one has $W_K(\rho) = \Gamma_K(\rho)$. In fact, in order to prove, in §3, that $W_K(\rho)$ does exist, we will only need this fact in the tamely ramified case. However, the proof is no more difficult in its general from, so we will prove it in that way. We will do this by computing two-dimensional, dihedral local root numbers, and this will be sufficient to allow us to apply §2.27 to compare W_K and Γ_K.

At the risk of irritating the assiduous reader, I will repeatedly be stating/proving results which are conditional upon the existence of $W_K(\rho)$. By this means I hope to convince everyone that the eventual proof of existence, given in §3, does not contain a circular argument.

We must evaluate what $W_K(\rho)$ must be for a continuous representation of the form

$$\rho: \Omega_K \to O_2(\mathbb{R}).$$

Since $\mathrm{im}(\rho)$ is a finite subgroup, one finds the following to be the only possibilities:

(2.29)

   (i) $\rho = \rho_1 \oplus \rho_2$ with $\rho_i: \Omega_K \to O_1(\mathbb{R}) \cong \{\pm 1\}$.

   (ii) $\mathrm{im}(\rho) \subset SO_2(\mathbb{R})$ is cyclic, and the complexification of ρ is the sum of a one-dimensional representation and its complex conjugate, $c(\rho) = \rho_1 + \bar{\rho}_1$.

   (iii) $\mathrm{im}(\rho)$ is isomorphic to the dihedral group, D_{2m}, for some $m \geq 3$.

(2.30) Theorem

A local root number, W_K, satisfying the conditions of (2.3) must satisfy the formula

$$SW_2(\rho)W_K(\det \rho) = W_K(\rho) \in S^1$$

for any two-dimensional, orthogonal Galois representation (i.e., as in (2.29) (i)–(iii)).

(2.31) Proposition

Let $\rho: \Omega_K \to O_2(\mathbb{R})$ in (2.29) (ii) be the realification of a one-dimensional complex representation, ρ_1, associated by class field theory to a character $\rho_1: K^* \to \mathbb{C}^*$. Then

$$W_K(\rho) = \rho_1(-1) = SW_2(\rho) \in \{\pm 1\}$$

(and, of course, $\det \rho$ is trivial).

Proof. Firstly we point out that $W_K(\rho)$ is actually a function of the complexification of ρ, which is $\rho_1 + \bar{\rho}_1$. That is, $W_K(—)$ is defined on complex representations so that

$$W_K(\rho) = W_K(\rho_1 + \bar{\rho}_1)$$

$$= W_K(\rho_1)W_K(\bar{\rho}_1)$$

$$= \rho_1(-1), \quad \text{by [Ta, Cor. 1 (ii), p. 109]}.$$

This last line comes from Gauss sum manipulations. Since $\det \rho$ is trivial, we have shown that

$$W_K(\rho)/W_K(\det \rho) = W_K(\rho) = \rho_1(-1).$$

Now consider the diagram of homomorphisms:

(2.32)

in which r is realification. We have the following isomorphisms:

$$H^*(BSO_2(\mathbb{R}); \mathbb{Z}/2) \cong H^*(BO_2(\mathbb{R}); \mathbb{Z}/2)/(SW_1)$$

$$\cong \mathbb{Z}/2[SW_1, SW_2]/(SW_1)$$

$$\cong \mathbb{Z}/2[SW_2] \qquad [\text{H}; \text{M-St}],$$

but $H^*(BU(1); \mathbb{Z})$ is the polynomial ring on the first Chern class, $c_1 \in H^2(BU(1); \mathbb{Z})$. Therefore we see that if $\tilde{c}_1$ is the mod 2 reduction of c_1, then

(2.33) $$SW_2[\rho] = \tilde{c}_1(\rho_1) \qquad (\text{mod } 2).$$

Consider the following commutative diagram of exact sequences:

$$
\begin{array}{ccccc}
\mathbb{Z} & \xrightarrow{(2\pi i \cdot ---)} & \mathbb{C} & \xrightarrow{\exp(---)} & \mathbb{C}^* \\
\downarrow & & \downarrow{\scriptstyle \exp(1/2(---))} & & \downarrow{\scriptstyle 1} \\
\{\pm 1\} \cong \mathbb{Z}/2 & \longrightarrow & \mathbb{C}^* & \xrightarrow{z^2} & \mathbb{C}^*
\end{array}
$$

Since $c_1(\rho)$ is the boundary of $\rho_1: K^* \to \mathbb{C}^*$ in the cohomology sequence associated to the upper row, we see that $\tilde{c}_1(\rho_1)$ is the boundary of ρ_1 in the sequence associated to the lower row. Consequently $\tilde{c}_1(\rho_1)$ is the obstruction to ρ_1, having (as a homomorphism) a square root, ρ_2. However, the diagram

$$
\begin{array}{ccc}
\{\pm 1\} \rightarrowtail K^* & \xrightarrow{z^2} & K^* \\
& {\scriptstyle \rho_1} \searrow & \downarrow{\scriptstyle \rho_2} \\
& & \mathbb{C}^*
\end{array}
$$

shows that $\tilde{c}_1(\rho_1)$ is trivial if and only if $\rho_1(-1) \in \{\pm 1\}$ is trivial. Hence, from (2.33),

$$SW_2[\rho] = \rho_1(-1)$$

as required. $\qquad\qquad\qquad\qquad\qquad\qquad\qquad\qquad\qquad\qquad\qquad\qquad\square$

(2.34) Suppose now that K is a non-Archimedean local field and that $G(E/K) \cong D_{2n}$. Hence we have a chain of extensions $E \supset L \supset K$ in which $G(E/L) \cong \mathbb{Z}/n$ and L/K is quadratic. Since $G(E/K)$ is dihedral, every $u \in G(E/K) - G(E/L)$ is of order 2. However, the transfer map

$$i^*: H_1(G(E/K)) = G(E/K)_{ab} \to \mathbb{Z}/n \cong H_1(G(E/L))$$

is given by

$$i^*(u) = u^2 = 1.$$

On the other hand, by local class field theory,

$$i^*: (\Omega_K)_{ab} \to (\Omega_L)_{ab}$$

is induced by the inclusion, $K^* \to L^*$. We have $(D_{2n})_{ab} \cong \mathbb{Z}/2$ or $\mathbb{Z}/2 \times \mathbb{Z}/2$, depending on the parity of n. The generators are u and possibly x, the generator of $\mathbb{Z}/n$. We have seen that $i^*(u) = 1$ and $i^*(x) = x(uxu) = xx^{-1} = 1$ also. Hence i^* is trivial—which means that if $\rho_1: G(E/L) \to S^1$ is a character corrresponding to $\rho_1: L^* \to \mathbb{C}^*$, then

$$K^* \rightarrowtail L^* \xrightarrow{\rho_1} \mathbb{C}^*$$

is trivial. Set $\rho = \mathrm{Ind}_{\mathbb{Z}/n}^{D_{2n}}(\rho_1)$.

(2.35) Proposition [F-Q, Thm 3; De §3.2; Ta, p. 121]

With the notation of §2.34, if $L = K(\delta)$ with $\delta^2 \in K^*$ and $\mathrm{Tr}_{L/K}(\delta) = 0$, then, if $W_K(\rho)$ satisfies the conditions of §2.3,

$$\rho_1(\delta) = W_L(\rho_1) = W_K(\rho)/(W_K(\det \rho)).$$

Proof. Let us establish the second equation first. By the conditions of §2.3,

$$W_K(\rho) = W_L(\rho_1) \cdot W_K(\mathrm{Ind}_{L/K}(1)),$$

but $\mathrm{Ind}_{L/K}(1) = \det(\rho) \oplus 1$, from which the result follows.

To prove the first equation, we follow [Ta, pp. 119–122]. If ρ_1 is nonramified then $\mathcal{O}_L = \mathcal{O}_K[\alpha]$ is the ring of integers. Let $f(X) \in \mathcal{O}_K[X]$ be the monic minimal polynomial of α over K. Then $f'(\alpha)^{-1} \in \ker(\mathrm{Tr}_{L/K})$ and $f'(\alpha)\mathcal{O}_L = \mathcal{D}_{L/K}$ so that, by definition,

$$W_L(\rho) = \rho_1(\mathcal{D}_{L/K}) = \bar{\rho}_1(f'(\alpha)^{-1}). \qquad \square$$

Now assume that ρ_1 is ramified, in which case the first equation will be a consequence of the next two results.

(2.36) Lemma

In the situation of §2.35, let $0 \neq y \in P_L$, the maximal ideal of $\mathcal{O}_L$; then

$$\sum_{\substack{x \in \mathcal{O}_L^* \\ \mathrm{mod}^\times f(\rho_1)}} \bar{\rho}_1(b^{-1}xy)\psi_L(b^{-1}xy) = 0,$$

where $f(\rho_1)$ is the conductor of ρ_1 and ψ_L is as in §§(1.55)–(1.60).

Proof. Observe that the sum is independent of the choices of representatives x of

$\mathcal{O}_L^*(\mathrm{mod}^{\times} f(\rho_1))$. Then

$$\sum_{\substack{x \in \mathcal{O}_L^* \\ \mathrm{mod}^{\times} f(\rho_1)}} \bar{\rho}_1(b^{-1}xy)\psi_L(b^{-1}xy)$$

$$= \sum_{\substack{x \in \mathcal{O}_L^* \\ \mathrm{mod}^{\times} \mathcal{A}}} \bar{\rho}_1(b^{-1}xy) \sum_{\substack{z \in (1+\mathcal{A}) \\ \mathrm{mod}^{\times} f(\rho_1)}} \bar{\rho}_1(z)\psi_L(b^{-1}xyz),$$

where $\mathcal{A} \lhd \mathcal{O}_L$, $f(\rho_1) = P_L\mathcal{A}$.

The map $(z \mapsto \psi_L(b^{-1}xyz))$ is constant on $1 + \mathcal{A}$, since $b^{-1}xy.\mathcal{A} \subset \mathcal{D}_L^{-1}$. But ρ_1 is a nontrivial character on the finite group $(1 + \mathcal{A})/(1 + f(\rho_1))$ so that the inner sum is zero, which yields the result. □

(2.37) Proposition

Let L/K be a finite extension of non-Archimedean local fields, and let ρ_1 be a character of finite order of L^* which is trivial on K^*. Then

$$W_L(\rho_1) = \sum_i \lambda_i \bar{\rho}_1(\delta_i)$$

for some real numbers $\lambda_i > 0$, and $\delta_i \in L^*$ such that $\mathrm{Tr}_{L/K}(\delta_i) = 0$.

Proof. Write $a \sim b$ if ab^{-1} is real and positive, $(a, b$ nonzero complex numbers). From the definition,

$$W_L(\rho_1) \sim \sum_{\substack{x \in \mathcal{O}_L^* \\ \mathrm{mod}^{\times} f(\rho_1)}} \bar{\rho}_1(b^{-1}x)\psi_L(b^{-1}x).$$

Let dy be a Haar measure on the locally compact additive group, K^+. The integrals that follow are really just finite sums over y which have been weighted:

$$W_L(\rho_1) \sim \int_{\mathcal{O}_K - (0)} \sum_x \bar{\rho}_1(b^{-1}xy)(\psi_L(b^{-1}xy)dy$$

by §2.36, since only y in $\mathcal{O}_K(\mathrm{mod}(P_L \cap \mathcal{O}_K))$ contributes, and each such y contributes the same. Since $(\rho_1|K^*)$ is trivial, and since $\bar{\rho}_1(b^{-1}xy) = \bar{\rho}_1(b^{-1}x)\bar{\rho}_1(y)$, the integral equals

$$\sum_x \bar{\rho}_1(b^{-1}x) \int_{\mathcal{O}_k} \psi_L(b^{-1}xy)dy$$

$$= \sum_x \bar{\rho}_1(b^{-1}x) \int_{\mathcal{O}_K} \psi_K(y \cdot \mathrm{Tr}_{L/K}(b^{-1}x))dy, \qquad \text{by definition of } \psi_L,$$

$$= \sum_{\substack{x \in \mathcal{O}_L^* \\ \mathrm{mod}^{\times} f(\rho_1) \\ \mathrm{Tr}_{L/K}(b^{-1}x) \in \mathcal{D}_K^{-1}}} \bar{\rho}_1(b^{-1}x) \int_{\mathcal{O}_K} \psi_K(y \cdot \mathrm{Tr}_{L/K}(b^{-1}x))dy,$$

because the integral vanishes if $\operatorname{Tr}_{L/K}(b^{-1}x) \notin \mathscr{D}_K^{-1}$. Otherwise, the integral is a positive real number, λ_x. But $\operatorname{Tr}_{L/K}(b^{-1}x) \in \mathscr{D}_K^{-1}$ if and only if $b^{-1}x = d_x + \delta_x$ for some $d_x \in \mathscr{D}_L^{-1}$ with $\operatorname{Tr}_{L/K}(\delta_x) = 0$. However, in that case $bd_x \in f(\rho_1)$ so that

$$\rho_1(b^{-1}x) = \rho_1(b^{-1}x - d_x)$$
$$= \rho_1(\delta_x)$$

and we obtain

$$W_L(\rho_1) \sim \sum_x \lambda_x \bar{\rho}_1(\delta_x), \qquad \text{as required.} \qquad \square$$

(2.38)   Before proceeding to the proof of Theorem 2.30 in the case of §2.29 (iii), we will need some preparatory facts about double coverings of $O_2(\mathbb{R})$.

Write $O_2(\mathbb{R}) = \mathbb{Z}/2 \ltimes S^1$ with $\mathbb{Z}/2 = \langle \tau \rangle$. We have a commutative diagram of central extensions.

(2.39)

$$
\begin{array}{ccccccc}
\mathbb{Z}/2 \cong \{\pm 1\} & \rightarrowtail & S^1 & \xrightarrow{z^2} & S^1 & \cong SO_2(\mathbb{R}) \\
\downarrow{\scriptstyle 1} & & \downarrow{\scriptstyle i} & & \downarrow{\scriptstyle i} & & \downarrow{\scriptstyle i} \\
\{\pm 1\} & \rightarrowtail & \mathbb{Z}/2 \ltimes S^1 & \xrightarrow{\mathbb{Z}/2 \ltimes (z^2)} & \mathbb{Z}/2 \ltimes S^1 & \cong O_2(\mathbb{R}).
\end{array}
$$

The central extensions of (2.39) correspond to cohomology classes [H-S]

(2.40) $\begin{cases} \chi \in H^2(BO_2(\mathbb{R}); \mathbb{Z}/2) = \langle SW_2, SW_1^2 \rangle \text{ and} \\ v \in H^2(BSO_2(\mathbb{R}); \mathbb{Z}/2) = \langle i^*(SW_2) \rangle, \text{ by (2.33).} \end{cases}$

If $j: \mathbb{Z}/2 = \langle \tau \rangle \to O_2(\mathbb{R})$ is the inclusion, $j^*(\chi) = 0$ since j lifts through $\mathbb{Z}/2 \ltimes (z^2)$ in the lower extension of (2.39).

Therefore we have shown that

(2.41) $\begin{cases} \chi = SW_2 \text{ in (2.3) and} \\ \rho: G \to O_2(\mathbb{R}) \text{ lifts in (2.39) if and only if} \\ SW_2[\rho] = 0. \end{cases}$

(2.42) Proof of Theorem (2.30) in Case §2.29 (iii)

First, we remark that $\delta \in F^*/K^*$ is the unique element of order 2 so that we have an exact sequence (although the squaring map is not necessarily onto)

(2.43) $$\mathbb{Z}/2 \cong \{\delta\} \rightarrowtail F^*/K^* \xrightarrow{z^2} F^*/K^*.$$

Suppose that $SW_2[\rho]$ is trivial so that $\rho: \Omega_K \to O_2(\mathbb{R})$ lifts, by (2.41), through

$\Sigma_2 \int z^2$ in the lower row of (2.39). The restriction $\rho_1 : \Omega_F \to SO_2(\mathbb{R}) \cong S^1$ lifts through z^2 in the upper row of (2.39). The lifting, ρ_2, of ρ_1 is the restriction of the (dihedral) lifting of ρ so that ρ_2 factors through F^*/K^*. Hence we have

$$\rho_1(\delta) = \rho_2(\delta(\mathrm{mod}\, K^*))^2$$

$$= \rho_2(\delta^2(\mathrm{mod}\, K^*))$$

$$= 1.$$

Conversely, suppose that $\rho_1(\delta) = 1$; then, by (2.43), we have a homomorphism $\rho_2 : \mathrm{im}\,(z^2) \to S^1$ such that $\rho_2(v^2) = \rho_1(v)$ $(v \in F^*/K^*)$. However, S^1 is injective, so we may extend ρ_2 to a homomorphism:

$$\rho_2 : F^*/K^* \to S^1$$

such that $\rho_1(v) = \rho_2(v)^2$. We may consider ρ_2 as a one-dimensional complex representation of Ω_F which lifts ρ_1 through the squaring map in the upper row of (2.39).

Now let $\Phi : \Omega_K \to \Sigma_2 \int \Omega_F$ be the homomorphism defined in Chapter 4, §1.10. We may form the composite

$$\psi : \Omega_K \xrightarrow{\Phi} \Sigma_2 \int \Omega_F \xrightarrow{\Sigma_2 \int \rho_2} \Sigma_2 \int S^1.$$

If $t \in \Omega_K - \Omega_F$, this composition is given by $(h \in \Omega_F)$,

(2.44)
$$\begin{cases} \psi(ht) = (\rho_2(h)\rho_2(t^2), \rho_2(t^{-1}ht))\tau \\ \psi(h) = (\rho_2(h), \rho_2(t^{-1}ht)). \end{cases}$$

However, $\rho_2(t^2) = 1$ since the image of t^2 in $(\Omega_F)_{ab} \cong \{\text{the closure of } F^*\}$ lies in K^*. Also if the image in F^* of h is $a + b\delta$ $(a, b \in K)$, then the image of $t^{-1}ht$ is $a - b\delta$ so that $\rho_2(h)\rho_2(t^{-1}ht) = \rho_2(a^2 - b^2\delta^2) = 1$. Therefore

$$\psi(ht^\varepsilon) = (\rho_2(h), \rho_2(h))\tau^\varepsilon \qquad (\varepsilon = 0, 1)$$

which lies in the image of the embedding $O_2(\mathbb{R}) = \mathbb{Z}/2 \ltimes S^1 \rightarrowtail \Sigma_2 \int S^1$. Hence ψ may be considered as a homomorphism

$$\psi : \Omega_K \to O_2(\mathbb{R}) \cong \mathbb{Z}/2 \ltimes S^1,$$

and clearly $(\mathbb{Z}/2 \ltimes z^2) \cdot \psi = \rho$ so that $SW_2[\rho] \in H^2(K; \mathbb{Z}/2) \cong \{\pm 1\}$ is trivial, by (2.41).

Thus, by (2.35), we have shown that $\rho_1(\delta) = SW_2[\rho]$, as required. $\qquad \square$

(2.45) Proof of Theorem 2.30 in Case §2.29 (i)

We may write $\rho_1 = l(a)$, $\rho_2 = l(b)$ with $a, b \in K^*/(K^*)^2$. If $l(a) = l(b)$, then

$$W_K(l(a))^2 = l(a)(-1), \qquad \text{by [Ta, p. 109]}$$

$$= (a, -1)$$

$$= (a, a), \qquad \text{as required.}$$

If $a \not\equiv b(\bmod (K^*)^2)$, then the compositum, L, of $K(\sqrt{a}), K(\sqrt{b}), K(\sqrt{ab})$ has Galois group, $G(L/K)$, isomorphic to D_8. We have a diagram of central extensions

$$
\begin{array}{ccccc}
\mathbb{Z}/2 & \rightarrowtail & \mathbb{Z}/4 & \rightarrow & \mathbb{Z}/2 \\
\cong \downarrow & & \downarrow & & \downarrow \\
\mathbb{Z}/2 & \rightarrowtail & D_8 & \rightarrow & \mathbb{Z}/2 \times \mathbb{Z}/2,
\end{array}
$$

where the groups are Galois groups. Suppose that $l(a)$ is a representation of the top right $\mathbb{Z}/2$ and that $l(a) \oplus l(b): D_8 \rightarrow \mathbb{Z}/2 \times \mathbb{Z}/2 \rightarrow O_2(\mathbb{R})$ is $\mathrm{Ind}_{\mathbb{Z}/4}^{D_8}(l(a))$; this can be arranged by the correct choice. Then, by the proof of Case §2.29 (iii), the result follows. It takes the form of (2.12), since

$$W_K(l(a) \oplus l(b)) = W_K \mathrm{Ind}_{\mathbb{Z}/4}^{D_8}(l(a))$$

$$= SW_2(l(a) + l(b)) W_K(1 \oplus l(ab))$$

$$= (a, b) W_K(l(ab)).$$

Combining §§2.27 and 2.30, we obtain the following result, which includes the result of [De], once we have established the existence of W_K in §3:　　□

(2.46) Corollary

If W_K exists satisfying the conditions of (2.3), then, on $RO(K)$,

$$W_K(\rho) = \Gamma_K(\rho) \qquad \text{of (2.24)}$$

and, in particular,

$$W_K(\rho) = SW_2(\rho) W_K(\det(\rho)) \in \mu_4.$$

(2.47)　We close this section with some discussion of characteristic classes on the Grothendieck-Witt ring of K. For this purpose we will assume that we have constructed the local root numbers, W_K.

　　The following topic is particularly appropriate because of the manner in which it interrelates the material of this section with the formulae of Chapter 3, (2.7) and (2.8).

Let $K/\mathbb{Q}_p$ be a local field. The Grothendieck-Witt ring is the set of equivalence classes of finite-dimensional, nonsingular, symmetric bilinear forms over K with addition and multiplication induced by (orthogonal) direct sum and tensor product. Denote the Grothendieck-Witt ring by $GW(K)$. The Witt ring, $W(K)$, is the quotient of $GW(K)$ by the free abelian group on the form

$$\langle 1 \rangle + \langle -1 \rangle = \left\{ \begin{bmatrix} 1 & 0 \\ 0 & -1 \end{bmatrix} \right\}.$$

Here, as usual, $\langle a \rangle \colon K \times K \to K$ sends (x, y) to axy. We have augmentation maps, given by rank and rank mod 2, and a commutative diagram

$$\begin{array}{ccc} GW(K) & \xrightarrow{\varepsilon} & \mathbb{Z} \\ \downarrow & & \downarrow \\ W(K) & \xrightarrow{\varepsilon} & \mathbb{Z}/2 \end{array}$$

so that $IGW(K) \cong IW(K)$; the augmentation ideals coincide.

Now consider the isomorphism of §2.20

$$\psi_K \colon IGW(K) \xrightarrow{\cong} Y_K.$$

(2.48) Proposition

The following diagram commutes:

$$\begin{array}{ccc} IGW(F) & \xrightarrow[\psi_F]{\cong} & Y_F \\ {\scriptstyle \mathrm{Tr}^S_{F/K}} \downarrow & & \downarrow {\scriptstyle \mathrm{Ind}_{F/K}} \\ IGW(K) & \xrightarrow[\psi_K]{\cong} & Y_K \end{array}$$

Proof. If $\mathrm{rank}(x) = 0$, $x \in IGW(F)$, then $\psi_K(\mathrm{Tr}^S_{F/K}(x))$ is detected by $\{SW_i(\psi_K(\mathrm{Tr}^S_{F/K}(x)))\}$ $(i = 1, 2)$, which, by definition of ψ_K, equal $\{HW_i(\mathrm{Tr}^S_{F/K}(x))\}$. By Chapter 3, (2.8), on the other hand, $SW_i(\mathrm{Ind}_{F/K}(\psi_F(x))) = HW_i(\mathrm{Tr}^S_{F/K}(x))$ $(i = 1, 2)$ since $\mathrm{rank}(x) = 0$. This proves the result. $\qquad \square$

(2.49) THE WEIL CHARACTER

The Weil character is a homomorphism

$$\gamma_K \colon GW(K) \to \mu_8 = \{\text{eighth roots of unity}\}$$

which actually factors through $W(K)$. It is defined as follows: Let (V, b) be a finite-dimensional, nonsingular, symmetric bilinear form over $\mathbb{Q}_p$. Let L be any $\mathbb{Z}_p$-lattice in V so that L is a finitely generated, free $\mathbb{Z}_p$-module in V such that $L \otimes_{\mathbb{Z}_p} \mathbb{Q}_p = V$ and $b(L, L) \subset \mathbb{Z}_p$. Introduce $L^{\#}$ such that

$$L^{\#} = \{x \in V \mid 2b(x, L) \in \mathbb{Z}_p\},$$

then $L^{\#} \supset L$ and $A = L^{\#}/L$ is a finite, abelian p-group. Define

$$(2.50) \qquad \begin{cases} q: A \to \mathbb{Q}_p/\mathbb{Z}_p \text{ by} \\ q(x + L) \equiv b(x, x) \,(\text{mod } \mathbb{Z}_p). \end{cases}$$

Set

$$(2.51) \qquad \gamma_{\mathbb{Q}_p}(V, b) = 1/(\sqrt{|A|}) \sum_{x \in A} \exp(-2\pi i q(x)),$$

and, in general, define γ_K by the composition

$$(2.52) \qquad \gamma_K: GW(K) \xrightarrow{\mathrm{Tr}^S_{K/\mathbb{Q}_p}} GW(\mathbb{Q}_p) \xrightarrow{\gamma_{\mathbb{Q}_p}} \mu_8.$$

By construction γ_K is invariant under the Scharlau transfer, $\mathrm{Tr}^S_{F/K}$.

(2.53) Lemma

With the preceding notation, if $IO(K)$ is as in (2.25), the homomorphism

$$IO(K) \twoheadrightarrow Y_K \xrightarrow{\psi_K^{-1}} IGW(K) \xrightarrow{\gamma_K} \mu_8$$

is invariant under induction.

Proof. This follows directly from §2.48. $\square$

(2.54)   THE FOURIER TRANSFORM OF W_K

For $a \in K^*$ define $\hat{r}_K(a)$ to be the Fourier transform of the local root number

$$(2.55) \qquad \hat{r}_K(a) = (1/\sqrt{(|K^*/(K^*)^2|)}) \sum_{x \in K^*/(K^*)^2} (a, x) W_K(l(x)) \in \mu_8,$$

where $(a, x) = l(a)l(x) \in H^2(K; \mathbb{Z}/2) \cong \{\pm 1\}$. Formula (2.55) extends to a well-defined homomorphism

$$\hat{r}_K: GW(K) \to \mu_8.$$

The following result was first proved in [C-Y]. The ambiguity of sign has been removed by B. Kahn [K2] and, independently, by P. E. Conner. They show that the correct sign is the positive one.

(2.56) Theorem

Let $K/\mathbb{Q}_p$ be a local field; then for $x \in GW(K)$,

$$\hat{r}_K(x) = \pm \overline{\gamma_K(x)}.$$

Proof. It suffices to show that, if $\mathrm{rank}(x) \equiv O(2)$, $\hat{r}_K(x) = \overline{\gamma_K(x)}$. Since both sides factor through $W(K)$, we may suppose that $\mathrm{rank}(x) = 0$. Now consider the two compositions

$$IO(K) \longrightarrow Y_K \xrightarrow{\cong} IGW(K) \xrightarrow{f} \mu_8$$

given by $f = \overline{\gamma_K(x)}$ and $f = \hat{r}_K$.

These are equal when $K = \mathbb{Q}_p$, by direct verification. In fact, $\hat{r}_{\mathbb{Q}_p}(l(a)) = \overline{\gamma_{\mathbb{Q}_p}(l(a))}$ for all $a \in \mathbb{Q}_p^*$. However, on $IO(K)$, both are invariant under induction, as the next formula shows, together with §§2.26 and 2.27. This completes the proof modulo the next result.   ☐

(2.57) Lemma

(i) Let $a, b \in K^*$, and let $V_{a,b}$ denote the orthogonal representation given by $V_{a,b} = l(a) \oplus i(b) = \psi_K(\langle a \rangle + \langle b \rangle)$. Then

$$\hat{r}_K(\langle a \rangle + \langle b \rangle) = SW_2(V_{a,b})W_K(l(-1)(\det V_{a,b})) \in \mu_4.$$

(ii) If $x \in GW(K)$, and $\mathrm{rank}(x) \equiv 0(2)$, then

$$\hat{r}_K(x) = W_K(\psi_K(x))W_K(l(-1))^{(\mathrm{rank}(x)/2)}(-1, SW_1(x)).$$

Proof. By definition, if $D = |K^*/(K^*)^2|$,

$$\hat{r}_K(\langle a \rangle + \langle b \rangle) = (1/D)\sum_{c,d}(a,c)(b,d)W_K(l(c))W_K(l(d))$$

$$= (1/D)\sum_{c,d}(a,c)(b,d)(c,d)W_K(l(cd)), \qquad \text{by (2.12).}$$

Now set $x = cd$, then $(c,d) = (c, cx) = (c, -x)(c, -c) = (c, -x)$ so that we obtain, by bilinearity of (α, β),

$$\hat{r}_K(\langle a \rangle + \langle b \rangle) = (1/D)\sum_{c,x}(b,x)(-abx,c)W_K(l(x)).$$

However, if $\{c \to (-abx, c)\}$ is not the trivial map, then the corresponding sum over c (for fixed x) is trivial. Therefore

$$\hat{t}_K(\langle a \rangle + \langle b \rangle) = (a, b) W_K(l(-ab))$$

as only $x = -ab$ contributes nontrivially. This proves part (i) of the result. Part (ii) follows from part (i) and §§2.26 and 2.27, or §2.46. $\square$

3.   EXISTENCE OF LOCAL ROOT NUMBERS

In this section we will use the presentation for $R(G)$, from Chapter 6, §13, together with §2, to construct local root numbers

$$(3.1) \qquad\qquad W_K : R(G(L/K)) \to S^1$$

(when L/K is a Galois extension of non-Archimedean local fields) satisfying the conditions of §2.3. Our proof will be fundamentally different from that of [De 2]. For example, we will construct W_K in such a way that the local/global factorization of (1.42) is a formal consequence of the construction. Formula (1.42) is usually derived as a consequence of the properties of (2.3), but our method will, in some cases, derive (2.3) from (1.42).

(3.2)   DEFINITION OF ω_K

Let $\mathscr{R}(G(L/K))$ denote the Grothendieck group of monomial homomorphisms as introduced in Chapter 6, §(3.23). Hence

$$\mathscr{R}(G(L/K)) = \left(\bigoplus_{n \geq 1} R_+ \left(G, \Sigma_n \int S^1 \right) \right) \Big/ \Lambda_\phi.$$

We will define a homomorphism,

$$(3.3) \qquad\qquad \omega_K : \mathscr{R}(G(L/K)) \to S^1,$$

by imitating the desired properties of the local root number, W_K, of (2.3). By Chapter 6, (2.6), $\mathscr{R}(G(L/K))$ is freely generated by monomial homomorphisms

$$v : G(L/K) \to \Sigma_n \int S^1$$

which are irreducible in the sense of Chapter 5, §1. Suppose that such a v is given by

$$(3.4) \qquad \begin{cases} v = \mathrm{Ind}_{F/K}(\phi), \text{ where } \phi : G(L/F) \to S^1 \\ \text{is defined up to inner automorphism.} \end{cases}$$

By (3.4) we may assign to v the number

$$\omega_K(v) = W_F(\phi)\Gamma_K(\mathrm{Ind}_{F/K}(1)) \in S^1,$$

where $W_F(\phi)$ is given by §§(1.55)–(1.60), and Γ_K is as in (2.24).
For a general monomial homomorphism

$$v: G(L/K) \to \Sigma_n \int S^1,$$

we may extend the preceding formula by setting

$$(3.5) \qquad \begin{cases} \omega_K(v) = \prod_\alpha W_{F_\alpha}(\rho_\alpha)\Gamma_K(\mathrm{Ind}_{F_\alpha/K}(1)), \\[2mm] \text{where } \rho(v): \{G(L/F_\alpha) \xrightarrow{\rho_\alpha} S^1\} \text{ is} \\[2mm] \text{the } \rho\text{-construction of Chapter 6, §3.28.} \end{cases}$$

In other words, our formula defines a homomorphism of the form

$$\hat{\omega}_K: R_+(G(L/K), S^1) \to S^1,$$

and we have defined ω_K as the composition

$$\omega_K: \mathscr{R}(G(L/K)) \xrightarrow{\rho} R_+(G(L/K), S^1) \xrightarrow{\hat{\omega}_K} S^1.$$

Formula (3.5) extends to give the homomorphism of (3.3), and then we may define, for a *representation*,

$$\alpha: G(L/K) \to U(n)$$

$$(3.6) \qquad W_K(\alpha) = \omega_K(\tau_{G(L/K)}(\alpha)) \in S^1.$$

Notice that, at first sight, $W_K(\alpha)$ as defined in (3.6) is only a well-defined *function of sets*:

$$(3.7) \qquad \begin{Bmatrix} \text{Representations} \\ \text{of } G(L/K) \end{Bmatrix} \xrightarrow{w_K} \begin{Bmatrix} \text{complex numbers} \\ \text{of unit norm} \end{Bmatrix}.$$

In fact, the situation is apparently worse. Although when $\dim(\alpha) = 1$, $\tau_G(\alpha) = \alpha$ and $W_K(\alpha)$ will be equal to the expression given in §§(1.55)–(1.60), there is at first sight no reason why, for a homomorphism, $\alpha: G(L/K) \to \Sigma_n \int S^1 \subset U(n)$ that $W_K(\alpha) = \omega_K(\alpha)$. Indeed, if this were evident, then, Theorem 3.22 of Chapter 6, in the form

$$B: \mathscr{R}(G)/\!\!\sim \xrightarrow{\cong} R(G),$$

where $\sim$ is the relation generated by $\{\tau_G(\alpha) - \alpha \,|\, \alpha: G \to \Sigma_n \int S^1\}$, would imply that W_K equals the homomorphism induced by ω_K. In this case it would be elementary to show that (2.3) is fulfilled.

To get around this problem, we prove (1.42) first. Notice that, up to this point, our methods have been entirely *local* (i.e., no appeal to Artin L-functions or analytic global Galois representation techniques).

(3.8) Proposition

Let L/K be a finite Galois extension of non-Archimedean local fields, and let F be an intermediate field. If $v: G(L/F) \to \Sigma_n \int S^1$ is a monomial homomorphism, then

$$\omega_K(\mathrm{Ind}_{F/K}(v)) = \omega_F(v)\omega_K(\mathrm{Ind}_{F/K}(1))^n \in S^1.$$

Proof. We may reduce to the case when $v = \mathrm{Ind}_{M/F}(\phi)$ for some $\phi: G(L/M) \to S^1$. Therefore

$$\omega_K(\mathrm{Ind}_{F/K}(v)) = \omega_K(\mathrm{Ind}_{M/K}(\phi))$$

$$= \omega_M(\phi)\Gamma_K(\mathrm{Ind}_{M/K}(1)), \qquad \text{by definition,}$$

$$= \omega_M(\phi)\Gamma_K(\mathrm{Ind}_{F/K}(\mathrm{Ind}_{M/F}(1))$$

$$= \omega_M(\phi)\Gamma_F(\mathrm{Ind}_{M/F}(1))\Gamma_K(\mathrm{Ind}_{F/K}(1))^n, \qquad \text{by §2.26 (i),}$$

$$= \omega_F(\mathrm{Ind}_{M/F}(\phi)) \cdot \omega_K(\mathrm{Ind}_{F/K}(1))^n, \qquad \text{as required.} \qquad \square$$

(3.9) Proposition

Let L/K be a finite, normal extension of number fields, and let $v: G(L/K) \to \Sigma_n \int S^1$ be a monomial homomorphism. Then, with the notation of (1.41) and (1.42) the Artin root number satisfies

$$W_K(B(v)) = \prod_{\substack{v \text{ place} \\ \text{of } K}} \omega_{K_v}(v_v) \in S^1.$$

Proof. We may consider each side of the preceding equation as a homomorphism on $\mathscr{R}(G(L/K))$, and, by the fact that (1.42) is true in the abelian case, we know that the result is true when $n = 1$.

On the other hand, the image of the ρ-construction applied to $\mathscr{R}(G(L/K))$ is generated by $\{v = \mathrm{Ind}_{F/K}(\phi); \phi: G(L/K) \to S^1\}$. For such a v we have

$$\prod_v \omega_{K_v}((\mathrm{Ind}_{F/K}(\phi))_v)$$

$$\prod_v \prod_{w|v} \omega_{K_v}(\mathrm{Ind}_{F_{w/K_v}}(\phi_w))$$

because, by the double-coset formula in $\mathscr{R}(G)$, or in $R_+(G, S^1)$ (see (3.27)),

$$\mathrm{Ind}_{F/K}(\phi)_v = \sum_{w|v} \mathrm{Ind}_{F_{w/K_v}}(\phi_w) \in \mathscr{R}(\Omega_{K_v}),$$

where w runs through places of F dividing v. By §3.8, this expression equals

$$\prod_v \prod_{w|v} \omega_{F_w}(\phi_w) \omega_{K_v}(\mathrm{Ind}_{F_{w/K_v}}(1)) = W_F(\phi) \prod_v \Gamma_{K_v}(\mathrm{Ind}_{F_{w/K_v}}(1)),$$

since $\dim(\phi) = 1$. However,

$$\prod_v \Gamma_{K_v}(x_v) = 1$$

if x is orthogonal (cf., [Ta, p. 130]). This follows from §2.26 (iii), the fact that it is true if $\dim(x) = 1$ by results of Hecke, and that $\Pi_v SW_2(x_v)$ is the image of $SW_2(x)$ under two steps in an exact sequence, namely,

$$H^2(K; \mathbb{Z}/2) \to \bigoplus_v H^2(K_v; \mathbb{Z}/2) \cong \bigoplus_v \{\pm 1\} \xrightarrow{\text{multiply}} \mu_2 = \{\pm 1\}.$$

Therefore the expression reduces to

$$W_K(\phi) = W_K(\mathrm{Ind}_{F/K}(\phi)) = W_K(B(v)), \qquad \text{as required.} \qquad \square$$

(3.10) Theorem

Let L/K be a finite, normal Galois extension of number fields. Let $\alpha: G(L/K) \to U(n)$ be a representation. Then, in the notation of (1.41) and (1.42),

$$W_K(\alpha) = \prod_{\substack{v \text{ place} \\ \text{of } K}} W_{K_v}(\alpha_v).$$

Proof. By (3.6), the definition of W_{K_v},

$$\prod_v W_{K_v}(\alpha_v) = \prod_v \omega_{K_v}(\tau_{G(L_u/K_v)}(\alpha_v)), \qquad \text{where } D_v = G(L_u/K_v) \text{ in (1.41)},$$

$$= \prod_v \omega_{K_v}((\tau_{G(L/K)}(\alpha))_v),$$

by naturality of τ_G (Chapter 6, §(3.25) and Theorem 3.13 (iv)) and of the ρ-construction (Chapter 6, §(3.37)),

$$= W_K(B(\tau_{G(L/K)}(\alpha)), \qquad \text{by §3.9,}$$

$$= W_K(\alpha), \qquad \text{by Chapter 6, §3.13 (ii).} \qquad \square$$

The following result is proved in [F-T] by local methods, and we will omit the proof. A local Galois extension is *tame* if the first wild ramification group, in (1.40), is trivial.

(3.11) Proposition

Let L/K be a finite, tame Galois extension of non-Archimedean local fields. Then the root number homomorphisms

$$\{W_F: R(G(L/F)) \to S^1, K \leq F \leq L\}$$

exist and satisfy the conditions of (2.3).
   In particular, on $RO(G(L/F))$, in this case $\Gamma_F = W_F$, by §2.46.

(3.12) Lemma

Let $E/\mathbb{Q}_p$ be a finite Galois extension with group $G = G(E/\mathbb{Q}_p)$. Then there exists a Galois of number fields, $\hat{E}/\hat{K}$ such that

   (i) *$\hat{E}$ is dense in E and $\hat{E} \subset E$ is the unique place over $\hat{K} \subset \mathbb{Q}_p$.*
   (ii) *$G(\hat{E}/\hat{K}) = G$.*
   (iii) *$\hat{E}/\hat{K}$ is tamely ramified at all finite places different from those lying above p.*

Proof. Let S denote the set of primes dividing $|G|$. Let $f(X) \in \mathbb{Z}_p[X]$ be a separable polynomial for which $E/\mathbb{Q}_p$ is a splitting field with $\deg(f) = n$. For each prime $l \in S$ different from p, choose distinct numbers

$$x_1(l), \ldots, x_n(l) \in \mathbb{Q}_l.$$

By the Chinese Remainder Theorem [La, p. 11], choose $\tilde{f}(X) \in \mathbb{Z}[X]$ of degree n such that

(i) $\tilde{f}(X) \equiv f(x) \pmod{p^{M(p)}}$.
(ii) $\tilde{f}(X) \equiv \prod_{i=1}^{n} (X - x_i(l)) \pmod{l^{M(l)}}$ for each $l \in S$, $l \neq p$.

By Newton's method [La, p. 42, Prop. 2], if we choose the integers $M(l)$ and $M(p)$ large enough, we may ensure that $\tilde{f}(X)$ is separable and splits in E and in $\mathbb{Q}_l$ for $l \in S$, $l \neq p$. Let $\hat{E}/\mathbb{Q}$ denote the splitting field for $\tilde{f}(X)$, and set $\hat{K} = \hat{E} \cap \mathbb{Q}_p$.
   Providing that $M(p)$ is large enough, $\hat{E}$ will be dense in E and $\hat{K}$ will be dense in $\mathbb{Q}_p$, which ensures that (ii) holds. Hence, as $\hat{E} \subset E$ is Galois invariant,

condition (i) also holds. Finally, (iii) holds because wild ramification can only occur at finite primes over $l \in S$, and at such primes the local extension is trivial if $M(l)$ is large enough, for then $\tilde{f}(X)$ splits in $\mathbb{Q}_l$. $\qquad \square$

Now we may describe one situation in which Theorem 3.10 implies that the root number satisfies (2.3). In local fields of *finite* characteristic the following method works in general.

(3.13) Theorem

Let $E/\mathbb{Q}_p$ be a finite, Galois extension of local fields. Suppose that $\hat{K} = \mathbb{Q}$, in the construction of §3.12, then (3.6) defines a homomorphism

$$W_K : R(G(E/\mathbb{Q}_p)) \to S^1$$

satisfying the conditions of (2.3).

Proof. By the discussion which follows (3.7), we have only to prove for every homomorphism

$$\rho : G(E/\mathbb{Q}_p) \to \Sigma_n \int S^1$$

that

$$\omega_{\mathbb{Q}_p}(\tau_{G(E/\mathbb{Q}_p)}(\rho)) = \omega_{\mathbb{Q}_p}(\rho) \in S^1.$$

Choose $\hat{E}/\hat{K}$ as in §3.12, and consider $\rho : G(\hat{E}/\hat{K}) \to \Sigma_n \int S^1$ as a global Galois representation. The Artin root number, $W_{\hat{K}}$, is well-defined on $R(G(\hat{E}/\hat{K}))$ so that

$$W_{\hat{K}}(\tau_{G(\hat{E}/\hat{K})}(\rho)) = W_{\hat{K}}(\rho) \in S^1.$$

But the same is true at the tame finite places, by §3.11. That is, if v is a place of $\hat{K}$ and if w is a place of $\hat{E}$ dividing v such that $\hat{E}_w/\hat{K}_v$ is tame, then

$$\omega_{\hat{K}_v}(\tau_{G(\hat{E}_w/\hat{K}_v)}(\rho_v)) = \omega_{\hat{K}_v}(\rho_v).$$

Similarly, all is well at the Archimedean places, because the Archimedean root numbers exist by §1.43.

Therefore, from the local/global factorizations of §3.9, we obtain, since $\hat{K} = \mathbb{Q} \subset \mathbb{Q}_p$ is the unique place over p,

$$\omega_{\mathbb{Q}_p}(\tau_{G(E/\mathbb{Q}_p)}(\rho)) = W_{\hat{K}}(\tau_{G(\hat{E}/\hat{K})}(\rho)) \left\{ \prod_{\hat{K}_v \neq \mathbb{Q}_p} \omega_{\hat{K}_v}(\tau_{G(\hat{E}_w/\hat{K}_v)}(\rho_v)) \right\}^{-1}$$

$$= W_{\hat{K}}(\rho) \left\{ \prod_{\hat{K}_v \neq \mathbb{Q}_p} \omega_{\hat{K}_v}(\rho_v) \right\}^{-1}$$

$$= \omega_{\mathbb{Q}_p}(\rho), \qquad \text{as required.} \qquad \square$$

(3.14) Exercise

Deligne's existence proof for the local root numbers is based on passing from the local extension to a global one, and therein twisting by (or tensoring with) a carefully chosen Hecke character. This method is described in [Ta, pp. 106–108].

Adapt Deligne's method to derive the properties of (2.3) for W_K from the results of §§3.9 and 3.10.

References

[A]   E. Artin. *Galois Theory*. Notre Dame Mathematical Lectures #2 (1965).

[A2]   E. Artin. Uber eine neue Art von *L*-Reihen. *Hamburg Abh.* 1 (1923): 89–108 (Collected papers #3).

[A3]   E. Artin. Zur Theorie der *L*-Reihen mit allgemeinen Gruppencharakteren. *Hamburg Abh.* 8 (1930): 292–306 (Collected papers #8).

[Ad]   J. F. Adams. *Lectures on Lie Groups*. Benjamin (1969).

[At]   M. F. Atiyah. Characters and Cohomology of Finite Groups. *Pub. Math. IHES (Paris)* 9 (1961): 23–64.

[At 2]   M. F. Atiyah. *K-Theory*. Benjamin (1968).

[A-B]   M. F. Atiyah and R. Bott. A Lefschetz Fixed Point Formula for Elliptic Complexes: I. *Annals of Math.* 86 (1967): 374–407.

[A-B-S] M. F. Atiyah, R. Bott, and A. Shapiro. Clifford Modules. *Topology* 3 (1964): 3–38.

[At-Se] M. F. Atiyah and G. B. Segal. Equivariant *K*-Theory and Completion. *J. Diff. Geom.* 3 (1969): 1–18.

[B-G]   J. C. Becker and C. H. Gottlieb. The Transfer Map and Fibre Bundles; *Topology* 14 (1975): 1–12.

[B-S]   Z. I. Borevich and I. R. Shafarevich. *Number Theory*. Academic Press (1966).

[B-Sch] J. C. Becker and R. E. Schultz. Equivariant Function Spaces and Stable Homotopy 1. *Comm. Math. Helv.* (49) 1 (1974): 1–34.

[Bra]   R. Brauer. Beziehungen zwischen Klassenzahlen von Teilkorpern eines galoischen Korpers. *Math. Nachr.* 4 (1951): 158–174.

[Bre]   G. E. Bredon. *Equivariant Cohomology Theories*. Lecture Notes in Mathematics #34. Springer-Verlag (1967).

[Br]   K. S. Brown. *Cohomology of Groups*. Springer-Verlag Graduate Texts in Mathematics #87 (1982).

[BK]   A. K. Bousfield and D. M. Kan. *Homotopy Limits, Completions and Localisations*. Springer-Verlag Lecture Notes in Math. #304 (1972).

[Ca]   G. Carlsson. Equivariant Stable Homotopy and Segal's Burnside Ring Conjecture; *Annals of Math.* 120 (1984): 189–224.

[Car]   P. Cartier. *La Conjecture locale de Langlands pour GL*(2) *et la demonstration de Ph. Kutzko*. Sem. Bourbaki (1979/80) #550 Lecture Notes in Mathematics #842, Springer-Verlag (1981).

[C-E]   H. Cartan and S. Eilenberg. *Homological Algebra*. Princeton University Press (1956).

[CN-T]   Ph. Cassou-Nogués and M. J. Taylor. Constante de l'équation fonctionelle L d'Artin d'une représentation symplectique et moderée. *Ann. Inst. Fourier Grenoble* (2) 33 (1983): 1–17.

[CN-T2]   Ph. Cassou-Nogués and M. J. Taylor. Local Root Numbers and Hermitian-Galois Module Structure of Rings of Integers. *Math. Ann.* 263 (1983): 251–261.

[CN-T3]   Ph. Cassou-Nogués and M. J. Taylor. Invariant de Clifford equivariant d'un caractère orthogonale. Preprint (1984).

[CLM]   F. R. Cohen, T. Lada, and J. P. May. *The Homology of Iterated Loop Spaces.* Springer-Verlag Lecture Notes in Math. #533 (1976).

[Co]   H. Cohn. *A Classical Invitation to Algebraic Numbers and Class Fields.* Springer-Verlag Universitext (1978).

[C-P]   P. E. Conner and R. Perlis. *A Survey of Trace Forms of Algebraic Number Fields.* Series in Pure Math. #2. World Scientific Publishing Co., Singapore (1984).

[C-Y]   P. E. Conner and N. Yui. The Additive Characters of the Witt Ring of an Algebraic Number Field. Preprint #MSRI 12608-85, Berkeley.

[C-R]   C. Curtis and I. Reiner. *Methods of Representation Theory.* Vol. 1. Wiley-Interscience (1981).

[D]   M. A. Delzant. Définition des classes de Stiefel-Whitney d'un module quadratique sur un corps de charactéristique différent de 2. *C.R. Acad. Sci. Paris* 255 (1962): 1366–1368.

[De]   P. Deligne. Les Constantes locales de l'equation fonctionelle de la fonction L d'Artin d'une représentation orthogonale. *Inventiones Math.* 35 (1976): 299–316.

[De2]   P. Deligne. *Les Constantes des équations fonctionelle des fonctions L.* Springer-Verlag Lecture Notes in Math. #349 (1973), 501–597.

[Dr]   A. Dress. *Contributions to the Theory of Induced Representations.* Springer-Verlag Lecture Notes in Math #342 (1973), 183–240.

[Dr 2]   A. Dress. A Characterisation of Solvable Groups. *Math Zeit.* 110 (1969): 213–217.

[Dw]   B. Dwork. On the Artin Root Number. *Amer. J. Math.* 78 (1956): 444–472.

[Fe]   M. Feshbach. The Transfer and Compact Lie Groups. Thesis Stanford University (1976). (*See also Bull. AMS* 83 (3) (1977): 372–374).

[FP]   Z. Fiedorowicz and S. B. Priddy. *Homology of Classical Groups over Finite Fields and Their Associated Infinite Loop Spaces.* Springer-Verlag Lecture Notes in Math. #674 (1978).

[F]   A. Frohlich. Orthogonal Representations of Galois Groups, Stiefel-Whitney Classes and Hasse-Witt invariants; *J.f. Reine ang math* 360 (1985): 85–123.

[F 2]   A. Frohlich. *Class Groups and Hermitian Modules.* Birkhauser, P.M. #48 (1984).

[F 3]   A. Frohlich. Orthogonal and Symplectic Representations of Groups. *Proc. L. M. Soc.* (3) 24 (1972): 470–506.

[F-Q]   A. Frohlich and J. Queyrut. On the Functional Equation of the Artin L-function for Characters of Real Representations; *Inventiones Math.* 20 (1973): 125–138.

[F-T]   A. Frohlich and M. J. Taylor. The Arithmetic Theory of Local Galois Gauss Sums for Characters. *Trans. Royal Soc.* A298 (1980): 141–181.

[Gel]   S. Gelbart. An Elementary Introduction to the Langlands Programme. *Bull. AMS* (10) 2 (1984): 177–219.

[H-Se]   B. Harris and G. B. Segal. *K*-groups of Rings of Algebraic Integers. *Annals of Math* (1) 101 (1975): 20–33.

[H]   D. Husemoller. *Fibre Bundles.* McGraw-Hill (1966), Springer-Verlag (1975).

[H-S]   P. Hilton and U. Stammbach. *A Course in Homological Algebra.* Springer-Verlag Graduate Texts in Mathematics #4 (1971).

[Ho-S]   G. Hochschild and J.-P. Serre. Cohomology of Group Extensions. *Trans. AMS* 74 (1953): 110–134.

[I-R]   K. Ireland and K. Rosen. *A Classical Introduction to Modern Number Theory.* Springer-Verlag, Graduate Texts in Mathematics #84 (1982).

[K]   B. Kahn. Classes de Stiefel-Whitney de formes quadratiques et de représentations Galoisiennes réelles. *Inventiones Math.* 78 (1984): 223–256.

[K2]   B. Kahn. Sommes de Gauss attachées aux charactères quadratiques—une conjecture de Pierre Conner. *Comm. Math. Helv.* 62 (1987): 532–541.

[K-P]   D. S. Kahn and S. B. Priddy. The Transfer and Stable Homotopy Theory. *Math. Proc. Cambr. Phil. Soc.* (83) 103 (1978): 103–111. (*See also* Applications of the transfer to stable homotopy theory; Bull. AMS 741 (1972): 981–987.)

[Ka]   M. Karoubi. Homology of the infinite orthogonal and symplectic groups over algebraically closed fields; *Inventiones Math.* 73 (1983): 247–250.

[Kn]   K.-H. Knapp. Some Applications of *K*-theory to Framed Bordism: *E*-Invariant and Transfer. *Habilitationschrift*, Univ. Bonn (1979).

[Kos]   A. Koslowski. The Evens-Kahn formula for the total Stiefel-Whitney Class; *Proc. AMS* (2) 91 (1984): 309–313.

[Kos2]   A. Koslowski. The Transfer in Segal's Cohomology. *Ill. J. Math.* (4) 27 (1983): 614–623.

[L]   T. Y. Lam. *Algebraic Theory of Quadratic Forms.* Benjamin (1973).

[La]   S. Lang. *Algebraic Number Theory.* Addison-Wesley (1970).

[LMM]   L. G. Lewis, J. P. May, and J. E. McClure. Classifying *G*-Spaces and the Segal Conjecture. *Current Trends in Algebraic Topology.* Can. Math. Soc. Conf. Proc. #2, Part 2 (1982), 165–179.

[Mac]   S. Maclane. *Homology.* Grund. Math. Wiss. Springer-Verlag (1963).

[Mar]   J. Martinet. Character Theory and Artin *L*-functions. *Algebraic Number Fields* (ed. A. Frohlich). Academic Press (1977).

[MM]   J. P. May and J. E. McClure. A Reduction of the Segal conjecture. *Current Trends in Algebraic Topology.* Can. Math. Soc. Conf. Proc. #2, Part 2 (1982), 209–222.

[M-Sn-Z] J. P. May, V. P. Snaith, and P. Zelewski. A further Generalisation of the Segal Conjecture. Univ. Chicago preprint (1987).

[Me]   A. S. Merkurjev. On the Torsion of K_2 of Local Fields. *Annals of Math.* (2) 118 (1983): 375–381.

[Me-S]   A. S. Merkurjev and A. A. Suslin. *K*-Cohomology of Severi-Brauer Varieties and the Norm Residue Homomorphism. *Izv. Acad. Sci. USSR*, forthcoming.

[Mil]   J.S. Milne. *Étale Cohomology.* Princeton University Series #33 (1980).

[M]   J. W. Milnor. Algebraic K-Theory and Quadratic Forms. *Inventiones Math.* 9 (1970). 318–344.

[M2]   J. W. Milnor. Introduction to Algebraic K-Theory. *Annals of Math.* Study #72, Princeton (1977).

[M-H]   J. W. Milnor and D. Husemoller. *Symmetric Bilinear Forms.* Ergebnisse Math. #73, Springer-Verlag (1973).

[M-St]   J. W. Milnor and J. D. Stasheff. *Characteristic Classes.* Ann. Math. Study #76, Princeton University Press (1974).

[O'M]   O. T. O'Meara. *Introduction to Quadratic Forms.* Grund. Math. Wiss. #117, Springer-Verlag (1963).

[Q]   D. G. Quillen. The Mod 2 Cohomology Ring of Extra-special 2-Groups, the Spinor Groups. *Math. Ann.* 94 (1971): 197–212.

[Q2]   D. G. Quillen. The Adams Conjecture. *Topology* (1) 10 (1971): 67–80.

[R]   E. G. Rees. *Notes on Geometry.* Universitext in Mathematics, Springer-Verlag (1983).

[Sch]   W. Scharlau. *Quadratic and Hermitian Forms.* Grund. Math. Wiss. #270, Springer-Verlag (1985).

[Seg]   G. B. Segal. The Multiplicative Group of Classical Cohomology; *Q. J. Math. Oxford* 26 (1975): 289–299.

[Ser]   J.-P. Serre. *Cohomologie Galoisienne.* Lecture Notes in Mathematics #5, Springer-Verlag (1966).

[Ser 2] J.-P. Serre. *Local Fields.* Graduate Texts in Mathematics #67, Springer-Verlag (1979).

[Ser 3] J.-P. Serre. L'Invariant de Witt de la forme $\mathrm{Tr}(x^2)$. *Comm. Math. Helv.* 59 (1984): 651–676.

[Ser 4] J.-P. Serre. *Extensions icosahedriques.* Sem. Théorie des Nombres, Bordeaux (1979/80), Exposé 19.

[Ser 5] J.-P. Serre. Sur la rationalité des représentations d'Artin. *Annals of Math.* (2) 72 (1960): 405–420.

[Ser 6] J.-P. Serre. *Linear Representations of Finite Groups*, Graduate Texts in Mathematics #42, Springer-Verlag (1977).

[Ser 7] J.-P. Serre: Conducteurs d'Artin des caracteres réels. *Inventiones Math.* 14 (1971): 173–183.

[Ser 8] J.-P. Serre. *A Course in Arithmetic.* Graduate Texts in Mathematics #7, Springer-Verlag (1974).

[Sou]   C. Soulé. K_2 et la groupe de Brauer. Sem. Bourbaki (1982) #601.

[Sn]   V. P. Snaith. Stiefel-Whitney Classes of Symmetric Bilinear Forms—A Formula of Serre. *Can. Bull. Math.* (2) 28 (1985). 218–222.

[Sn 1]   V. P. Snaith. A Descent Theorem for Hermitian Algebraic K-Theory. *Can. J. Math.* (4) 39 (1987): 835–847.

[Sn 2]   V. P. Snaith. On the Classifying Spaces of Galois Groups. Proc. Can. Math. Soc. Conf. on algebraic topology, St. John's, Nfld. (1983), *AMS Contemp. Math. Series.* 37 (1985): 145–148.

[Sn 3]   V. P. Snaith. *Algebraic Cobordism and K-Theory.* Mem. AMS #221 (1979).

[Sn 4]   V. P. Snaith. *Explicit Brauer Induction.* Inventiones Math., forthcoming.

[Sn 5]   V. P. Snaith. A Construction of the Deligne-Langlands Local Root Numbers of Orthogonal Galois Representations. *Topology* (2) 27 (1988) 119–127.

[Sn-Z]   V. P. Snaith and P. Zelewski. Stable Maps into the Classifying Spaces of Compact Lie Groups. UWO preprint (1986).

[Spa]   E. H. Spanier. *Algebraic Topology*. McGraw-Hill (1966).

[Sp]   T. A. Springer. On the Equivalence of Quadratic Forms. *Proc. Akad. Amsterdam* 62 (1959). 241–253.

[Sp2]   T. A. Springer. *Invariant Theory*. Lecture Notes in Mathematics #585, Springer-Verlag (1977).

[S-E]   N. E. Steenrod (written by D. B. A. Epstein). *Cohomology Operations. Annals of Math.* Study #50 (1962).

[St]   M. Stein. Generators, Relations and Coverings of Chevalley Groups over Commutative Rings. *Amer. J. Math.* 43 (1971): 965–1004.

[Su]   A. A. Suslin. On the K-Theory of Algebraically Closed Fields. *Inventiones Math.* 73 (1983): 243–249.

[Su 1]   A. A. Suslin. On the K-Theory of Local Fields. *J. Pure and App. Alg.* 34 (1984): 301–318.

[Ta]   J. T. Tate (with C. J. Bushnell and M. Taylor). Local constants. *Algebraic Number Fields* (ed. A. Frohlich), Academic Press (1977).

[T]   M. J. Taylor. On Frohlich's Conjecture for Rings of Integers of Tame Extensions. *Inventiones Math.* 63 (1981): 41–79.

[T 1]   R. W. Thomason. Algebraic K-Theory and Étale Cohomology. MIT preprint (1980), revised (1984).

[T 2]   R. W. Thomason. The Lichtenbaum-Quillen Conjecture for $K/l[1/\beta]$. *Current Trends in Algebraic Topology*. Proc. Can. Math. Soc. Conf. Series, Vol. 2, Part 1 (1982): 117–140.

[Wa]   C. D. Walter. Brauer's Class Number Relation. *Acta Arith.* 35 (1979): 33–40.

[W]   E. Witt. Die algebraische Struktur des Gruppenringes einer endlichen Gruppe über einem Zahlenkörper. *J. Reine Angew. Math.* 190 (1952): 231–245.

Index

Mathematics–Bestsellers

HANDBOOK OF MATHEMATICAL FUNCTIONS: with Formulas, Graphs, and Mathematical Tables, Edited by Milton Abramowitz and Irene A. Stegun. A classic resource for working with special functions, standard trig, and exponential logarithmic definitions and extensions, it features 29 sets of tables, some to as high as 20 places. 1046pp. 8 x 10 1/2. 0-486-61272-4

ABSTRACT AND CONCRETE CATEGORIES: The Joy of Cats, Jiri Adamek, Horst Herrlich, and George E. Strecker. This up-to-date introductory treatment employs category theory to explore the theory of structures. Its unique approach stresses concrete categories and presents a systematic view of factorization structures. Numerous examples. 1990 edition, updated 2004. 528pp. 6 1/8 x 9 1/4. 0-486-46934-4

MATHEMATICS: Its Content, Methods and Meaning, A. D. Aleksandrov, A. N. Kolmogorov, and M. A. Lavrent'ev. Major survey offers comprehensive, coherent discussions of analytic geometry, algebra, differential equations, calculus of variations, functions of a complex variable, prime numbers, linear and non-Euclidean geometry, topology, functional analysis, more. 1963 edition. 1120pp. 5 3/8 x 8 1/2. 0-486-40916-3

INTRODUCTION TO VECTORS AND TENSORS: Second Edition--Two Volumes Bound as One, Ray M. Bowen and C.-C. Wang. Convenient single-volume compilation of two texts offers both introduction and in-depth survey. Geared toward engineering and science students rather than mathematicians, it focuses on physics and engineering applications. 1976 edition. 560pp. 6 1/2 x 9 1/4. 0-486-46914-X

AN INTRODUCTION TO ORTHOGONAL POLYNOMIALS, Theodore S. Chihara. Concise introduction covers general elementary theory, including the representation theorem and distribution functions, continued fractions and chain sequences, the recurrence formula, special functions, and some specific systems. 1978 edition. 272pp. 5 3/8 x 8 1/2. 0-486-47929-3

ADVANCED MATHEMATICS FOR ENGINEERS AND SCIENTISTS, Paul DuChateau. This primary text and supplemental reference focuses on linear algebra, calculus, and ordinary differential equations. Additional topics include partial differential equations and approximation methods. Includes solved problems. 1992 edition. 400pp. 7 1/2 x 9 1/4. 0-486-47930-7

PARTIAL DIFFERENTIAL EQUATIONS FOR SCIENTISTS AND ENGINEERS, Stanley J. Farlow. Practical text shows how to formulate and solve partial differential equations. Coverage of diffusion-type problems, hyperbolic-type problems, elliptic-type problems, numerical and approximate methods. Solution guide available upon request. 1982 edition. 414pp. 6 1/8 x 9 1/4. 0-486-67620-X

VARIATIONAL PRINCIPLES AND FREE-BOUNDARY PROBLEMS, Avner Friedman. Advanced graduate-level text examines variational methods in partial differential equations and illustrates their applications to free-boundary problems. Features detailed statements of standard theory of elliptic and parabolic operators. 1982 edition. 720pp. 6 1/8 x 9 1/4. 0-486-47853-X

LINEAR ANALYSIS AND REPRESENTATION THEORY, Steven A. Gaal. Unified treatment covers topics from the theory of operators and operator algebras on Hilbert spaces; integration and representation theory for topological groups; and the theory of Lie algebras, Lie groups, and transform groups. 1973 edition. 704pp. 6 1/8 x 9 1/4. 0-486-47851-3

Browse over 9,000 books at www.doverpublications.com

A SURVEY OF INDUSTRIAL MATHEMATICS, Charles R. MacCluer. Students learn how to solve problems they'll encounter in their professional lives with this concise single-volume treatment. It employs MATLAB and other strategies to explore typical industrial problems. 2000 edition. 384pp. 5 3/8 x 8 1/2.   0-486-47702-9

NUMBER SYSTEMS AND THE FOUNDATIONS OF ANALYSIS, Elliott Mendelson. Geared toward undergraduate and beginning graduate students, this study explores natural numbers, integers, rational numbers, real numbers, and complex numbers. Numerous exercises and appendixes supplement the text. 1973 edition. 368pp. 5 3/8 x 8 1/2.   0-486-45792-3

A FIRST LOOK AT NUMERICAL FUNCTIONAL ANALYSIS, W. W. Sawyer. Text by renowned educator shows how problems in numerical analysis lead to concepts of functional analysis. Topics include Banach and Hilbert spaces, contraction mappings, convergence, differentiation and integration, and Euclidean space. 1978 edition. 208pp. 5 3/8 x 8 1/2.   0-486-47882-3

FRACTALS, CHAOS, POWER LAWS: Minutes from an Infinite Paradise, Manfred Schroeder. A fascinating exploration of the connections between chaos theory, physics, biology, and mathematics, this book abounds in award-winning computer graphics, optical illusions, and games that clarify memorable insights into self-similarity. 1992 edition. 448pp. 6 1/8 x 9 1/4.   0-486-47204-3

SET THEORY AND THE CONTINUUM PROBLEM, Raymond M. Smullyan and Melvin Fitting. A lucid, elegant, and complete survey of set theory, this three-part treatment explores axiomatic set theory, the consistency of the continuum hypothesis, and forcing and independence results. 1996 edition. 336pp. 6 x 9.   0-486-47484-4

DYNAMICAL SYSTEMS, Shlomo Sternberg. A pioneer in the field of dynamical systems discusses one-dimensional dynamics, differential equations, random walks, iterated function systems, symbolic dynamics, and Markov chains. Supplementary materials include PowerPoint slides and MATLAB exercises. 2010 edition. 272pp. 6 1/8 x 9 1/4.   0-486-47705-3

ORDINARY DIFFERENTIAL EQUATIONS, Morris Tenenbaum and Harry Pollard. Skillfully organized introductory text examines origin of differential equations, then defines basic terms and outlines general solution of a differential equation. Explores integrating factors; dilution and accretion problems; Laplace Transforms; Newton's Interpolation Formulas, more. 818pp. 5 3/8 x 8 1/2.   0-486-64940-7

MATROID THEORY, D. J. A. Welsh. Text by a noted expert describes standard examples and investigation results, using elementary proofs to develop basic matroid properties before advancing to a more sophisticated treatment. Includes numerous exercises. 1976 edition. 448pp. 5 3/8 x 8 1/2.   0-486-47439-9

THE CONCEPT OF A RIEMANN SURFACE, Hermann Weyl. This classic on the general history of functions combines function theory and geometry, forming the basis of the modern approach to analysis, geometry, and topology. 1955 edition. 208pp. 5 3/8 x 8 1/2.   0-486-47004-0

THE LAPLACE TRANSFORM, David Vernon Widder. This volume focuses on the Laplace and Stieltjes transforms, offering a highly theoretical treatment. Topics include fundamental formulas, the moment problem, monotonic functions, and Tauberian theorems. 1941 edition. 416pp. 5 3/8 x 8 1/2.   0-486-47755-X

Browse over 9,000 books at www.doverpublications.com

Mathematics–Algebra and Calculus

VECTOR CALCULUS, Peter Baxandall and Hans Liebeck. This introductory text offers a rigorous, comprehensive treatment. Classical theorems of vector calculus are amply illustrated with figures, worked examples, physical applications, and exercises with hints and answers. 1986 edition. 560pp. 5 3/8 x 8 1/2. 0-486-46620-5

ADVANCED CALCULUS: An Introduction to Classical Analysis, Louis Brand. A course in analysis that focuses on the functions of a real variable, this text introduces the basic concepts in their simplest setting and illustrates its teachings with numerous examples, theorems, and proofs. 1955 edition. 592pp. 5 3/8 x 8 1/2. 0-486-44548-8

ADVANCED CALCULUS, Avner Friedman. Intended for students who have already completed a one-year course in elementary calculus, this two-part treatment advances from functions of one variable to those of several variables. Solutions. 1971 edition. 432pp. 5 3/8 x 8 1/2. 0-486-45795-8

METHODS OF MATHEMATICS APPLIED TO CALCULUS, PROBABILITY, AND STATISTICS, Richard W. Hamming. This 4-part treatment begins with algebra and analytic geometry and proceeds to an exploration of the calculus of algebraic functions and transcendental functions and applications. 1985 edition. Includes 310 figures and 18 tables. 880pp. 6 1/2 x 9 1/4. 0-486-43945-3

BASIC ALGEBRA I: Second Edition, Nathan Jacobson. A classic text and standard reference for a generation, this volume covers all undergraduate algebra topics, including groups, rings, modules, Galois theory, polynomials, linear algebra, and associative algebra. 1985 edition. 528pp. 6 1/8 x 9 1/4. 0-486-47189-6

BASIC ALGEBRA II: Second Edition, Nathan Jacobson. This classic text and standard reference comprises all subjects of a first-year graduate-level course, including in-depth coverage of groups and polynomials and extensive use of categories and functors. 1989 edition. 704pp. 6 1/8 x 9 1/4. 0-486-47187-X

CALCULUS: An Intuitive and Physical Approach (Second Edition), Morris Kline. Application-oriented introduction relates the subject as closely as possible to science with explorations of the derivative; differentiation and integration of the powers of x; theorems on differentiation, antidifferentiation; the chain rule; trigonometric functions; more. Examples. 1967 edition. 960pp. 6 1/2 x 9 1/4. 0-486-40453-6

ABSTRACT ALGEBRA AND SOLUTION BY RADICALS, John E. Maxfield and Margaret W. Maxfield. Accessible advanced undergraduate-level text starts with groups, rings, fields, and polynomials and advances to Galois theory, radicals and roots of unity, and solution by radicals. Numerous examples, illustrations, exercises, appendixes. 1971 edition. 224pp. 6 1/8 x 9 1/4. 0-486-47723-1

AN INTRODUCTION TO THE THEORY OF LINEAR SPACES, Georgi E. Shilov. Translated by Richard A. Silverman. Introductory treatment offers a clear exposition of algebra, geometry, and analysis as parts of an integrated whole rather than separate subjects. Numerous examples illustrate many different fields, and problems include hints or answers. 1961 edition. 320pp. 5 3/8 x 8 1/2. 0-486-63070-6

LINEAR ALGEBRA, Georgi E. Shilov. Covers determinants, linear spaces, systems of linear equations, linear functions of a vector argument, coordinate transformations, the canonical form of the matrix of a linear operator, bilinear and quadratic forms, and more. 387pp. 5 3/8 x 8 1/2. 0-486-63518-X

Browse over 9,000 books at www.doverpublications.com

Mathematics–Geometry and Topology

PROBLEMS AND SOLUTIONS IN EUCLIDEAN GEOMETRY, M. N. Aref and William Wernick. Based on classical principles, this book is intended for a second course in Euclidean geometry and can be used as a refresher. More than 200 problems include hints and solutions. 1968 edition. 272pp. 5 3/8 x 8 1/2.   0-486-47720-7

TOPOLOGY OF 3-MANIFOLDS AND RELATED TOPICS, Edited by M. K. Fort, Jr. With a New Introduction by Daniel Silver. Summaries and full reports from a 1961 conference discuss decompositions and subsets of 3-space; n-manifolds; knot theory; the Poincaré conjecture; and periodic maps and isotopies. Familiarity with algebraic topology required. 1962 edition. 272pp. 6 1/8 x 9 1/4.   0-486-47753-3

POINT SET TOPOLOGY, Steven A. Gaal. Suitable for a complete course in topology, this text also functions as a self-contained treatment for independent study. Additional enrichment materials make it equally valuable as a reference. 1964 edition. 336pp. 5 3/8 x 8 1/2.   0-486-47222-1

INVITATION TO GEOMETRY, Z. A. Melzak. Intended for students of many different backgrounds with only a modest knowledge of mathematics, this text features self-contained chapters that can be adapted to several types of geometry courses. 1983 edition. 240pp. 5 3/8 x 8 1/2.   0-486-46626-4

TOPOLOGY AND GEOMETRY FOR PHYSICISTS, Charles Nash and Siddhartha Sen. Written by physicists for physics students, this text assumes no detailed background in topology or geometry. Topics include differential forms, homotopy, homology, cohomology, fiber bundles, connection and covariant derivatives, and Morse theory. 1983 edition. 320pp. 5 3/8 x 8 1/2.   0-486-47852-1

BEYOND GEOMETRY: Classic Papers from Riemann to Einstein, Edited with an Introduction and Notes by Peter Pesic. This is the only English-language collection of these 8 accessible essays. They trace seminal ideas about the foundations of geometry that led to Einstein's general theory of relativity. 224pp. 6 1/8 x 9 1/4. 0-486-45350-2

GEOMETRY FROM EUCLID TO KNOTS, Saul Stahl. This text provides a historical perspective on plane geometry and covers non-neutral Euclidean geometry, circles and regular polygons, projective geometry, symmetries, inversions, informal topology, and more. Includes 1,000 practice problems. Solutions available. 2003 edition. 480pp. 6 1/8 x 9 1/4.   0-486-47459-3

TOPOLOGICAL VECTOR SPACES, DISTRIBUTIONS AND KERNELS, François Trèves. Extending beyond the boundaries of Hilbert and Banach space theory, this text focuses on key aspects of functional analysis, particularly in regard to solving partial differential equations. 1967 edition. 592pp. 5 3/8 x 8 1/2.
0-486-45352-9

INTRODUCTION TO PROJECTIVE GEOMETRY, C. R. Wylie, Jr. This introductory volume offers strong reinforcement for its teachings, with detailed examples and numerous theorems, proofs, and exercises, plus complete answers to all odd-numbered end-of-chapter problems. 1970 edition. 576pp. 6 1/8 x 9 1/4.   0-486-46895-X

FOUNDATIONS OF GEOMETRY, C. R. Wylie, Jr. Geared toward students preparing to teach high school mathematics, this text explores the principles of Euclidean and non-Euclidean geometry and covers both generalities and specifics of the axiomatic method. 1964 edition. 352pp. 6 x 9.   0-486-47214-0

Browse over 9,000 books at www.doverpublications.com